T0183384

Lecture Notes in Artificial Intelligence 8816

Subseries of Lecture Notes in Computer Science

More information about this series at http://www.springer.com/series/1244

Fernando Bobillo · Rommel N. Carvalho
Paulo C.G. Costa · Claudia d'Amato
Nicola Fanizzi · Kathryn B. Laskey
Kenneth J. Laskey · Thomas Lukasiewicz
Matthias Nickles · Michael Pool (Eds.)

Uncertainty Reasoning for the Semantic Web III

ISWC International Workshops
URSW 2011–2013
Revised Selected Papers

 Springer

Editors

Fernando Bobillo
University of Zaragoza
Zaragoza
Spain

Rommel N. Carvalho
Universidade de Brasília
Brasília
Brazil

Paulo C.G. Costa
George Mason University
Fairfax, VA
USA

Claudia d'Amato
Università degli Studi di Bari
Bari
Italy

Nicola Fanizzi
Università degli Studi di Bari
Bari
Italy

Kathryn B. Laskey
George Mason University
Fairfax, VA
USA

Kenneth J. Laskey
MITRE Corporation
McLean, VA
USA

Thomas Lukasiewicz
University of Oxford
Oxford
UK

Matthias Nickles
National University of Ireland
Galway
Ireland

Michael Pool
Goldman Sachs
Washington, DC
USA

ISSN 0302-9743   ISSN 1611-3349 (electronic)
Lecture Notes in Artificial Intelligence
ISBN 978-3-319-13412-3   ISBN 978-3-319-13413-0 (eBook)
DOI 10.1007/978-3-319-13413-0

Library of Congress Control Number: 2014956507

LNCS Sublibrary: SL7 – Artificial Intelligence

Springer Cham Heidelberg New York Dordrecht London

Printed on acid-free paper

Springer International Publishing AG Switzerland is part of Springer Science+Business Media
(www.springer.com)

Preface

This is the third volume on Uncertainty Reasoning for the Semantic Web, containing revised and significantly extended versions of selected workshop papers presented at three workshops on Uncertainty Reasoning for the Semantic Web (URSW), collocated with the International Semantic Web Conferences (ISWC) in 2011, 2012, and 2013. The first volume contained the proceedings of the first three workshops on URSW at ISWC in 2005, 2006, and 2007, while the second volume included revised versions of papers presented at the 2008, 2009, and 2010 editions of URSW or at the First International Workshop on Uncertainty in Description Logics (UniDL), held in 2010.

These three volumes together represent a comprehensive compilation of state-of-the-art research approaches to uncertainty reasoning in the context of the Semantic Web, capturing different models of uncertainty and approaches to deductive as well as inductive reasoning with uncertain formal knowledge.

The World Wide Web community envisions effortless interaction between humans and computers, seamless interoperability and information exchange among Web applications, and rapid and accurate identification and invocation of appropriate Web services. As work with semantics and services grows more ambitious, there is increasing appreciation of the need for principled approaches to the formal representation of and reasoning under uncertainty. The term *uncertainty* is intended here to encompass a variety of forms of incomplete knowledge, including incompleteness, inconclusiveness, vagueness, ambiguity, and others. The term *uncertainty reasoning* is meant to denote the full range of methods designed for representing and reasoning with knowledge when Boolean truth values are unknown, unknowable, or inapplicable. Commonly applied approaches to uncertainty reasoning include probability theory, Dempster-Shafer theory, fuzzy logic and possibility theory, and numerous other methodologies.

A few Web-relevant challenges that are addressed by reasoning under uncertainty include:

Uncertainty of available information: Much information on the World Wide Web is uncertain. Examples include weather forecasts or gambling odds. Canonical methods for representing and integrating such information are necessary for communicating it in a seamless fashion.

Information incompleteness: Information extracted from large information networks such as the World Wide Web is typically incomplete. The ability to exploit partial information is very useful for identifying sources of service or information. For example, that an online service deals with greeting cards may be evidence that it also sells stationery. It is clear that search effectiveness could be improved by appropriate use of technologies for handling uncertainty.

Information incorrectness: Web information is also often incorrect or only partially correct, raising issues related to trust or credibility. Uncertainty representation and reasoning helps to resolve tension among information sources having different confidence and trust levels, and can facilitate the merging of controversial information obtained from multiple sources.

Uncertain ontology mappings: The Semantic Web vision implies that numerous distinct but conceptually overlapping ontologies will coexist and interoperate. It is likely that in such scenarios, ontology mapping will benefit from the ability to represent degrees of membership and/or likelihoods of membership in categories of a target ontology, given information about class membership in the source ontologies.

Indefinite information about Web services: Dynamic composability of Web services will require runtime identification of processing and data resources and resolution of policy objectives. Uncertainty reasoning techniques may be necessary to resolve situations in which existing information is not definitive.

Uncertainty is thus an intrinsic feature of many important tasks on the Web and the Semantic Web, and a full realization of the World Wide Web as a source of processable data and services demands formalisms capable of representing and reasoning under uncertainty. Unfortunately, none of these needs can be addressed in a principled way by current Web standards. Although it is to some degree possible to use semantic markup languages such as OWL or RDF(S) to represent qualitative and quantitative information about uncertainty, there is no established foundation for doing so, and feasible approaches are severely limited. Furthermore, there are ancillary issues such as how to balance representational power versus simplicity of uncertainty representations, which uncertainty representation techniques address uses such as the examples listed above, how to ensure the consistency of representational formalisms and ontologies, etc.

In response to these pressing demands, in recent years, several promising approaches to uncertainty reasoning on the Semantic Web have been proposed. The present volume covers a representative cross section of these approaches, from extensions to existing Web-related logics for the representation of uncertainty to approaches to inductive reasoning under uncertainty on the Web.

In order to reflect the diversity of the presented approaches and to relate them to their underlying models of uncertainty, the contributions to this volume are grouped as follows:

Probabilistic and Dempster-Shafer Models

Probability theory provides a mathematically sound representation language and formal calculus for rational degrees of belief, which gives different agents the freedom to have different beliefs about a given hypothesis. As this provides a compelling framework for representing uncertain, imperfect knowledge that can come from diverse agents, there are many distinct approaches using probability in the context of the Semantic Web. Classes of probabilistic models covered with the present volume are Bayesian networks, probabilistic extensions to description and first-order logics, and models based on the Dempster-Shafer theory (a generalization of the classical Bayesian approach).

Fuzzy and Possibilistic Models

Fuzzy formalisms allow for representing and processing degrees of truth about vague (or imprecise) pieces of information. In fuzzy description logics and ontology languages, concept assertions, role assertions, concept inclusions, and role inclusions have a degree of truth rather than a binary truth value. The present volume presents various approaches that exploit fuzzy logic and possibility theory in the context of the Semantic Web.

Inductive Reasoning and Machine Learning

Machine learning is supposed to play an increasingly important role in the context of the Semantic Web by providing various tasks, such as the learning of ontologies from incomplete data or the (semi-)automatic annotation of data on the Web. Results obtained by machine learning approaches are typically uncertain. As a logic-based approach to machine learning, inductive reasoning provides means for inducing general propositions from observations (example facts). Papers in this volume exploit the power of inductive reasoning for the purpose of ontology learning, and project future directions for the use of machine learning on the Semantic Web.

Hybrid Approaches

This volume segment contains papers that either combine approaches from two or more of the previous segments, or that do not rely on any specific classical approach to uncertainty reasoning.

Acknowledgments. We would like to express our gratitude to the authors of this volume for their contributions and to the workshop participants for inspiring discussions, as well as to the members of the workshop Program Committees and the additional reviewers for their reviews and for their overall support.

October 2014

Fernando Bobillo
Rommel N. Carvalho
Paulo C.G. Costa
Claudia d'Amato
Nicola Fanizzi
Kathryn B. Laskey
Kenneth J. Laskey
Thomas Lukasiewicz
Matthias Nickles
Michael Pool

Organization

Reviewers of Chapters for URSW III

M. David Allen	MITRE Corporation, USA
Fernando Bobillo	University of Zaragoza, Spain
Christopher Burnett	University of Aberdeen, UK
Silvia Calegari	University of Milano-Bicocca, Italy
Rommel N. Carvalho	Universidade de Brasília and Office
	of the Comptroller General, Brazil
Davide Ceolin	VU University Amsterdam, The Netherlands
Paulo C.G. Costa	George Mason University, USA
Fabio G. Cozman	Universidade de São Paulo, Brazil
Claudia d'Amato	University of Bari, Italy
Nicola Fanizzi	University of Bari, Italy
Jhonatan Garci	University of Aberdeen, UK
Huan Gao	Southeast University, China
Marcelo Ladeira	Universidade de Brasília, Brazil
Kathryn B. Laskey	George Mason University, USA
Thomas Lukasiewicz	University of Oxford, UK
Trevor Martin	University of Bristol, UK
Matthias Nickles	Technische Universität München, Germany
Rafael Peñaloza	Technische Universität Dresden, Germany
Michael Pool	Convera Technologies, Inc., USA
David Robertson	University of Edinburgh, UK
Thomas Scharrenbach	University of Zurich, Switzerland
Giorgos Stoilos	National Technical University of Athens, Greece
Umberto Straccia	ISTI-CNR, Italy
Merwyn Taylor	MITRE Corporation, USA
Matthias Thimm	Universität Koblenz-Landau, Germany
Peter Vojtáš	Charles University in Prague, Czech Republic

Seventh International Workshop on Uncertainty Reasoning for the Semantic Web (URSW 2011)

Organizing Committee

Fernando Bobillo	University of Zaragoza, Spain
Rommel N. Carvalho	George Mason University, USA
Paulo C.G. Costa	George Mason University, USA
Claudia d'Amato	University of Bari, Italy
Nicola Fanizzi	University of Bari, Italy
Kathryn B. Laskey	George Mason University, USA
Kenneth J. Laskey	MITRE Corporation, USA
Thomas Lukasiewicz	University of Oxford, UK
Trevor Martin	University of Bristol, UK
Matthias Nickles	University of Bath, UK
Michael Pool	Vertical Search Works, Inc., USA

Program Committee

Fernando Bobillo	University of Zaragoza, Spain
Silvia Calegari	University of Milano-Bicocca, Italy
Rommel N. Carvalho	George Mason University, USA
Paulo C.G. Costa	George Mason University, USA
Fabio G. Cozman	Universidade de São Paulo, Brazil
Claudia d'Amato	University of Bari, Italy
Nicola Fanizzi	University of Bari, Italy
Marcelo Ladeira	Universidade de Brasília, Brazil
Kathryn B. Laskey	George Mason University, USA
Kenneth J. Laskey	MITRE Corporation, USA
Thomas Lukasiewicz	University of Oxford, UK
Trevor Martin	University of Bristol, UK
Matthias Nickles	University of Bath, UK
Jeff Z. Pan	University of Aberdeen, UK
Rafael Peñaloza	Technische Universität Dresden, Germany
Michael Pool	Vertical Search Works, Inc., USA
Livia Predoiu	University of Mannheim, Germany
Guilin Qi	Southeast University, China
Celia Ghedini Ralha	Universidade de Brasília, Brazil
David Robertson	University of Edinburgh, UK
Daniel Sánchez	University of Granada, Spain
Thomas Scharrenbach	University of Zurich, Switzerland

Sergej Sizov	University of Koblenz-Landau, Germany
Giorgos Stoilos	University of Oxford, UK
Umberto Straccia	ISTI-CNR, Italy
Andreas Tolk	Old Dominion University, USA
Peter Vojtáš	Charles University in Prague, Czech Republic

Eighth International Workshop on Uncertainty Reasoning for the Semantic Web (URSW 2012)

Organizing Committee

Fernando Bobillo   University of Zaragoza, Spain
Rommel N. Carvalho   George Mason University, USA
Paulo C.G. Costa   George Mason University, USA
Nicola Fanizzi   University of Bari, Italy
Kathryn B. Laskey   George Mason University, USA
Kenneth J. Laskey   MITRE Corporation, USA
Thomas Lukasiewicz   University of Oxford, UK
Trevor Martin   University of Bristol, UK
Matthias Nickles   Technical University of Munich, Germany
Michael Pool   Goldman Sachs, USA

Program Committee

Fernando Bobillo   University of Zaragoza, Spain
Silvia Calegari   University of Milano-Bicocca, Italy
Rommel N. Carvalho   George Mason University, USA
Davide Ceolin   VU University Amsterdam, The Netherlands
Paulo C.G. Costa   George Mason University, USA
Fabio G. Cozman   Universidade de São Paulo, Brazil
Nicola Fanizzi   University of Bari, Italy
Marcelo Ladeira   Universidade de Brasília, Brazil
Kathryn B. Laskey   George Mason University, USA
Kenneth J. Laskey   MITRE Corporation, USA
Thomas Lukasiewicz   University of Oxford, UK
Trevor Martin   University of Bristol, UK
Matthias Nickles   Technical University of Munich, Germany
Jeff Z. Pan   University of Aberdeen, UK
Rafael Peñaloza   Technische Universität Dresden, Germany
Michael Pool   Goldman Sachs, USA
Livia Predoiu   University of Mannheim, Germany
Guilin Qi   Southeast University, China
Celia Ghedini Ralha   Universidade de Brasília, Brazil
David Robertson   University of Edinburgh, UK
Daniel Sánchez   University of Granada, Spain
Sergej Sizov   University of Koblenz-Landau, Germany
Giorgos Stoilos   University of Oxford, UK

Umberto Straccia	ISTI-CNR, Italy
Matthias Thimm	Universität Koblenz-Landau, Germany
Andreas Tolk	Old Dominion University, USA
Peter Vojtáš	Charles University in Prague, Czech Republic

Additional Reviewers

Christopher Burnett
Yuqing Tang

Ninth International Workshop on Uncertainty Reasoning for the Semantic Web (URSW 2013)

Organizing Committee

Fernando Bobillo   University of Zaragoza, Spain
Rommel N. Carvalho   George Mason University, USA
Paulo C.G. Costa   George Mason University, USA
Claudia d'Amato   University of Bari, Italy
Nicola Fanizzi   University of Bari, Italy
Kathryn B. Laskey   George Mason University, USA
Kenneth J. Laskey   MITRE Corporation, USA
Thomas Lukasiewicz   University of Oxford, UK
Trevor Martin   University of Bristol, UK
Matthias Nickles   National University of Ireland, Ireland
Michael Pool   Goldman Sachs, USA

Program Committee

Fernando Bobillo   University of Zaragoza, Spain
Silvia Calegari   University of Milano-Bicocca, Italy
Rommel N. Carvalho   George Mason University, USA
Davide Ceolin   VU University Amsterdam, The Netherlands
Paulo C.G. Costa   George Mason University, USA
Fabio G. Cozman   Universidade de São Paulo, Brazil
Claudia d'Amato   University of Bari, Italy
Nicola Fanizzi   University of Bari, Italy
Marcelo Ladeira   Universidade de Brasília, Brazil
Kathryn B. Laskey   George Mason University, USA
Kenneth J. Laskey   MITRE Corporation, USA
Thomas Lukasiewicz   University of Oxford, UK
Trevor Martin   University of Bristol, UK
Alessandra Mileo   National University of Ireland, Ireland
Matthias Nickles   National University of Ireland, Ireland
Jeff Z. Pan   University of Aberdeen, UK
Rafael Peñaloza   Technische Universität Dresden, Germany
Michael Pool   Goldman Sachs, USA
Livia Predoiu   University of Mannheim, Germany
Guilin Qi   Southeast University, China
Celia Ghedini Ralha   Universidade de Brasília, Brazil
David Robertson   University of Edinburgh, UK

Contents

UMP-ST Plug-in: Documenting, Maintaining and Evolving Probabilistic Ontologies Using UnBBayes Framework

Rommel N. Carvalho[1], Laécio L. dos Santos[2][⊠], Marcelo Ladeira[2],
Henrique A. da Rocha[1], and Gilson L. Mendes[1]

[1] Department of Research and Strategic Information (DIE),
Brazilian Office of the Comptroller General (CGU), SAS, Quadra 01, Bloco A,
Edifício Darcy Ribeiro, Brasília, Distrito Federal, Brazil
{rommel.carvalho,henrique.rocha,liborio}@cgu.gov.br
http://www.cgu.gov.br

[2] Department of Computer Science (CIC), University of Brasília (UnB),
Campus Universitário Darcy Ribeiro, Brasília, Distrito Federal, Brazil
laecio@gmail.com, mladeira@unb.br
http://www.cic.unb.br

Abstract. Several approaches have been proposed for dealing with uncertainty in the Semantic Web (SW). Although probabilistic ontologies (PO) is one of the most promising approach to model uncertainty in ontologies, no support has been offered to ontological engineers on how to create this more complex type of ontologies. This task has proven to be extremely difficult and hard, which motivated the creation of the Uncertainty Modeling Process for Semantic Technologies (UMP-ST), a process that guides users in modeling POs. This paper presents the UMP-ST plug-in, a tool that implements this process and shows how the plug-in, implemented in UnBBayes Framework, overcomes the main problems on modeling probabilistic ontologies: the complexity in creating; the difficulty in maintaining and evolving; and the lack of a centralized tool for documenting these ontologies. The probabilistic ontology for Procurement Fraud Detection and Prevention in Brazil is used to show how the UMP-ST plug-in overcomes these problems. This probabilistic ontology is a proof-of-concept use case created as part of a research project at the Brazilian Office of the Comptroller General (CGU). (A short version of this paper was presented on the URSW 2013 [3]).

Keywords: Uncertainty Modeling Process · Semantic Web · UMP-ST · POMC · Probabilistic ontology · Fraud detection · MEBN · UnBBayes

1 Introduction

In the last decade there has been a significant increase in formalisms that integrate uncertainty representation into ontology languages. This was motivated by the need for representation and inference in domains with uncertainty, since

© Springer International Publishing Switzerland 2014
F. Bobillo et al. (Eds.): URSW 2011-2013, LNAI 8816, pp. 1–20, 2014.
DOI: 10.1007/978-3-319-13413-0_1

OWL, the standard Web Ontology Language, supports only deterministic ontologies. This has given birth to several new languages like: PR-OWL [8–10], PR-OWL 2 [4,5], OntoBayes [26], BayesOWL [11], and probabilistic extensions of SHIF(**D**) and SHOIN(**D**) [20].

However, the increase of expressive power that these languages have provided did not come without its drawbacks. In order to express more, the user is also expected to deal with more complex representations. This increase in complexity has been a major obstacle to making these languages more popular and used more often in real world problems.

While there is a robust literature on ontology engineering [1,14] and knowledge engineering for Bayesian networks [16,19], the literature contains little guidance on how to model a probabilistic ontology.

To fill the gap, Carvalho [5] proposed the Uncertainty Modeling Process for Semantic Technologies (UMP-ST), a methodology based on the Unified Process, which describes the main tasks involved in creating probabilistic ontologies incrementally and iteratively.

Nevertheless, the UMP-ST is only a guideline on things you should think about and things you should do, but it does not provide a tool for doing so. In this paper we present the UMP-ST plug-in for UnBBayes. This plug-in has the objective of dealing with three main problems: the complexity in creating probabilistic ontologies; the difficulty in maintaining and evolving existing probabilistic ontologies; and the lack of a centralized tool for documenting probabilistic ontologies.

This paper is organized as follows. Section 2 introduces the UMP-ST process and the Probabilistic Ontology Modeling Cycle (POMC). Section 3 presents the Probabilistic Web Ontology Language (PR-OWL) and the Multi-Entity Bayesian Network (MEBN), semantic technologies that motivated the creation of the UMP-ST. Section 4 presents UnBBayes and its plug-in framework. Then, Sect. 5 describes UMP-ST plug-in, which is the main contribution of this paper. Section 6 illustrates how this tool can be used to create a probabilistic ontology for procurement fraud detection and prevention. Finally, Sect. 7 presents some concluding remarks.

2   UMP-ST

The UMP-ST was proposed by Carvalho [5] as a methodology to build probabilistic ontologies. The UMP-ST is based on the Unified Process (UP), a framework that describes the activities that a team performs to transform a set of requirements into a software system [23]. Like the UP, the UMP-ST uses an iterative and incremental approach, building the ontology through several deliveries, each adding new requirements to the previous ones.

The UMP-ST divides the construction of a PO in four phases: Inception, where goals are defined; Elaboration, where the ontology will be modeled; Construction, where the ontology will be implemented; and Transition, where a new version of the ontology will be available. Inside each phase, four major disciplines guide the modeler: Requirements, Analysis & Design, Implementation and Test.

Figure 1 depicts the intensity of each discipline during the UMP-ST, which is iterative and incremental. The basic idea behind iterative enhancement is to model the domain incrementally, allowing the modeler to take advantage of what is learned during earlier iterations of the model. Learning comes from discovering new rules, entities, and relations that were not obvious previously. Some times it is possible to test some of the rules defined during the Analysis & Design stage even before having implemented the ontology. This is usually done by creating simple probabilistic models to evaluate whether the model will behave as expected before creating the more complex first-order probabilistic models. That is why some testing occurs during the first iteration (I1) of the inception phase, prior to the start of the implementation phase.

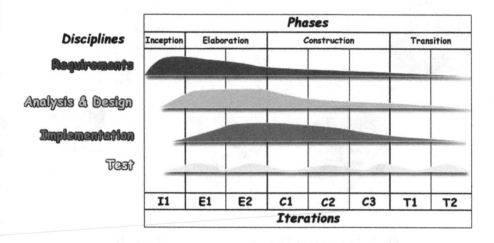

Fig. 1. Uncertainty Modeling Process for Semantic Technologies (UMP-ST).

Figure 2 presents the Probabilistic Ontology Modeling Cycle (POMC). This cycle depicts the major outputs from each discipline and the natural order in which the outputs are produced. Unlike the waterfall model [22], the POMC cycles through the steps iteratively, using what is learned in one iteration to improve the result of the next. The arrows reflect the typical progression, but are not intended as hard constraints. Indeed, it is possible to have interactions between any pair of disciplines. For instance, it is not uncommon to discover a problem in the rules defined in the Analysis & Design discipline during the activities in the Test discipline. As a result, the engineer might go directly from Test to Analysis & Design in order to correct the problem.

The **Requirements** discipline defines the goals that should be achieved by reasoning with the semantics provided by our model.

The **Analysis & Design** discipline describes classes of entities, their attributes, how they relate, and what rules apply to them in our domain. This definition is independent of the language used to implement the model.

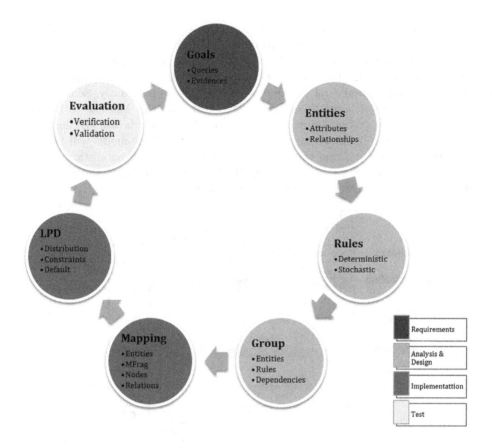

Fig. 2. Probabilistic Ontology Modeling Cycle (POMC)

The **Implementation** discipline maps our design to a specific language that allows uncertainty in semantic technologies (ST). In Fig. 2, the mapping is to PR-OWL where the elements defined during Analysis & Design will be mapping to constructions like entities, random variables, arcs and MFrags. In this discipline, the Local Probability Distribution (LPD) will also be defined. When mapping to other technologies (e.g., OntoBayes), it will be necessary to use different constructions (e.g., L-Nodes and arcs).

Finally, the **Test** discipline is responsible for evaluating whether the model developed during the Implementation discipline is behaving as expected from the rules defined during Analysis & Design and whether they achieve the goals elicited during the Requirements discipline. As noted previously, it is a good idea to test some rules and assumptions even before the implementation. This is a crucial step to mitigate risk by identifying problems before wasting time in developing an inappropriate complex model.

An important aspect of the UMP-ST process is defining traceability of requirements. Gotel and Finkelstein [15] define requirements traceability as:

Requirements traceability refers to the ability to describe and follow the life of a requirement, in both forward and backward directions.

To provide traceability, requirements should be arranged in a specification tree, so that each requirement is linked to its "parent" requirement, allowing a fast visualization of the dependencies between the requirements. In the UMP-ST model, each item of evidence is linked to a query it supports, which in turn is linked to its higher level goal. This linkage supports requirements traceability.

In addition to the hierarchical decomposition of the specification tree, requirements should also be linked to work products of other disciplines, such as the rules in the Analysis & Design discipline, probability distributions defined in the Implementation discipline, and goals, queries, and evidence elicited in the Requirements discipline. These links provide traceability that is essential to validation and management of change.

This kind of link between work products of different disciplines is typically done via a Requirements Traceability Matrix (RTM) [24,25]. Although useful and very important to guarantee the goals are met, the RTM is extremely hard to keep track without a proper tool. Therefore, this was a crucial feature that we incorporated into the UMP-ST plug-in.

3 PR-OWL and MEBN

OWL, the standard language for creating ontologies in the Semantic Web, lacks a proper support for uncertainty representation. The great quantity of domains involving uncertainty has made urgent the creation of a language able to represent them. In this context, Paulo Costa, created the PR-OWL (Probabilistic OWL) language in 2005, extending OWL with statements that allow the creation of probabilistic ontologies [8].

PR-OWL uses the MEBN formalism to represent uncertainty in standard OWL ontologies. This is done by modeling the domain into a MEBN Theory (MTheory), a set of MEBN Fragments (MFrags) composed of nodes that represent the random variables of the model. MEBN provides a probabilistic inference based on first-order logic and Bayesian networks[1]. Figure 3 shows the main elements of PR-OWL.

MEBN is a language for representing a probabilistic knowledge based on Bayesian networks and First-Order Logic (FOL) [18]. MEBN increases the power of Bayesian networks to add the expressive power of FOL. Moreover, it extends FOL by adding a way to specify probabilistic distributions. MEBN solves the main limitation of Bayesian networks: the inability to represent situations where the number of random variables involved are unknown in advance. This limitation makes it impossible to use Bayesian networks for domains that involve recursion.

MEBN represents the domain knowledge with a MEBN Theory (MTheory). A MTheory is composed by a set of MFrags, each of which representing probability information about a group of related random variables [18]. This set of

---

[1] PR-OWL requires a MEBN inference engine to process the additional syntax.

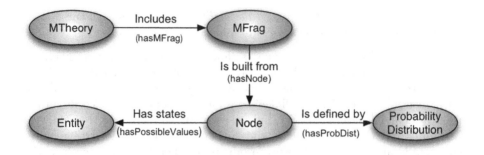

Fig. 3. PR-OWL main elements (reproduced with permission from Costa [8])

MFrags satisfies consistency constraints that guarantee the existence of a unique joint distribution over its random variables.

An MFrag is composed of resident nodes, input nodes and context nodes. Resident nodes and input nodes represent properties and relationships of entities, and have arguments (ordinary variables) that will be filled with entities during the instantiation of the model. A resident node has its LPD defined into the MFrag, while an input node is a reference to a resident node in a different MFrag. Oriented edges represent dependencies between the nodes. Context nodes define the constraints that should be observed for the probabilistic relations defined in its MFrag to be valid.

An MTheory works like a template, where MFrags will be instantiated from entities, relationships and existing evidence (findings). This instantiation will result in a Situation-Specific Bayesian Network (SSBN), where nodes of MFrags become standard Bayesian network nodes. After instantiating the SSBN, a Bayesian network inference algorithm can be used to calculate the distribution of a node of interest.

In 2011, Carvalho proposed PR-OWL 2, extending PR-OWL to provide a formal mapping between OWL concepts and PR-OWL random variables and to make the types already present in OWL compatible with PR-OWL [5]. PR-OWL 2 made it easier to construct hybrid ontologies containing probabilistic and deterministic statements.

4   UnBBayes Plug-in Architecture

UnBBayes is an open-source Java[TM] application developed by the Artificial Intelligence Group from the Computer Science Department at the University of Brasilia in Brazil that provides a framework for building probabilistic graphical models and performing plausible reasoning. It features a graphical user interface (GUI), an application programming interface (API), as well as plug-in support for unforeseen extensions. It offers a comprehensive programming model that supports the exploitation of probabilistic reasoning and provides a high degree of scalability [17, 21]. Figure 4 shows a screenshot of the tool with several plug-ins opened as different internal windows.

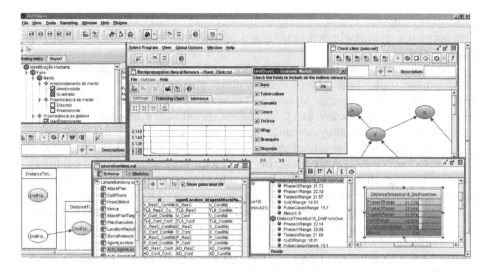

Fig. 4. Screenshot of the framework UnBBayes with several plug-ins.

Unlike APIs, plug-ins offer a means to run new code inside the UnBBayes' runtime environment. A plug-in is a program that interacts with a host application (a core) to provide a given function (usually very specific) "on demand". The binding between a plug-in and a core application usually happens at loading time (when the application starts up) or at runtime.

In UnBBayes, a plug-in is implemented as a folder, a *ZIP* or a *JAR* file containing the following elements: (a) **a plug-in descriptor file**[2] (a XML file containing meta-data about the plug-in itself), (b) **classes** (the Java program itself - it can be a set of ".*class*" files or a packaged *JAR* file), and (c) **resources** (e.g. images, icons, message files, mark-up text).

UnBBayes currently relies on Java Plug-in Framework[3] (JPF) version 1.5.1 to provide a flexible plug-in environment. JPF is an open source plug-in infrastructure framework for building scalable Java projects, providing a runtime engine that can dynamically discover and load plug-ins on-the-fly. The activation process (i.e. the class loading process) is done in a lazy manner, so plug-in classes are loaded into memory only when they are needed.

One specific type of plug-in that can be added to UnBBayes is the module plug-in. Module plug-ins provide a means to create a relatively self-sufficient feature in UnBBayes (e.g. new formalisms or completely new applications). In UnBBayes vocabulary, *modules* are basically new internal frames that are initialized when tool bars or menu buttons are activated. Those internal frames do not need to be always visible, so one can create modules that add new functionalities to the application without displaying any actual "internal" frame

---

[2] A plug-in descriptor file is both the main and the minimal content of a UnBBayes plug-in, thus one can create a plug-in composed only by a sole descriptor file.

[3] http://jpf.sourceforge.net/

(wizards or pop-ups can be emulated this way). The UMP-ST tool presented in this paper is a completely new application, since it was implemented as a module plug-in.

Figure 5 illustrates the main classes of a module plug-in. `UnBBayesModule` is the most important class of a module plug-in and it is an internal frame (thus, it is a subclass of *swing* `JInternalFrame`). Classes implementing `IPersistenceAwareWindow` are GUI classes containing a reference to an I/O class, and because `UnBBayesModule` implements `IPersistenceAwareWindow`, a module should be aware of what kind of files it can handle (so that UnBBayes can consistently delegate I/O requests to the right modules). `NewModuleplug-in` and `NewModuleplug-inBuilder` are just placeholders representing classes that should be provided by plug-ins. The builder is necessary only if `NewModuleplug-in` does not provide a default constructor with no parameters. For more information on UnBBayes plug-in framework see [21].

UnBBayes provides plug-ins for various formalisms based on Bayesian Networks, including Influence Diagram (ID), Multiply-Sectioned Bayesian Network (MSBN), Hybrid Bayesian Network (HBN), Object-Oriented Bayesian Network (OOBN), Probabilistic Relational Model (PRM), and Multi-Entity Bayesian Network (MEBN).

UnBBayes was the first tool to implement MEBN. It has a GUI for creating the model graphically, a knowledge base for representation and reasoning in FOL, a language for specifying the LPD's, and an algorithm for generating the SSBN from a set of queries and findings [6]. In this first version, UnBBayes uses PR-OWL format to persist the model, and PowerLoom [7] as the Knowledge Base system. Recently, a plug-in for PR-OWL 2 was implemented, integrating Protégé [13], a free, open-source ontology editor, to UnBBayes, allowing the user to build both the deterministic and the probabilistic part of a ontology using the UnBBayes.

5   UMP-ST Plug-in

The UMP-ST tool was implemented by the Artificial Intelligence Group of the University of Brasilia as a plug-in for UnBBayes. As seen in Sect. 2, the UMP-ST process consists of four major disciplines: Requirements, Analysis & Design, Implementation, and Test. Nevertheless, the UMP-ST plug-in focuses only on the Requirements and Analysis & Design disciplines, since they are the only language independent disciplines. As seen in Sect. 4, UnBBayes has plug-ins for building probabilistic ontologies in PR-OWL and PR-OWL 2, which can be used in the Implementation and Test disciplines.

The objective of the UMP-ST plug-in is overcoming three main problems:

1. The complexity in creating probabilistic ontologies;
2. The difficulty in maintaining and evolving existing probabilistic ontologies; and
3. The lack of a centralized tool for documenting probabilistic ontologies.

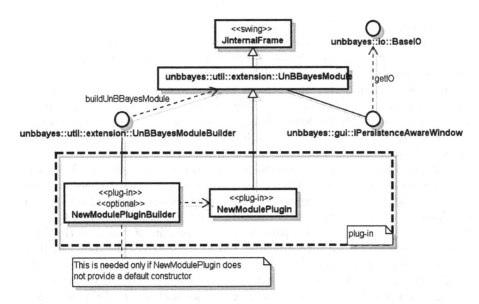

Fig. 5. Class diagram of classes that must be extended to create a module plug-in.

The UMP-ST plug-in is almost like a wizard tool that guides the user in each and every step of the Requirements and Analysis & Design disciplines. The user begins defining the goals that should be achieved by the probabilistic ontology (PO) as well as the queries that should be answered by the PO in order to achieve that goal and the evidence needed in order to answer these queries. Then the user is allowed to move to the next phase of the process which is defining the entities, with their attributes and relationships. After this, he should define rules, both probabilistic and deterministic. Finally he should define the groups related to the defined goals, queries, and evidence (see Fig. 2).

Respecting this order of steps defined in the process allows the tool to incorporate an important aspect which is traceability. In every step of the way, the user is required to associate which working product previously defined requires the definition of this new element. For instance, when defining a new query, the user has to say which goal that query helps achieve. We call this feature backtracking. This feature allows, for instance, the user to identify which goals are being achieved by the implementation of a specific group. This feature provides an easy and friendly way of maintaining the RTM matrix, defined previously.

The step by step guidance provided by the tool allows the user to overcome the complexity in creating POs (first problem). Moreover, the plug-in also solves the third problem, since all the documentation related to the PO being designed is centralized in the tool and can be saved for future use. This documentation is essential in every project, since it allows other users to quickly capture the solution and the reasons that led it to be built that way. The plug-in allows the user to enter comments for each editable element and storing information about the author and creation date.

Finally, the difficulty in maintaining and evolving existing POs (second problem) is addressed mainly by the traceability feature. When editing any element (e.g., a goal, an entity, a rule, etc.), two panels are always present. On the one hand, the backtracking panel shows every element from previous steps of the process associated with the element being edited. On the other hand, the forwardtracking panel shows every element created in the following steps of the process associated with the element being edited. This provides a constant attention to where and what changes might impact, which facilitates maintainability and evolution of existing POs.

The Fig. 6 shows the initial pane for editing entities in the plug-in. As shown, the GUI opens inside the UnBBayes window as a new frame. There are four tabs available: to edit the Requirements, the user utilizes the tab Goals; to edit the Analysis & Design, the user utilizes the tabs Entities, Rules and Groups. Selecting a tab initially opens a panel listing all elements of that type (in Fig. 6 all entities are listed, since the Entities tab was selected). The user can then choose to edit or to delete an existing element, or to add new elements. In the tab Entities there is also a button that opens the panel for editing relationships.

Figure 7 presents the panel for editing entities with some of the main features of the UMP-ST plug-in. Note the backtracking panel that in this case shows which goals and hypothesis originated the creation of the entity. The fowardtracking panel lists all elements created from this entity (attributes, relationships, rules and groups).

The UMP-ST tool was implemented with the Java programming language, using the Swing API for developing the user interface. The plug-in is distributed under GPL license, and it is available on sourceforge[4]. Since it is a module plug-in in UnBBayes, the `UmpstModule` class extends the `UnBBayesModule` class. The UMP-ST plug-in is mostly structured in a Model-View-Controller (MVC[5]) design pattern[6], which explicitly separates the program's elements into three distinct roles, in order to provide separation of concern (i.e. the software is separated into three different set of classes with minimum overlap of functionality). The View is implemented by the `umpst.GUI` package, the Controller by the `umpst.Controller` package, and the Model by the `umpst.IO` and `umpst.Model` packages. In this current beta version of the plug-in, the project is saved using the serialization method, in which an image of Java objects is saved into a file, allowing it to be reassembled during the load process.

---

[4] http://sourceforge.net/projects/unbbayes/

[5] An MVC design isolates logic and data from the user interface, by separating the components into three independent categories: *Model* (data and operations), *View* (user interface) and *Controller* (mostly, a mediator, scheduler, or moderator of other classes) [2].

[6] Design patterns are a set of generic approaches aiming to avoid known problems in software engineering [12].

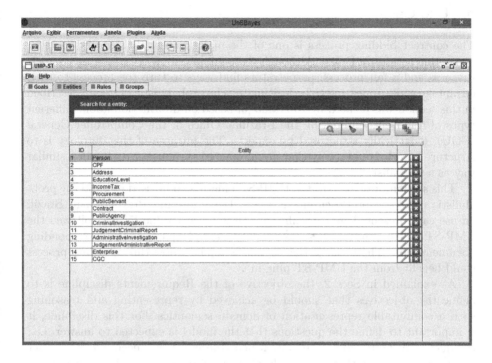

Fig. 6. Initial Panel of the Tab for editing Entities

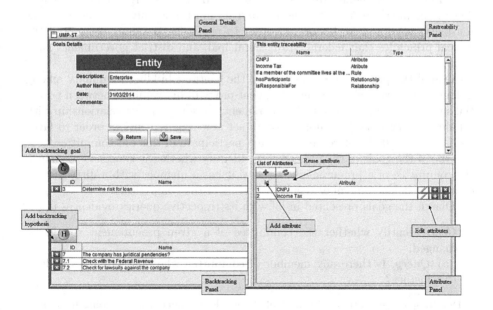

Fig. 7. Panel for editing an entity with a few features highlighted.

6   Use Case

The contract bidding process is one of the main process that can be used for corruption actions in Brazil. Although there are laws that try to ensure competitiveness and a fair process, perpetrators find ways to turn the process to their advantage while appearing to be legitimate. The probabilistic ontology described in this section was created by Carvalho [5] based on information about different types of fraud encountered by the Brazilian Office of the Comptroller General (CGU), didactically structured by experts. The purpose of this ontology is to structure knowledge in a way that an automated system can reason in a similar way as a specialist.

This section presents how the UMP-ST plug-in can be used to build the probabilistic ontology for Procurement Fraud Detection and Prevention in Brazil, an use case presented by Carvalho [5]. Although Carvalho [5] has followed the UMP-ST process, there was no tool at the time to help create the corresponding documentation. The focus of this section is to show how this modeling process could benefit from the UMP-ST plug-in[7].

As explained in Sect. 2, the objective of the Requirements discipline is to define the objectives that should be achieved by representing and reasoning with a computable representation of domain semantics. For this discipline, it is important to define the questions that the model is expected to answer, i.e., the queries to be posed to the system being designed. For each question, a set of information items that might help answer the question (evidence) should be defined.

Public procurements are very complex and involve large sums of money. Therefore, members forming the committee must not only be prepared, but must also have a clean history, in order to maximize the morality, one of the ethical principles that federal, state, municipal, and district governments must adopt.

One of the principles established for the contract bidding is equality among bidders. This principle prohibits the agent proxy to discriminate between potential suppliers. If an agent of the procurement has some form or relationship with a bidder, he can provide information or set new requirements in order to favor that bidder, and therefore he should not participate in this bid committee.

One of the goals of the use case is to identify whether it is needed to change the committee of a procurement, because it has a member with a dirty history or because it has a member related with any of the competitors.

One of the goals presented in [5] with its respective queries/evidences is:

1. *Goal*: Identify whether the committee of a given procurement should be changed.
   (a) *Query*: Is there any member of committee who does not have a clean history?

---

[7] Due to space limitation, only part of the whole documentation is going to be presented in this paper. The focus will be on presenting several features available in the UMP-ST plug-in.

     i. *Evidence*: Committee member has criminal history;
    ii. *Evidence*: Committee member has been subject to administrative investigation.
(b) *Query*: Is there any relation between members of the committee and the enterprises that participated in previous procurements?
     i. *Evidence*: Member and responsible person of an enterprise are relatives (mother, father, brother, or sister);
    ii. *Evidence*: Member and responsible person of an enterprise live at the same address.

Figure 8 presents how this goal and its corresponding queries and evidence would be displayed in the UMP-ST plug-in. Note that both query and evidence are considered hypothesis in our tool. The idea is to generalize, since an evidence for a query could be another query, and an evidence can be supported by sub-evidences. Therefore, we decided to call them both hypothesis, which, defined hierarchically will assemble an implicit specification tree.

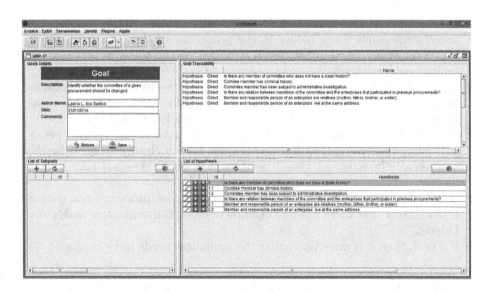

Fig. 8. Panel for displaying the hypothesis (queries and evidence) for a goal.

Note that for the definition of the requirements, the user does not need have knowledge about probabilistic ontologies, or even about semantic technologies. Thus, these can be defined by the domain expert himself. In fact, the next discipline, Analysis & Design can be done by someone with minimal knowledge about ontologies, leaving only the implementation and testing disciplines for the expert in the chosen technology.

The next step in the POMC model is to define the entities, attributes, and relationships by looking on the set of goals/queries/evidence defined in the previous step. For instance, from the evidence that says "responsible person of an enterprise" we need to define the entities Person and Enterprise, besides the relationship isResponsibleFor.

Figure 7 presents the entity Enterprise with its attributes, goals and hypothesis defined as backtraking elements, as well as traceability panel with its forward-tracking elements (attributes, rules, relationships, groups, etc.). Relationships are edited under its own panel, where the user must name it and set which entities are involved.

Once the entities, its attributes, and relationships are defined, we are able to define the rules for our PO. The panel for editing rules are really similar to the panel for editing entities. The difference is that we can define what type of rule it is (deterministic or stochastic). Moreover, the backtraking panel allows the user to add elements from the previous step in the POMC cycle, i.e., entities, attributes, and relationships, as well as elements in the current step, i.e., other rules. Thus, the forwardtracking panel only allows elements from the current and future steps in the process, i.e., other rules and groups.

The rules presented in [5] for the goal previously described are:

1. If a member of the committee has a relative (mother, father, brother, or sister) responsible for a bidder in the procurement, then it is more likely that a relationship exists between the committee and the enterprises, which inhibits competition.
2. If a member of the committee lives at the same address as a person responsible for a bidder in the procurement, then it is more likely that a relationship exists between the committee and the enterprises, which lowers competition.
3. If a member of the committee has been convicted of a crime or has been penalized administratively, then he/she does not have a clean history. If he/she was recently investigated, then it is likely that he/she does not have a clean history.
4. If the relation defined in 1 and 2 is found in previous procurements, then it is more likely that there will be a relation between this committee and future bidders.
5. If 3 or 4, then it is more likely that the committee needs to be changed.

As it can be seen, rules 4 and 5 illustrate how a rule can be defined from others rules. Thus, we would add rules 1 and 2 in tracking list of rule 4 and rules 3 and 4 in the tracking list of rule 5.

Figure 9 shows the panel for editing rule 2. The backtracking panel shows that this rule involves the entities Person, Address, Enterprise and Procurement, which in turn are used in the relationships livesAt, isResponsibleFor, and hasPartipants. This information will be useful on the implementation of this rule.

Finally, once the rules are defined, the user can go to the final step of the Analysis & Design discipline, which is to define the groups, which will facilitate the implementation of the PO. The panel for creating groups is similar to the panel for editing rules. The difference is that the forwardtracking panel shows only other groups. Figure 10 presents a list of groups created.

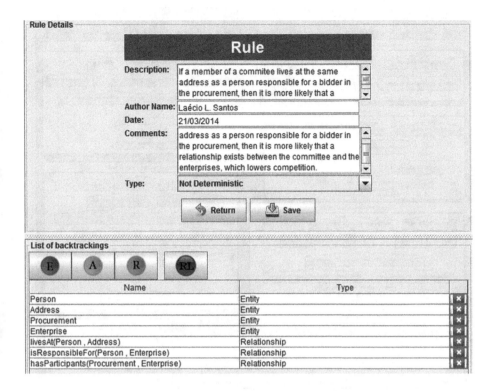

Fig. 9. Rules panel.

ID	Group		
1	Personal Information		×
2	Procurement Information		×
3	Enterprise Information		×
4	Criminal Judgement Information		×
5	Admnistrative Judgement Information		×
6	Front of Enterprise		×
7	Exists Front in Enterprise		×
8	Related Enterprises		×
9	Member Related to Participant		×
10	Competition Compromised		×
11	Related to Previous Participant		×
12	Suspicious Committee		×
13	Owns Suspended Enterprise		×
14	Suspicious Procurement		×

Fig. 10. Panel displaying some groups.

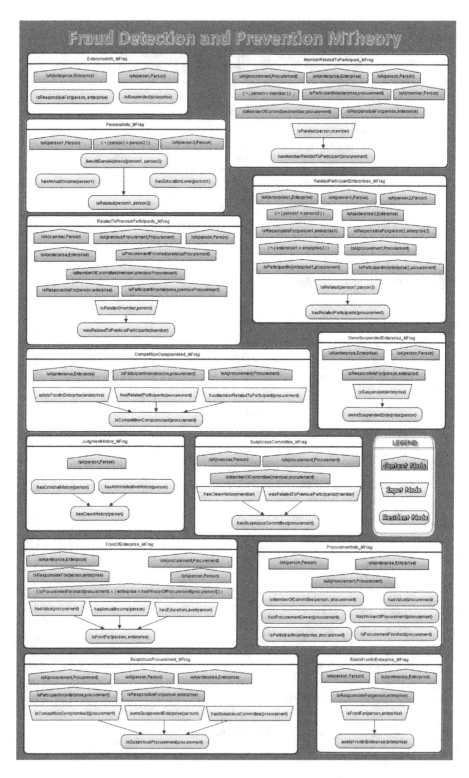

Fig. 11. Implementation of the PO in UnBBayes-MEBN.

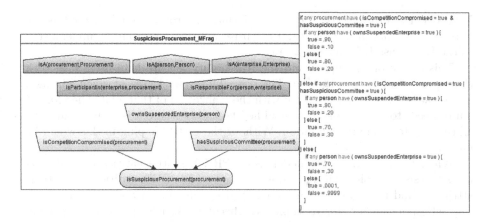

Fig. 12. LPD of the node `isRelated`.

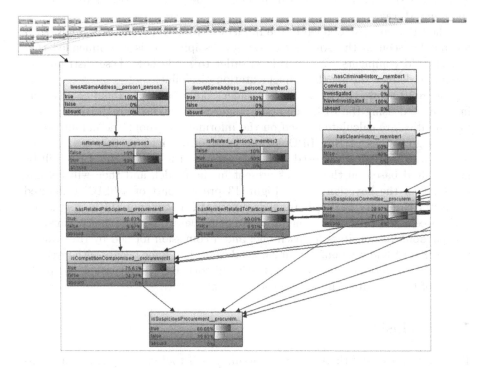

Fig. 13. SSBN for the query `IsSuspiciousProcurement(procurement1)`.

Figure 11 shows the MTheory created in the implementation of this ontology in MEBN[8], suggested in [5]. Note that there is pretty much a one-to-one correspondence between the groups defined in UMP-ST and MFrags created in

---

[8] Avaiable in https://sourceforge.net/projects/unbbayes/files/examples/.

MTheory. For instance, the Personal Information group is implemented as the Personal Information MFrag, the Enterprise Information group is implemented as the Enterprise Information MFrag, etc.

This one-to-one mapping and the traceability feature help users deal with change and evolution of the PO. The traceability panel present when editing a goal shows all elements associated with the realization of that goal. Therefore, if a user needs to change a specific goal he/she knows where it is going to impact, all the way to the implementation. Without the UMP-ST plug-in this would be infeasible.

When implementing the model using PR-OWL, the entities defined during Analysis & Design will be simply created as PR-OWL entities, while their attributes and relationships will be mapped to random variables. The relationship isResponsibleFor, for example, will become the isResponsibleFor(person, enterprise) resident node into the MFrag Enterprise Information. The rules define the dependencies between the nodes and also the context nodes that will set the constraints that must be followed in order to be able to instantiate its corresponding MFrag.

The LPDs are also defined in the Implementation discipline. Although LPDs are not foreseen in the Analysis & Design discipline, it is recommended that experts add comments to each rule in order to guide how these distributions should be defined during the Implementation discipline.

Figure 12 shows the LPD for the resident node isSuspiciousProcurement implemented using the LPD definition language sintax used in UnBBayes. The probabilities are calculated based on the informations about the parent nodes. The assign values of this LPD are notional only, since in this use case no real data or statistics was used. When generating the SSBN, this LPD will be instanciated based on the entities present in the model, and then will became the CPT of the bayesian nodes. Figure 13 present part of a SSBN generated to answer the query isSuspiciousProcurement(procurement1) in a situation with three enterprises competing for the procurement1, with person1, person2, and person3 responsibles for each one, and a comission formed by three members. Person1 and person2 live at same address. The same occurs with member2 and person2. Member1 have administrative history. For this situation, the probability of the procurement be suspicious is high (60 %).

7   Conclusion

This paper presented the UMP-ST plug-in, a GUI tool for designing, maintaining, and evolving POs. To the best of our knowledge, this is the first tool in the world to support the design of POs.

The UMP-ST plug-in provides a step by step guidance in designing POs, which allows the user to overcome the complexity in creating POs. Moreover, the plug-in also provides a centralized tool for documenting POs, whereas before the documentation was spread in different documents (word documents with requirements, UML diagrams with entities, attributes, and relations, etc.).

Finally, the difficulty in maintaining and evolving existing POs is addressed mainly by the traceability feature. The implementation of both forwardtracking and backtracking provide a constant attention to where and what changes might impact, which facilitates maintainability and evolution of existing POs. Although this traceability can be achieved by a simple implementation of RTM in tools like spreadsheets, as the PO becomes larger this manual traceability becomes infeasible and error prone.

The UMP-ST plug-in is still in beta phase. Some of the features that should be included in the future are: saving the project as an xml file; exporting all documentation to a single PDF of HTML file; allowing the user to edit the rules in a more visual way; and generating MTheories automatically based on the entities, attributes, relationships and groups defined in the Analysis & Design discipline, in order to facilitate the creation of a MEBN model (i.e., PR-OWL PO) during the Implementation discipline.

Acknowledgments. The authors gratefully acknowledge full support from the Brazilian Office of the Comptroller General (CGU) for the research reported in this paper.

References

1. Allemang, D., Hendler, J.A.: Semantic Web for the Working Ontologist. Morgan Kaufmann, San Francisco (2008)
2. Buschmann, F., Meunier, R., Rohnert, H., Sommerlad, P., Stal, M.: Pattern-Oriented Software Architecture. A System of Patterns, vol. 1. Wiley, Chichester (1996)
3. Carvalho, R.N., Ladeira, M., de Souza, R.M., Matsumoto, S., Da Rocha, H.A., Mendes, G.L.: UMP-ST plug-in: a Tool for documenting, maintaining, and evolving probabilistic ontologies. In: Bobillo, F., Carvalho, R.N., da Costa, P.C.G., d'Amato, C., Fanizzi, N., Laskey, K.B., Laskey, K.J., Lukasiewicz, T., Martin, T., Nickles, M., Pool, M. (eds.) Proceedings of the 9th International Workshop on Uncertainty Reasoning for the Semantic Web (URSW 2013). CEUR Workshop Proceedings, vol. 1073, pp. 15–26 (2013). CEUR-WS.org
4. Carvalho, R.N., Laskey, K.B., Costa, P.C.G.: PR-OWL 2.0 – bridging the gap to OWL semantics. In: Bobillo, F., Costa, P.C.G., d'Amato, C., Fanizzi, N., Laskey, K.B., Laskey, K.J., Lukasiewicz, T., Nickles, M., Pool, M. (eds.) URSW 2008-2010/UniDL 2010. LNCS, vol. 7123, pp. 1–18. Springer, Heidelberg (2013)
5. Carvalho, R.N.: Probabilistic ontology: representation and modeling methodology. Ph.D., George Mason University, Fairfax, VA, USA (2011)
6. Carvalho, R.N., Santos, L.L., Ladeira, M., Costa, P.C.G.: A GUI tool for plausible reasoning in the semantic web using MEBN. In: Proceedings of the Seventh International Conference on Intelligent Systems Design and Applications, ISDA '07, Los Alamitos, CA, USA, pp. 381–386. IEEE Computer Society, October 2007
7. Chalupsky, H., MacGregor, R.M., Russ, T.: Powerloom manual (2010)
8. Costa, P.C.G.: Bayesian semantics for the semantic Web. Ph.D., George Mason University, Fairfax, VA, USA (2005)
9. Costa, P.C.G., Laskey, K.B., Laskey, K.J.: PR-OWL: a Bayesian framework for the semantic Web. In: Proceedings of the First Workshop on Uncertainty Reasoning for the Semantic Web (URSW 2005), Galway, Ireland, November 2005

10. da Costa, P.C.G., Laskey, K.B., Laskey, K.J.: PR-OWL: a Bayesian ontology language for the semantic Web. In: da Costa, P.C.G., d'Amato, C., Fanizzi, N., Laskey, K.B., Laskey, K.J., Lukasiewicz, T., Nickles, M., Pool, M. (eds.) URSW 2005 - 2007. LNCS (LNAI), vol. 5327, pp. 88–107. Springer, Heidelberg (2008)
11. Ding, Z., Peng, Y., Pan, R.: BayesOWL: uncertainty modeling in semantic web ontologies. In: Ma, Z. (ed.) Soft Computing in Ontologies and Semantic Web. Studies in Fuzziness and Soft Computing, vol. 204, pp. 3–29. Springer, Heidelberg (2006). doi:10.1007/978-3-540-33473-6_1
12. Gamma, E., Helm, R., Johnson, R., Vlissides, J.M.: Design Patterns: Elements of Reusable Object-Oriented Software. Addison-Wesley, New York (1994)
13. Gennari, J.H., Musen, M.A., Fergerson, R.W., Grosso, W.E., Crubzy, M., Eriksson, H., Noy, N.F., Tu, S.W.: The evolution of protégé: an environment for knowledge-based systems development. Int. J. Hum.-Comput. Stud. **58**(1), 89–123 (2003)
14. Gomez-Perez, A., Corcho, O., Fernandez-Lopez, M.: Ontological Engineering: With Examples from the Areas of Knowledge Management, e-Commerce and the Semantic Web, 1st edn. Springer, Heidelberg (2004)
15. Gotel, O.C.Z., Finkelstein, C.W.: An analysis of the requirements traceability problem. In: 1994 Proceedings of the First International Conference on Requirements Engineering, pp. 94–101 (1994)
16. Korb, K.B., Nicholson, A.E.: Bayesian Artificial Intelligence. Chapman & Hall/CRC, Boca Raton (2003)
17. Ladeira, M., da Silva, D., Vieira, M., Onishi, M., Carvalho, R.N., da Silva, W.: Platform independent and open tool for probabilistic networks. In: Proceedings of the IV Artificial Intelligence National Meeting (ENIA 2003) on the XXIII Congress of the Brazilian Computer Society (SBC 2003), Unicamp, Campinas, Brazil, August 2003
18. Laskey, K.B.: MEBN: a language for first-order Bayesian knowledge bases. Artif. Intell. **172**(2–3), 140–178 (2008)
19. Mahoney, M.: Network engineering for agile belief network models. IEEE Trans. Knowl. Data Eng. **12**(4), 487–498 (2000)
20. Lukasiewicz, T.: Expressive probabilistic description logics. Artif. Intell. **172**(6–7), 852–883 (2008)
21. Matsumoto, S., Carvalho, R.N., Ladeira, M., da Costa, P.C.G., Santos, L.L., Silva, D., Onishi, M., Machado, E.: UnBBayes: a java framework for probabilistic models in AI. In: Cai, K. (ed.) Java in Academia and Research. iConcept Press, Annerley (2011)
22. Royce, W.W.: Managing the development of large software systems: concepts and techniques. In: Proceedings of IEEE WESTCON, pp. 1–9 (1970). Reprinted in Proceedings of the Ninth International Conference on Software Engineering, pp. 328–338, March 1987
23. Scott, K.: The Unified Process Explained. Addison-Wesley Longman Publishing Co Inc., Boston (2002)
24. Sommerville, I.: Software Engineering, 9th edn. Addison Wesley, Boston (2010)
25. Wiegers, K.E.: Software Requirements, 2nd edn. Microsoft Press, Redmond (2003)
26. Yang, Y., Calmet, J.: OntoBayes: an ontology-driven uncertainty model. In: Proceedings of the International Conference on Computational Intelligence for Modelling, Control and Automation and International Conference on Intelligent Agents, Web Technologies and Internet Commerce (CIMCA-IAWTIC'06), vol. 1, pp. 457–463. IEEE Computer Society (2005)

Computing Inferences for Relational Bayesian Networks Based on $\mathcal{ALC}$ Constructs

Fabio G. Cozman[1(✉)], Rodrigo B. Polastro[1], Felipe I. Takiyama[1], and Kate C. Revoredo[2]

[1] Universidade de São Paulo, Av. Prof. Mello Moraes, 2231, São Paulo–SP, Brazil
fgcozman@usp.br, rodrigopolastro@gmail.com, felipe.takiyama@usp.br
[2] Departamento de Informática Aplicada, UNIRIO, Av. Pasteur, 458,
Rio de Janeiro–RJ, Brazil
katerevoredo@gmail.com

Abstract. Credal $\mathcal{ALC}$ combines the constructs of the well-known $\mathcal{ALC}$ logic with probabilistic assessments, so as to let terminologies convey uncertainty about concepts and roles. We present a restricted version of Credal $\mathcal{ALC}$ that can be viewed as a description language for a class of relational Bayesian networks. The resulting "CR$\mathcal{ALC}$ networks" offer a simplified and illuminating route both to Credal $\mathcal{ALC}$ and to relational Bayesian networks. We then describe the implementation, in freely available packages, of approximate variational and lifted exact inference algorithms.

1 Introduction

This paper focuses on a probabilistic description logic, called Credal $\mathcal{ALC}$ [7], that adds probabilistic operators to the well-known description logic $\mathcal{ALC}$ [2]. Credal $\mathcal{ALC}$ lets terminologies convey uncertainty about concepts and roles. The idea is to adopt all constructs of acyclic $\mathcal{ALC}$ terminologies, plus probabilistic assessments such as $\mathbb{P}(C|D) \in [\underline{\alpha}, \overline{\alpha}]$, where C and D are concepts, and $\mathbb{P}(r) \in [\underline{\beta}, \overline{\beta}]$, where r is a role. These probabilities are supposed elicited from experts, or learned from data.

The semantics of Credal $\mathcal{ALC}$ is based on probabilities over interpretations, with implicit independence assumptions that are encoded through a Markov condition. Given a domain, a Credal $\mathcal{ALC}$ terminology can be grounded into a set of Bayesian networks. Instead of the usual satisfiability or subsumption problems that are studied in description logics [2], here the focus is on probabilistic *inference*: given a domain and a set of assertions, compute the conditional probability of some assertion.

While the syntax of Credal $\mathcal{ALC}$ is relatively simple to grasp, the semantics is quite complex. The adopted Markov condition is far from obvious, and one needs several assumptions to guarantee that any well-formed terminology specifies a single probability measure over all interpretations.

In this paper we present a reformulation of Credal $\mathcal{ALC}$, such that any well-formed set of formulas can be directly translated into a *relational* Bayesian network. The semantics is then inherited from the theory of relational Bayesian

© Springer International Publishing Switzerland 2014
F. Bobillo et al. (Eds.): URSW 2011-2013, LNAI 8816, pp. 21–40, 2014.
DOI: 10.1007/978-3-319-13413-0_2

networks [28,29]. This reformulation simplifies the development of probabilistic terminologies, and leads to insights concerning inference algorithms. Additionally, the syntax offers a new way of specifying relational Bayesian networks. Indeed, a profitable way to understand Credal $\mathcal{ALC}$ is to take it as a description language for a restricted but useful class of relational Bayesian networks, a class that can be valuable in specifying terminologies containing uncertainty.

We then move to inference algorithms; that is, algorithms that compute probabilities given a set of sentences and assertions in the language. One advantage of connecting Credal $\mathcal{ALC}$ with Bayesian networks is that algorithms for the latter formalism can be applied to the former. We present an implementation of variational message-passing algorithms for inference, in particular algorithms that exploit symmetries amongst individuals in a domain. Such symmetries allow us to cluster variables together, and to approximate inferences solely by exchanging messages between such clusters. We then present an implementation of exact inference using *lifted* algorithms; that is, algorithms that again treat sets of variables together. We examine the use of *aggregation parfactors* in exact lifted inference.

The paper is organized as follows. Basic definitions and notation, as well as related literature, are reviewed in Sect. 2. Credal $\mathcal{ALC}$ is presented as a syntax for relational Bayesian networks in Sect. 3. Sections 4 and 5 respectively describe our implementation of variational and lifted algorithms.

2 $\mathcal{ALC}$ and (Relational) Bayesian Networks

In this section we review some necessary notions, mostly related to knowledge representation formalisms. We are interested in description logics and in relational Bayesian networks, respectively as representation for deterministic and probabilistic relationships between objects.

2.1 The Description Logic $\mathcal{ALC}$

In the popular description logic $\mathcal{ALC}$ [66] we have *individuals*, *concepts*, and *roles*, to be understood as constants, unary relations, and binary relations. Throughout, $a, b, a_1, a_2, \ldots$ are individuals; $C, D, C_1, C_2, \ldots$ are concepts; and $r, r_1, r_2, \ldots$ are roles. Concepts and roles can combined to form new concepts using a set of *constructors*: *intersection* $(C \sqcap D)$, *union* $(C \sqcup D)$, *complement* $(\neg C)$, *existential restriction* $(\exists r.C)$, and *value restriction* $(\forall r.C)$. *Concept inclusions/definitions* are denoted respectively by $C \sqsubseteq D$ and $C \equiv D$, where C and D are concepts. Concept $C \sqcup \neg C$ is denoted by $\top$, and concept $C \sqcap \neg C$ is denoted by $\bot$. Restrictions $\exists r.\top$ and $\forall r.\top$ are abbreviated by $\exists r$ and $\forall r$ respectively. A set of concept inclusions and definitions is a *terminology*. If an inclusion/definition contains a concept C in its left hand side and a concept D in its right hand side, C *directly uses* D. Indicate the transitive closure of *directly uses* by *uses*. A terminology is *acyclic* if it is a set of concept inclusions/definitions such that no concept in the terminology uses itself [2].

A terminology may be associated with *assertions* about individuals or pairs of individuals; for instance, Fruit(appleFromJohn) and buyFrom(houseBob, John). Intuitively, an assertion is the grounding of a unary/binary relation. A set of assertions is called an *Abox*.

The semantics of $\mathcal{ALC}$ is given by a nonempty set $\mathcal{D}$, the *domain*, and a mapping $\mathcal{I}$, the *interpretation*. An interpretation $\mathcal{I}$ maps each individual to an element of the domain, each concept name to a subset of the domain, each role name to a binary relation on $\mathcal{D} \times \mathcal{D}$. An interpretation is extended to other concepts as follows: $(\neg C)^{\mathcal{I}} = \mathcal{D} \backslash (C)^{\mathcal{I}}$, $(C \sqcap D)^{\mathcal{I}} = (C)^{\mathcal{I}} \cap (D)^{\mathcal{I}}$, $(C \sqcup D)^{\mathcal{I}} = (C)^{\mathcal{I}} \cup (D)^{\mathcal{I}}$, $(\exists r.C)^{\mathcal{I}} = \{x \in \mathcal{D} | \exists y \in \mathcal{D} : (x,y) \in (r)^{\mathcal{I}} \wedge y \in (C)^{\mathcal{I}}\}$, $(\forall r.C)^{\mathcal{I}} = \{x \in \mathcal{D} | \forall y \in \mathcal{D} : (x,y) \in (r)^{\mathcal{I}} \rightarrow y \in (C)^{\mathcal{I}}\}$. We have $C \sqsubseteq D$ if and only if $(C)^{\mathcal{I}} \subseteq (D)^{\mathcal{I}}$; and $C \equiv D$ if and only if $(C)^{\mathcal{I}} = (D)^{\mathcal{I}}$.

Most description logics have direct translations into multi-modal logics [65] and fragments of first-order logic [4]. We often treat a concept C as a unary predicate $C(x)$, and a role r as a binary predicate $r(x,y)$.

2.2 Bayesian Networks

Now consider Bayesian networks, a popular representation for probability distributions. A Bayesian network consists of a directed acyclic graph $\widehat{\mathbf{G}}$ where each node is a random variable V_i and where the following Markov condition is assumed [53]: every random variable V_i is independent of its nondescendants nonparents given its parents. For categorial variables $V_1, \ldots, V_n$, this Markov condition implies the following factorization for joint probabilities:

$$\mathbb{P}(V_1 = v_1, \ldots, V_n = v_n) = \prod_{i=1}^{n} \mathbb{P}(V_i = v_i | \mathsf{pa}(V_i) = \pi_i), \qquad (1)$$

where $\mathsf{pa}(V_i)$ denotes the parents of V_i in the graph, and π_i denotes the configuration of parents of random variable V_i. Note that if a random variable V_i has no parents, then the unconditional probability $\mathbb{P}(V_i = x_i)$ is used in Expression (1). We say that $\mathbb{P}$ factorizes according to $\widehat{\mathbf{G}}$ if $\mathbb{P}$ satisfies Expression (1).

2.3 Probabilistic Description Logics

There has been considerable interest in languages that mix probability assessments and constructs employed in description logics [44,61]. Early proposals by Heinsohn [24], Jaeger [27] and Sebastiani [67] adopt probabilistic inclusion axioms with a *domain-based* semantics; that is, probabilities are assigned to subsets of the domain. Proposals in the literature variously adopt a domain-based semantics [13,14,20,35,37,42,76], or an interpretation-based semantics where probabilities are assigned to sets of interpretations [6,21,43,45,67].

Several probabilistic description logics rely on graphs to encode stochastic independence relations. The first language to resort to Bayesian networks, P-CLASSIC, enlarges the logic CLASSIC with a set of Bayesian networks so as to specify a single probability measure over the domain [37]. A limitation is that P-CLASSIC does not handle assertions. Other logics that combine terminologies with Bayesian networks are Yelland's Tiny Description Logic [76], Ding and Peng's BayesOWL language [13], and Staker's logic [69] (none can handle assertions). Costa and Laskey's PR-OWL language [6] adopts an interpretation-based semantics inherited from multi-entity Bayesian networks (MEBNs) [5]. Another path is to consider undirected models, for instance based on Markov logic [49].

For the purposes of this paper, a particularly interesting class of languages has been produced by combining Poole's choice variables [57] with description logics [8,43,63].

Besides the literature just reviewed, there is a large body of work on knowledge databases [26,59] and on fuzzy description logics [44]; also notable is Nottelmann and Fuhr's probabilistic version of the OWL language [51].

2.4   Relational Bayesian Networks

Combinations of logic, probabilities and independence assumptions are not limited to description logics. They range from simple template languages [46,71,74], to rule-based languages akin to Prolog [48,56,64], and to more sophisticated languages such as multi-entity Bayesian networks [39] and Markov logic [62]. Research on probabilistic logics sometimes emphasizes automated learning [19,60]. The term *Probabilistic Relational Model* (PRM) is frequently associated with languages that combine Bayesian networks with relational logic [16,18,38]. Overall, these languages move beyond older probabilistic logics [3,9,50] by explicitly considering Markov conditions. For our purposes, relational Bayesian networks [28–30] offer the most relevant language, which we now discuss.

A *relational Bayesian network* is a compact, graph-based representation for a joint distribution over a set of random variables specified via relations and their groundings over a domain [28,29]. We start with a vocabulary $\mathcal{S}$ containing finitely many relations. We wish to specify a probability measure over the set of interpretations for these relations. To do so, we specify a directed acyclic graph **G** where each node is a relation in $\mathcal{S}$. Each relation s is then associated with a *probability formula* F_s. To understand these formulas, we must understand the intended semantics.

To define the semantics, consider a domain $\mathcal{D}$ (a set with individuals). An *interpretation* $\mathcal{I}$ is a function that takes each k-ary relation to a set of k-tuples of elements of $\mathcal{D}$. Now given a k-ary relation $s \in \mathcal{S}$ and a k-tuple $\mathbf{a} \in \mathcal{D}^k$, associate with the grounding $s(\mathbf{a})$ the indicator function

$$\mathbb{1}_{s(\mathbf{a})}(\mathcal{I}) = \begin{cases} 1 \text{ if } \mathbf{a} \in (s)^{\mathcal{I}}, \\ 0 \text{ otherwise.} \end{cases}$$

Note: to emphasize the connection between interpretations in $\mathcal{ALC}$ and in relational Bayesian networks, we used the notation $(s)^{\mathcal{I}}$ in this expression.

We extend this notation to any formula ϕ, indicating by $\mathbb{1}_\phi(\mathcal{I})$ the function that yields 1 if ϕ holds in $\mathcal{I}$, and 0 otherwise. We wish to specify a probability

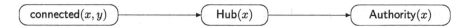

distribution over the set of all indicator variables $\mathbb{1}_\phi$. Probability formulas allow us to do so: a probability formula F_s specifies how to compute the distribution of $\mathbb{1}_{s(\mathbf{a})}$ for any appropriate tuple $\mathbf{a}$ of elements of $\mathcal{D}$.

Now return to the syntax. Jaeger restricted probability formulas to four constructs, defined recursively as follows [28,29]. First, a real number in $[0,1]$ is a probability formula. Second, a parameterized indicator function $\mathbb{1}_{s(\mathbf{x})}$, where $\mathbf{x}$ is a tuple of logical variables, is a probability formula. Third, $F_1 \times F_2 + (1 - F_1) \times F_3$, where F_1, F_2, and F_3 are probability formulas, is a probability formula.

Example 1. Consider the graph in Fig. 1. The assessment

$$\forall (x, y) \in \mathcal{D} \times \mathcal{D} : \mathbb{P}\big(\mathbb{1}_{\mathsf{connected}(x,y)} = 1\big) = 0.2$$

corresponds to the concise probability formula $F_{\mathsf{connected}}(x, y) = 0.2$. Consider also the assessments: for $\eta \in \{0, 1\}$,

$$\forall x \in \mathcal{D} : \mathbb{P}\big(\mathbb{1}_{\mathsf{Authority}(x)} = 1 \mid \mathbb{1}_{\mathsf{Hub}(x)} = \eta\big) = 0.9\eta + 0.2(1 - \eta).$$

These assessments correspond to the probability formula:

$$F_{\mathsf{Authority}}(x) = 0.9 \times \mathbb{1}_{\mathsf{Hub}(x)} + 0.2 \times (1 - \mathbb{1}_{\mathsf{Hub}(x)}).$$

Finally, consider the assessments: for $\eta \in \{0, 1\}$,

$$\forall x \in \mathcal{D} : \mathbb{P}\big(\mathbb{1}_{\mathsf{Hub}(x)} = 1 \mid \mathbb{1}_\phi = \eta\big) = 0.3\eta + 0.01(1 - \eta),$$

where ϕ is the first-order formula $\exists y \in \mathcal{D} : \mathsf{connected}(x, y)$. To encode this into a probability formula, we need to somehow quantify over y. To do so, Jaeger introduced a fourth construct, as we now describe. □

A *combination function* is any function that takes a tuple of numbers in $[0, 1]$ and returns a number in $[0, 1]$. Examples are:

$$\text{Noisy-OR}(A) = 1 - \prod_{p \in A}(1 - p), \qquad \text{Mean}(A) = \sum_{p \in A} p/|A|.$$

Tuples are specified as follows. Denote by $c(\mathbf{x}, \mathbf{y})$ a set of equality constraints containing logical variables in tuples $\mathbf{x}$ and $\mathbf{y}$. Denote by $\langle \mathbf{y} : c(\mathbf{x}, \mathbf{y}) \rangle$ the set of all groundings of $\mathbf{y}$ that satisfy $c(\mathbf{x}, \mathbf{y})$ for fixed $\mathbf{x}$. For each tuple $\mathbf{x}$, generate the set $\langle \mathbf{y} : c(\mathbf{x}, \mathbf{y}) \rangle$; now for each tuple in this set, evaluate $F_1, \ldots, F_k$. So if there are m tuples in $\langle \mathbf{y} : c(\mathbf{x}, \mathbf{y}) \rangle$ for fixed $\mathbf{x}$, then there are $k \times m$ elements in the resulting tuple. Denote by $\{F_1(\mathbf{x}, \mathbf{y}), \ldots, F_k(\mathbf{x}, \mathbf{y}); \langle \mathbf{y} : c(\mathbf{x}, \mathbf{y}) \rangle\}$ the tuple; note that not necessarily all variables in $(\mathbf{x}, \mathbf{y})$ appear in all probability formulas F_i.

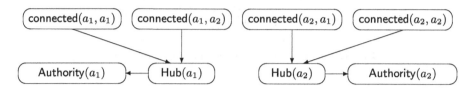

Fig. 2. Grounded graph $\widehat{\mathbf{G}}$ for Example 1.

Returning to Example 1, we can write the last assessment as:

$$F_{\mathsf{Hub}}(x) = \text{Noisy-OR}(\{\mathsf{connected}(x, y); \langle y : y = y\rangle\}).$$

Combination functions are quite powerful, but as a knowledge representation tool they have somewhat difficult syntax, and may be hard to understand.

We can take the graph $\mathbf{G}$ and the associated probability formulas, and generate a grounded graph $\widehat{\mathbf{G}}$. The nodes of $\widehat{\mathbf{G}}$ are indicator functions $\mathbb{1}_{s(\mathbf{a})}$ for all s and all appropriate $\mathbf{a}$. An edge is added to $\widehat{\mathbf{G}}$ from node $s_1(\mathbf{a}_1)$ to node $s_2(\mathbf{a}_2)$ if the probability formula for s_2 mentions the relation s_1. The Markov condition for Bayesian networks is then assumed for $\widehat{\mathbf{G}}$, and hence we obtain a factorization for the joint distribution of all indicator functions $\mathbb{1}_{s(\mathbf{a})}$.

Returning again to Example 1, Fig. 2 shows the grounded graph $\widehat{\mathbf{G}}$ for domain $\{a_1, a_2\}$. To simplify the figure, every indicator function $\mathbb{1}_{s(\mathbf{a})}$ is denoted simply by $s(\mathbf{a})$.

In this paper we do not consider *recursive relational Bayesian networks*, defined by Jaeger to allow typed relations and temporal evolution [28,29].

3   Credal $\mathcal{ALC}$

Credal $\mathcal{ALC}$ was proposed [7] as a syntactically simple probabilistic extension of $\mathcal{ALC}$. The syntax and associated semantics allows one to specify sets of probability measures over the set of interpretations for a given vocabulary.

In Sect. 3.1 we summarize the properties of the most flexible and general version of the language. Then in Sect. 3.2 we reformulate Credal $\mathcal{ALC}$ as a specification language for (a class of) relational Bayesian networks.

3.1   Credal $\mathcal{ALC}$ as a Flexible Mix of $\mathcal{ALC}$ and Probabilities

The language consists of well-formed $\mathcal{ALC}$ sentences, plus assessments

$$\mathbb{P}(C|D) \in [\underline{\alpha}, \overline{\alpha}], \qquad \mathbb{P}(r) \in [\underline{\beta}, \overline{\beta}],$$

where C is a concept name, D is a concept, and r is a role name. If the concept D is equal to $\top$ in the first assessment, we just write $\mathbb{P}(C) \in [\underline{\alpha}, \overline{\alpha}]$. Hence we have a subset of many existing probabilistic logics such as Lukasiewicz's conditional constraints [41]. The idea behind Credal $\mathcal{ALC}$ is to impose additional

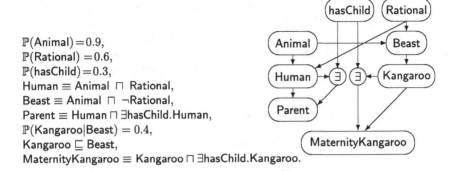

P(Animal)=0.9,
P(Rational) =0.6,
P(hasChild)=0.3,
Human ≡ Animal ⊓ Rational,
Beast ≡ Animal ⊓ ¬Rational,
Parent ≡ Human ⊓ ∃hasChild.Human,
P(Kangaroo|Beast) = 0.4,
Kangaroo ⊑ Beast,
MaternityKangaroo ≡ Kangaroo ⊓ ∃hasChild.Kangaroo.

Fig. 3. The Kangaroo network. Note that, as a visual aid, existential restrictions are explicitly shown as nodes in the graph, even though these nodes do not correspond directly to concepts in the terminology.

assumptions so that each consistent set of sentences can be grounded into a set of Bayesian networks, all sharing the same graph. Note that a set of probability distributions is often called a *credal set* [1], hence the name Credal $\mathcal{ALC}$.

We first need to have an appropriate concept of acyclicity. We adopt the relations *directly uses* and *uses* from $\mathcal{ALC}$, and extend them as follows. Given the assessment $\mathbb{P}(C|D) \in [\underline{\alpha}, \overline{\alpha}]$, we say that C *directly uses* D. And *uses* is the transitive closure of *directly uses*. Now for any concept C in the terminology, denote by $\mathsf{pa}(C)$ the set of concepts directly used by C, and by $\mathsf{nd}(C)$ the set of concepts that do not use C. Say that an assertion directly uses another one if the corresponding concepts satisfy the *directly uses* relation. Again, a terminology is acyclic if no concept in the terminology uses itself.

Given an acyclic terminology, we can draw a directed acyclic graph where nodes are concept and role names, and arrows encode the *directly uses* relation. For instance, Fig. 3 shows a probabilistic version of the Kangaroo ontology and the associated graph.[1]

We assume that every terminology is acyclic. By adopting acyclicity, we can define an appropriate Markov condition; to so, we need to examine the semantics.

First, all $\mathcal{ALC}$ constructs have the usual semantics, based on a domain and interpretations. As in many probabilistic logics [22,23], we adopt an assumption of *rigidity*: the interpretation of an individual does not change across interpretations. We also adopt throughout the *unique name assumption*: distinct individual names refer to distinct elements of the domain. Thus we can equate individuals with elements of the domain.

To define the semantics of probabilistic assessments, take a domain $\mathcal{D}$. Then define, for any individual a and concept C, a set of interpretations

$$\langle\!\langle C(a) \rangle\!\rangle = \{\mathcal{I} : a \in (C)^{\mathcal{I}}\}.$$

---

[1] The Kangaroo ontology is distributed with the CEL System at the site http://lat. inf.tu-dresden.de/systems/cel/.

Here $\langle\!\langle C(a)\rangle\!\rangle$ depends on $\mathcal{D}$, but we leave this dependency implicit to avoid burdening the notation (the domain of interest can be inferred from the context). Similarly, define the set of interpretations $\langle\!\langle r(a,b)\rangle\!\rangle = \{\mathcal{I} : (a,b) \in (r)^{\mathcal{I}}\}$, for any pair of individuals (a,b) and any role r. The semantics of a set of assessments is given by a probability measure $\mathbb{P}$ over the set of interpretations: each assessment $\mathbb{P}(C|D) \in [\underline{\alpha}, \overline{\alpha}]$ means

$$\forall x \in \mathcal{D} : \mathbb{P}(\langle\!\langle C(x)\rangle\!\rangle | \langle\!\langle D(x)\rangle\!\rangle) \in [\underline{\alpha}, \overline{\alpha}],$$

and each assessment $\mathbb{P}(r) \in [\underline{\beta}, \overline{\beta}]$ means

$$\forall (x,y) \in \mathcal{D} \times \mathcal{D} : \mathbb{P}(\langle\!\langle r(x,y)\rangle\!\rangle) \in [\underline{\beta}, \overline{\beta}].$$

Note that we abuse notation by using the same symbol $\mathbb{P}$ in the syntax and the semantics.

Given a terminology and a domain, we can always construct a directed acyclic graph $\widehat{\mathbf{G}}$ where nodes are all possible assertions, and where arrows encode the *directly uses* relation. As in Fig. 3, we add a node to $\widehat{\mathbf{G}}$ for each existential (or universal) quantifier in the terminology. Thus $\widehat{\mathbf{G}}$ is the grounding of the terminology for the given domain.

For example, consider a terminology with the following axioms and assessments (based on Ref. [7]): $\mathbb{P}(A) = 0.9$, $B \sqsubseteq A$, $D \equiv \forall r.A$, $C \equiv B \sqcup \exists r.D$, $\mathbb{P}(r) = 0.3$. The terminology can be drawn as a directed acyclic graph as in Fig. 4. Now suppose we have a domain with just two individuals, a and b. The grounded graph $\widehat{\mathbf{G}}$ is also shown in Fig. 4.

With this, we can state a *Markov condition:*[2] $\langle\!\langle C(a)\rangle\!\rangle$ is independent of all $\langle\!\langle ND\rangle\!\rangle$ where ND is a nondescendant nonparent of $C(a)$ in $\widehat{\mathbf{G}}$, given all $\langle\!\langle PA\rangle\!\rangle$ where PA is a parent of $C(a)$ in $\widehat{\mathbf{G}}$. That is, we just have the usual Markov condition for the grounded directed acyclic graph.

In practice it may be useful to adopt a number of assumptions that imply that a terminology can always be grounded into a single probability measure over interpretations (for instance, each assessment collapses to a single number). Indeed, Cozman and Polastro [7] have identified a number of assumptions that together guarantee uniqueness; some of these assumptions are easy to grasp, while others are quite convoluted.

We present in the next section a syntax, and a set of associated assumptions, that guarantees that any set of well-formed sentences can be grounded into a single Bayesian network given a finite domain. We do so by framing Credal $\mathcal{ALC}$ as a language for specification of relational Bayesian networks.

---

[2] We use the following concept of independence: an event E is independent of a set of events $\{F_i\}_i$ given a set of events $\{G_j\}_j$ if $\mathbb{P}(E \cap H'|H'') = \mathbb{P}(E|H'')\mathbb{P}(H'|H'')$ for any $H' = \cap_{i \in I} F_i$ and any nonempty $H'' = (\cap_{j \in J} G_j) \cap (\cap_{k \in K} G_k^c)$, for any subsets of indexes I, J, K.

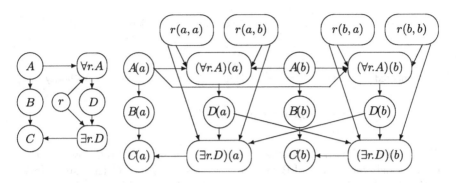

Fig. 4. Left: directed acyclic graph representing terminology. Right: grounding of the terminology for $\mathcal{D} = \{a, b\}$.

3.2 Credal $\mathcal{ALC}$ as Specification Language for Relational Bayesian Networks

As before, consider a vocabulary $\mathcal{S}$ containing individuals, concepts, and roles. A CR$\mathcal{ALC}$ *network* consists of a directed acyclic graph **G** where each node is either a concept name or a relation name, and where each node is associated either with

- a direct assessment: $\mathbb{P}(C) = \alpha$ if the node is a concept C, or $\mathbb{P}(r) = \alpha$ if the node is a role r, for $\alpha \in [0, 1]$; or
- a definition $C \equiv \phi$, if the node is a concept C, that must be a well-formed definition in $\mathcal{ALC}$ whose right-hand side only refers to parents of C.

To establish semantics for CR$\mathcal{ALC}$ networks, we translate this syntax directly into relational Bayesian networks. Consider a domain $\mathcal{D}$. Concepts and roles are viewed as unary and binary relations, and the semantics is given by a probability measure over $\mathcal{I}$, the set of interpretations of these relations. Additionally:

- A direct assessment $\mathbb{P}(C) = \alpha$, where C is a concept, is interpreted just as the probability formula $F_C(x) = \alpha$; that is, as

$$\forall x \in \mathcal{D} : \mathbb{P}(\mathbb{1}_{C(x)} = 1) = \alpha.$$

- A direct assessment $\mathbb{P}(r) = \alpha$, where r is a role, is interpreted just as the probability formula $F_r(x, y) = \alpha$; that is, as

$$\forall (x, y) \in \mathcal{D} \times \mathcal{D} : \mathbb{P}(\mathbb{1}_{r(x,y)} = 1) = \alpha.$$

The semantics of a definition $C \equiv \phi$ is immediate: for all $x \in \mathcal{D}$, every interpretation satisfies $C(x) \leftrightarrow \phi(x)$, where $\phi(x)$ is the translation of ϕ to first-order logic, mapping intersection to conjunction, complement to negation, and so on. As a digression, note that such definitions can be expressed through probability functions, as any first-order logic formula can be encoded through Jaeger's probability formulas [28, Lemma 2.4].

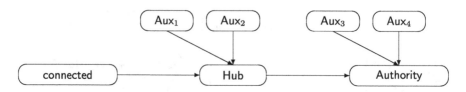

Fig. 5. Simple CR$\mathcal{ALC}$ network.

Now consider the graph $\widehat{\mathbf{G}}$ where each node is a grounded relation, exactly as done previously for relational Bayesian networks. Attach to $\widehat{\mathbf{G}}$ the usual Markov condition for Bayesian networks. The semantics of a CR$\mathcal{ALC}$ network is given by a probability measure that factorizes in accordance with this Markov condition.

Example 2. Fig. 5 shows a CR$\mathcal{ALC}$ network with seven variables, four of which are auxiliary variables Aux_i, where the same probabilities in Example (1) can be obtained as follows:

$$\mathbb{P}(\mathsf{connected}) = 0.2,$$

$$\mathsf{Hub} \equiv (\mathsf{Aux}_1 \sqcap \exists\mathsf{connected}) \sqcup (\mathsf{Aux}_2 \sqcap \neg\exists\mathsf{connected}),$$

$$\mathsf{Authority} \equiv (\mathsf{Aux}_3 \sqcap \mathsf{Hub}) \sqcup (\mathsf{Aux}_4 \sqcap \neg\mathsf{Hub}),$$

$$\mathbb{P}(\mathsf{Aux}_1) = 0.3, \quad \mathbb{P}(\mathsf{Aux}_2) = 0.01, \quad \mathbb{P}(\mathsf{Aux}_3) = 0.9, \quad \mathbb{P}(\mathsf{Aux}_4) = 0.2.$$

The grounded graph in Fig. 2 is again obtained for domain $\{a_1, a_2\}$. □

This example shows that, if properly combined, direct assessments and definitions can be used to build conditional probabilities. In fact, suppose we have a CR$\mathcal{ALC}$ network where C and D are concepts, D is the only parent of C, and we wish to express that, for all $x \in \mathcal{D}$,

$$\mathbb{P}\big(\mathbb{1}_{C(x)} = 1 | \mathbb{1}_{D(x)} = 1\big) = \alpha \quad \text{and} \quad \mathbb{P}\big(\mathbb{1}_{C(x)} = 1 | \mathbb{1}_{D(x)} = 0\big) = \beta.$$

Introduce two fresh concept names C_1 and C_2, add them as parents of C, but leave them without parents; then specify:

$$C \equiv (C_1 \sqcap D) \sqcup (C_2 \sqcap \neg D), \qquad \mathbb{P}(C_1) = \alpha, \qquad \mathbb{P}(C_2) = \beta. \qquad (2)$$

Note that, as desired, $\mathbb{P}\big(\mathbb{1}_{C(x)} = 1 | \mathbb{1}_{D(x)} = 1\big)$ is equal to

$$\sum_{\mathbb{1}_{C_1(x)}, \mathbb{1}_{C_2(x)}} \mathbb{P}\big(\mathbb{1}_{C(x)} = 1 | \mathbb{1}_{C_1(x)}, \mathbb{1}_{C_2(x)}, \mathbb{1}_{D(x)} = 1\big) \mathbb{P}\big(\mathbb{1}_{C_1(x)}\big) \mathbb{P}\big(\mathbb{1}_{C_2(x)}\big)$$

$$= \sum_{\mathbb{1}_{C_2(x)}} \mathbb{P}\big(\mathbb{1}_{C_1(x)} = 1\big) \mathbb{P}\big(\mathbb{1}_{C_2(x)}\big) = \alpha.$$

By similar reasoning, we see that $\mathbb{P}\big(\mathbb{1}_{C(x)} = 1 | \mathbb{1}_{D(x)} = 0\big) = \beta$. This sort of device let us specify conditional probability tables for a concept conditional on arbitrarily many concepts. In fact we can even introduce some syntactic sugar by writing the set of constructs in Expression (2) directly as $C \equiv \alpha D \sqcup \beta(\neg D)$. Moreover, there is no need to draw auxiliary variables in our graphs, as the next example shows.

Fig. 6. Simple CR*ALC* network without auxiliary variables.

Example 3. Consider again the CR*ALC* network in Example 2. The graph can be drawn as in Fig. 6, with additional concise assessments:

$$\mathbb{P}(\text{connected}) = 0.2, \quad \begin{matrix} \text{Hub} \equiv 0.3\ (\exists\text{connected}) \sqcup \\ 0.01\ (\neg\exists\text{connected}), \end{matrix} \quad \begin{matrix} \text{Authority} \equiv 0.9\ (\text{Hub}) \sqcup \\ 0.2\ (\neg\text{Hub}). \end{matrix}$$

Again, for domain $\{a_1, a_2\}$, the grounded graph is depicted in Fig. 2. □

Another example is given by the Kangaroo network in Fig. 3. We can replace both the assessment $\mathbb{P}(\text{Kangaroo}|\text{Beast}) = 0.4$ and the inclusion Kangaroo $\sqsubseteq$ Beast by

$$\text{Kangaroo} \equiv 0.4\text{Beast}.$$

Note that auxiliary variables resemble *choice* construct in ICL [57], but the semantics for the latter employs different assumptions regarding negation. Similarly, auxiliary variables resemble *switches* in PRISM [64]; the latter are more general and take parameters. Finally, auxiliary variables resemble the *exogenous variables* used in structural equations [54]; of course the latter are couched in terms of algebraic modeling.

We can even extend the previous constructs if we have three concepts C_1, C_2 and C_3. We can then use the definition

$$C \equiv (C_1 \sqcap C_2) \sqcup ((\neg C_1) \sqcap C_3),$$

similarly to Jaeger's probability formula $F_{C_1} F_{C_2} + (1 - F_{C_1}) F_{C_3}$.

Now consider the final construct in Jaeger's relational Bayesian networks; that is, combination functions. We do not have them in full generality here. For a fixed and finite domain, the node $\exists r.C$ is in fact a Noisy-OR: the indicator function $\mathbb{1}_{(\exists r.C)(x)}$ is a disjunction of all conjunctions of nodes $C(y)$ with inhibitor nodes $r(x, y)$, for all $y \in \mathcal{D}$. Likewise a node $\forall r.C$ can be written as a conjunction of implications.

Hence, as far as finite fixed domains are concerned, the syntax for CR*ALC* networks presented here can be viewed as syntactic sugar for relational Bayesian networks that only contain unary and binary relations, where binary relations have no parents, and where combination functions are restricted to Noisy-OR. What reasons can we offer to study such a subset of relational Bayesian networks? At the risk of repeating arguments already stated, we offer the following answers. First, the syntax and semantics of CR*ALC* networks are much easier to grasp and to use than general Credal *ALC*, and also easier to grasp than general relational Bayesian networks — in our experience, the syntax and semantics of Jaeger's probability functions are somewhat difficult for the novice. Second, techniques honed by current research for both description logics and relational Bayesian networks can be used. Third, theoretical results about description logics can lead to novel interesting results about relational Bayesian networks and related languages. We do not focus on the latter topic, as it has been examined before [7].

3.3   A Few Applications

We now briefly present examples of CR$\mathcal{ALC}$ networks that have been described in previous publications.

First, Polastro and Correa [55] have examined the use of CR$\mathcal{ALC}$ networks in robot navigation, a task where high level reasoning is important [17,25]. Deterministic facts were encoded in sentences such as

Office $\equiv$ Room $\sqcap$ $\exists$contains.Desk $\sqcap$ $\exists$contains.Cabinet $\sqcap$ $\exists$contains.Monitor

together with probabilistic assessments obtained by experimental analysis. During operation, images were collected and inferences were run; for instance, an image containing 3 chairs, 1 table, 1 monitor, 1 cabinet and 1 door was taken and the inference $\mathbb{P}(\text{Office}|\text{detected objects}) = 0.4278$ was computed. A similar application of CR$\mathcal{ALC}$ networks in spatial reasoning has been reported by Fenelon et al. [15].

A second example is link prediction, where the goal is to predict whether there is a link between two nodes in a social network [40]. Description logics are particularly well suited to handle social networks, because they deal with concepts (applied to single nodes) and roles (applied to pairs of nodes). Ochoa-Luna, Revoredo and Cozman [52] employed a CR$\mathcal{ALC}$ network to predict links in a social network consisting of researchers, where links indicate co-authorship. Approximate variational inference was used to compute the probability of various links between individuals. The combination of topological features and probabilistic reasoning led to accuracy of 89.5 % in link prediction, compared to 85.6 % accuracy with topological features only.

4   Approximate Variational Inference with CR$\mathcal{ALC}$ Networks

Given a CR$\mathcal{ALC}$ network, a finite domain $\mathcal{D}$, and an Abox $\mathcal{A}$, consider the query $Q = \mathbb{P}(C(a)|\mathcal{A})$. We refer to the computation of Q as an *inference*. One obvious idea is to ground the CR$\mathcal{ALC}$ network into a Bayesian network, and compute Q there by any existing algorithm [16,74]. However, the size of the grounded Bayesian network may become too large as the size of the domain grows. A solution is to approximate Q. One can ground the CR$\mathcal{ALC}$ network and run approximate inference there. It is not even necessary to generate the whole grounded Bayesian network, but only the groundings that are relevant to the query. However, for large domains it may be difficult to even generate the grounded network. A promising alternative is to avoid grounding concepts and roles, by noting that many of these grounding lead to identical computations.

So, consider running belief propagation [36] in the ground network. As many messages are actually identical, it makes sense to group them [32,68]. An additional step is to group nodes and run exact inference within each group, and to treat each group as a single variable as far as messages are concerned. That is, we have belief propagation among groups, and exact inference within groups; the messages can be derived using a variational scheme [75].

The idea then is to take the CR$\mathcal{ALC}$ network and to divide the ground Bayesian network in *slices*. Each slice congregates groundings with respect to a single individual: assertion $C(a)$ clearly belongs to the slice of a, and assertion $r(a, y)$ belongs to the slice of a for any $y \in \mathcal{D}$. For instance, in Fig. 1 we have two slices, one for individual a and another for individual b. We now consider messages in a variational scheme where each slice is a group [75]; to do so, we introduce a node for each function in the Bayesian network, and connect this node to the variables that affect the function (that is, we create the *factor graph*). Now we have messages sent to and from these function-nodes; for instance, suppose that node $A(a)$ is to send a message to the function-node that represents $f(c)$ where c is some individual that appears in the evidence. This message is $\prod_g m_{g,A(a)}(A(a))$, where $m_{g,A(a)}$ are the messages sent to $A(a)$ from function-nodes g, as g sweeps through the set of function-nodes connected to $A(a)$ except f. Similar messages are then sent to $A(a)$, and so on. Note that as we are grouping variables in each slice, and assuming that exact inference is run inside a slice, only messages between slices must be exchanged.

Now if all messages were exchanged amongst all possible slices until convergence, we would have a variant of belief propagation run in the fully grounded Bayesian network. But note that many slices are exchangeable. In particular, every element of the domain for which there is no evidence leads to an identical slice. Thus we can put together all such slices: everytime a message is to be sent from this "combined" slice, we simple compute the message that would be sent by a single slice, and raise it to the number of slices in the "combined" slice.

To summarize, approximate inferences are produced by generating a set of grounded Bayesian networks, one for each slice mentioned in the query and in the evidence, plus an additional Bayesian network for a "generic" individual; a detailed description of this algorithm can be found in Ref. [7]. Exact Bayesian network inference is performed in each one of these networks and messages are exchanged between the networks. The whole scheme resembles the RCR framework [73]. Experiments indicate that this approach leads to fast and accurate approximations [7].

A package implementing this algorithm has been coded by the second author using the Java language (version 6.0), and can work either from the command prompt or through a graphical user interface.[3] Once the package is downloaded and uncompressed, it must be run together with the jar file JavaBayes.jar (in the folder Proj01\libs). The distribution comes with the file CrALC.java; this file must be compiled and run.

---

[3] The package is freely available, in compressed form, at the site http://sites.poli.usp. br/pmr/ltd/Software/CRALC/inf-cralc-v21may2012.zip.

The package can be run from the command prompt with some additional parameters:

- gui {on|off}: loads (or not) the graphical user interface (default is on);
- i input_file N: loads terminology in input_file and sets domain size to N;
- p output_file: saves a ground Bayesian network into output_file;
- e query_file: loads query and evidence in query_file;
- r name: saves inference in file name.txt, and each slice in auxiliary file.

To describe CR$\mathcal{ALC}$ networks as input, we have chosen to adapt the Knowledge Representation System Specification (KRSS).[4] We use the following constructs: (and C1...Cn) for conjunction; (or C1...Cn) for disjunction; (not C) for complement; (all r C) to indicate the quantifier $\forall r.C$; (some r C) to indicate the quantifier $\exists r.C$; (define-concept C D) for $C \equiv D$. The package also allows (define-primitive-concept C D) for $C \sqsubseteq D$, in case the user wishes to use this construct (note that doing so is somewhat risky as it may generate a terminology that cannot be grounded into a unique Bayesian network). Probabilistic assessments are specified as (probability B α), denoting $P(B) = \alpha$. The package also allows (conditional-probability B A α) for $P(A|B) = \alpha$, even though such an assessment is not strictly necessary in CR$\mathcal{ALC}$ networks.

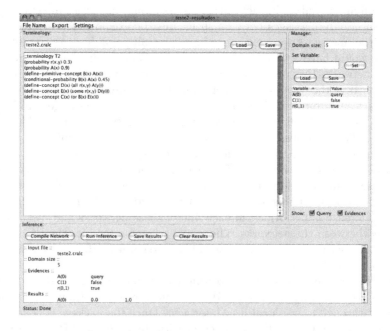

Fig. 7. Terminologies are written in the larger panel, while assertions are set in the right panel; the lower panel reports on inferences.

---

[4] The standard specification of KRSS can be found at http://dl.kr.org/krss-spec.ps.

An example of valid input file is:

```
;; This is a comment...
(probability A(x) 0.7)
(probability B(x) 0.4)
(define-concept C(x) (and A(x) (not B(x))))
```

Assertions can be inserted through simple files as well; for instance:

```
A(1)    query
B(0)    true
r(0,1)  false
```

The graphical user interface depicted in Fig. 7 can be used to load/save files, to specify the size of the domain and the assertions, to ask for inferences, and to check results. The interface lets the user insert domain size and assertions. Each assertion is inserted either as $C(i)$, where C is a concept and i is an integer, or as $r(i, j)$, where r is a role and i and j are integers.

5   Exact Lifted Inference with CR$\mathcal{ALC}$ Networks

Because any CR$\mathcal{ALC}$ network is a relational Bayesian network, we can use any algorithm that runs *lifted* inference in relational Bayesian networks. Here lifted means that inference is performed without completely grounding all concepts and roles [58]. There has been significant effort in exact lifted inference [10–12,31,47,70,72]. In our case we are interested in lifted inference in the presence of "aggregation parfactors" [33]; that is, in the presence of terms that aggregate the effect of many random variables, such as quantifiers.

Kisynski and Poole's AC-FOVE algorithm, an extension of first-order variable elimination [33,34], is currently the state-of-art for lifted inference with aggregation parfactors. We still focus on finite domains, and in this case each construct in a CR$\mathcal{ALC}$ network can be translated either into parameterized functions or into aggregation parfactors. For instance, an existential quantifier $\exists r.C$ can be encoded into an aggregation parfactor that yields 1 when the number of instances of $r(x, y) \wedge C(y)$ is larger than one, and 0 otherwise. AC-FOVE then applies a set of rules to the functions and parfactors. Each rule transforms a function or parfactor until only the query is present. The basic rule is *lifted elimination*, where all instances of a relation (concept or role, in our setting) are eliminated at once, without any actual grounding. For lifted elimination to be applied, several conditions must be met. When these conditions do not hold, AC-FOVE has several options: it can split groundings into groups, unify several groundings, multiply functions after unification, or exponentiate probabilities so as to account for exchangeable elements of the domain. Finally, AC-FOVE can use *counting formulas* [47]; that is, it can use random variables whose values indicate how many individuals satisfy a given condition (rather than dealing with all individuals separately). AC-FOVE is a greedy algorithm that chooses, at each step, one of these operations, resorting to grounding as a last resource.

A package implementing the AC-FOVE algorithm has been coded by the third author using the Java language (version 6.0). The package consists of an API that must be called as needed. Not only the AC-FOVE algorithm is implemented, but also the variable elimination and the C-FOVE algorithms.[5]

To illustrate, consider a few code fragments. First, the construction of a domain:

```
// Creates a list of individuals
List<Constant> individuals = new ArrayList<Constant>();
individuals.add(Constant.getInstance("a1"));
individuals.add(Constant.getInstance("a2"));
individuals.add(Constant.getInstance("a3"));
// Creates a population based on the list
Population popul  = Population.getInstance(individuals);
// Creates a logical variable bound to the population
LogicalVariable x = StdLogicalVariable.getInstance("x", popul);
```

An existential quantifier leads to an aggregation parfactor as follows:

```
Parfactor g = new AggParfactorBuilder(p, c, Or.OR).build();
```

Assertions can be inserted as evidence:

```
List<BigDecimal> fEvidence = TestUtils.toBigDecimalList(0.0, 1.0);
Parfactor evidence = new StdParfactorBuilder().variables(sprinkler).
values(fEvidence).build();
```

Finally, to run the algorithm:

```
ACFOVE acfove = new ACFOVE(input);
Parfactor result = acfove.run();
```

6   Conclusion

In this paper we have presented a syntax for CR$\mathcal{ALC}$ networks, with two goals in mind. First, it simplifies knowledge representation when probabilistic assessments must be coupled with the $\mathcal{ALC}$ logic. Second, it offers a description language for a useful class of relational Bayesian networks. The resulting language avoids many complexities of CR$\mathcal{ALC}$ and of general relational Bayesian networks and can be easily grasped by a user.

We also described freely available packages that implement approximate and exact inference for CR$\mathcal{ALC}$ networks. The software packages we have presented still require substantial development, but they are steps in a direction we feel has not received enough attention. As efficient inference is a key to combinations of uncertainty and semantic information, we hope that these efforts may be useful in future applications. Clearly there are many paths for future work; for instance, the study of open-world reasoning, infinite domains, and interval probabilities.

---

[5] The package is freely available at https://github.com/ftakiyama/AC-FOVE, where source code and examples can be found.

Acknowledgements. The first author was partially supported by CNPq. The second author was supported by FAPESP. The work reported here has received substantial support by FAPESP grant 2008/03995-5.

References

1. Augustin, T., Coolen, F.P.A., de Cooman, G., Troffaes, M.C.M.: Introduction to Imprecise Probabilities. Wiley, Chichester (2014)
2. Baader, F., Calvanese, D., McGuinness, D.L., Nardi, D., Patel-Schneider, P.F.: Description Logic Handbook. Cambridge University Press, Cambridge (2002)
3. Boole, G.: The Laws of Thought. Dover edition, New York (1958)
4. Borgida, A.: On the relative expressiveness of description logics and predicate logics. Artif. Intell. **82**(1–2), 353–367 (1996)
5. Costa, P.C.G., Laskey, K.B.: Of Klingons and starships: Bayesian logic for the 23rd century. In: Conference on Uncertainty in Artificial Intelligence (2005)
6. Costa, P.C.G., Laskey, K.B.: PR-OWL: a framework for probabilistic ontologies. In: Conference on Formal Ontology in Information Systems (2006)
7. Cozman, F.G., Polastro, R.B.: Complexity analysis and variational inference for interpretation-based probabilistic description logics. In: Conference on Uncertainty in Artificial Intelligence, pp. 117–125. AUAI Press (2009)
8. d'Amato, C., Fanizzi, N., Lukasiewicz, T.: Tractable reasoning with Bayesian description logics. In: Greco, S., Lukasiewicz, T. (eds.) SUM 2008. LNCS (LNAI), vol. 5291, pp. 146–159. Springer, Heidelberg (2008)
9. de Finetti, B.: Theory of Probability, vol. 1–2. Wiley, New York (1974)
10. de Salvo Braz, R., Amir, E., Roth, D.: Lifted first-order probabilistic inference. In: International Joint Conference in Artificial Intelligence (2006)
11. de Salvo Braz, R., Amir, E., Roth, D.: Lifted first-order probabilistic inference. In: Getoor, L., Taskar, B. (eds.) An Introduction to Statistical Relational Learning, pp. 433–451. MIT Press, Cambridge (2007)
12. de Salvo Braz, R., Amir, E., Roth, D.: A survey of first-order probabilistic models. In: Holmes, D.E., Jain, L.C. (eds.) Innovations in Bayesian Networks. Studies in Computational Intelligence, pp. 289–317. Springer, Heidelberg (2008)
13. Ding, Z., Peng, Y., Pan, R.: BayesOWL: uncertainty modeling in semantic web ontologies. In: Ma, Z. (ed.) Soft Computing in Ontologies and Semantic Web, pp. 3–29. Springer, Heidelberg (2006)
14. Dürig, M., Studer, T.: Probabilistic ABox reasoning: preliminary results. In: Description Logics, pp. 104–111 (2005)
15. Fenelon, V., Hummel, B., Santos, P.E., Cozman, F.G.: Encoding spatial domains with relational Bayesian networks. In: Spatio-temporal Dynamics Workshop, pp. 49–54 (2010)
16. Friedman, N., Getoor, L., Koller, D., Pfeffer, A.: Learning probabilistic relational models. In: International Joint Conference on Artificial Intelligence, pp. 1300–1309 (1999)
17. Galindo, C., Fernandez-Madrigal, J.-A., Gonzalez, J., Saffiotti, A.: Robot task planning using semantic maps. Robot. Auton. Syst. **11**, 955–966 (2008)
18. Getoor, L., Friedman, N., Koller, D., Taskar, B.: Learning probabilistic models of relational structure. In: International Conference on Machine Learning, pp. 170–177 (2001)
19. Getoor, L., Taskar, B.: Introduction to Statistical Relational Learning. MIT Press, Cambridge (2007)

20. Giugno, R., Lukasiewicz, T.: P-SHOQ(D): a probabilistic extension of SHOQ(D) for probabilistic ontologies in the semantic web. In: Flesca, S., Greco, S., Leone, N., Lanni, G. (eds.) European Conference on Logics in Artificial Intelligence, pp. 86–97 (2002)

21. Gutierrez-Basulto, V., Jung, J.C., Lutz, C., Schröder, L.: A closer look at the probabilistic description logic prob-$\mathcal{EL}$. In: Burgard, W., Roth, D. (eds.) Conference on Artificial Intelligence, pp. 197–202 (2011)

22. Halpern, J.Y.: An analysis of first-order logics of probability. Artif. Intell. **46**, 311–350 (1990)

23. Halpern, J.Y.: Reasoning about Uncertainty. MIT Press, Cambridge (2003)

24. Heinsohn, J.: Probabilistic description logics. In: Conference on Uncertainty in Artificial Intelligence, pp. 311–318 (1994)

25. Hertzberg, J., Saffiotti, A.: Using semantic knowledge in robotics. Robot. Auton. Syst. **56**, 875–877 (2008)

26. Hung, E., Getoor, L., Subrahmanian, V.S.: Probabilistic interval XML. ACM Trans. Comput. Logic **8**(4), 1–38 (2007)

27. Jaeger, M.: Probabilistic reasoning in terminological logics. In: Principles of Knowledge Representation, pp. 461–472 (1994)

28. Jaeger, M.: Relational Bayesian networks. In: Geiger, D., Shenoy, P.P. (eds.) Conference on Uncertainty in Artificial Intelligence, pp. 266–273. Morgan Kaufmann (1997)

29. Jaeger, M.: Complex probabilistic modeling with recursive relational Bayesian networks. Ann. Math. Artif. Intell. **32**, 179–220 (2001)

30. Jaeger, M.: Relational Bayesian networks: a survey. Linkoping Electronic Articles in Computer and Information Science, 6 (2002)

31. Kersting, K.: Lifted probabilistic inference. In: De Raedt, L., Bessiere, C., Dubois, D., Doherty, P., Frasconi, P., Heintz, F., Lucas, P. (eds.) European Conference on Artificial Intelligence. IOS Press (2012)

32. Kersting, K., Ahmadi, B., Natarajan, S.: Counting belief propagation. In: Conference on Uncertainty in Artificial Intelligence. AUAI Press (2009)

33. Kisynski, J.J., Poole, D.: Lifted aggregation in directed first-order probabilistic models. In: International Joint Conference on Artificial Intelligence, pp. 1922–1929 (2009)

34. Kisynski, J.J.: Aggregation and constraint processing in lifted probabilistic inference. Ph.D. thesis, Computer Science, University of British Columbia (2010)

35. Klinov, P., Parsia, B.: A hybrid method for probabilistic satisfiability. In: Bjørner, N., Sofronie-Stokkermans, V. (eds.) CADE 2011. LNCS, vol. 6803, pp. 354–368. Springer, Heidelberg (2011)

36. Koller, D., Friedman, N.: Probabilistic Graphical Models: Principles and Techniques. MIT Press, Cambridge (2009)

37. Koller, D., Pfeffer, A.: Object-oriented Bayesian networks. In: Conference on Uncertainty in Artificial Intelligence, pp. 302–313 (1997)

38. Koller, D., Pfeffer, A.: Probabilistic frame-based systems. In: AAAI, pp. 580–587 (1998)

39. Laskey, K.B.: MEBN: a language for first-order Bayesian knowledge bases. Artif. Intell. **172**(2–3), 140–178 (2008)

40. Liben-Nowell, D., Kleinberg, J.: The link prediction problem for social networks. J. Am. Soc. Inform. Sci. Technol. **7**(58), 1019–1031 (2007)

41. Lukasiewicz, T.: Probabilistic description logic programs. Int. J. Approx. Reason. **45**(2), 288–307 (2007)

42. Lukasiewicz, T.: Expressive probabilistic description logics. Artif. Intell. **172**(6–7), 852–883 (2008)
43. Lukasiewicz, T., Predoiu, L., Stuckenschmidt, H.: Tightly integrated probabilistic description logic programs for representing ontology mappings. Ann. Math. Artif. Intell. **63**(3/4), 385–425 (2011)
44. Lukasiewicz, T., Straccia, U.: Managing uncertainty and vagueness in description logics for the semantic web. J. Web Semant. **6**, 291–308 (2008)
45. Lutz, C., Schröder, L.: Probabilistic description logics for subjective uncertainty. In: Principles of Knowledge Representation and Reasoning, pp. 393–403. AAAI Press (2010)
46. Mahoney, S., Laskey, K.B.: Network engineering for complex belief networks. In: Conference on Uncertainty in Artificial Intelligence (1996)
47. Milch, B., Zettlemoyer, L.S., Kersting, K., Haimes, M., Kaelbling, L.P.: Lifted probabilistic inference with counting formulas. In: AAAI, pp. 1062–1068 (2008)
48. Ngo, L., Haddawy, P.: Answering queries from context-sensitive probabilistic knowledge bases. Theor. Comput. Sci. **171**(1–2), 147–177 (1997)
49. Niepert, M., Noessner, J., Stuckenschmidt, H.: Log-linear description logics. In: International Joint Conference on Artificial Intelligence (2011)
50. Nilsson, N.J.: Probabilistic logic. Artif. Intell. **28**, 71–87 (1986)
51. Nottelmann, H., Fuhr, N.: Adding probabilities and rules to OWL lite subsets based on probabilistic datalog. Int. J. Uncertain. Fuzziness Knowl. Based Syst. **14**(1), 17–42 (2006)
52. Ochoa-Luna, J.E., Revoredo, K.C., Cozman, F.G.: An experimental evaluation of a scalable probabilistic description logics approach for semantic link prediction. In: International Workshop on Uncertainty Reasoning for the Semantic Web, Shangai, China, pp. 63–74 (2012). http://ceur-ws.org
53. Pearl, J.: Probabilistic Reasoning in Intelligent Systems: Networks of Plausible Inference. Morgan Kaufmann, San Mateo (1988)
54. Pearl, J.: Causality: Models, Reasoning, and Inference. Cambridge University Press, New York (2000)
55. Polastro, R., Corrêa, F., Cozman, F., Okamoto Jr., J.: Semantic mapping with a probabilistic description logic. In: da Rocha Costa, A.C., Vicari, R.M., Tonidandel, F. (eds.) SBIA 2010. LNCS (LNAI), vol. 6404, pp. 62–71. Springer, Heidelberg (2010)
56. Poole, D.: Probabilistic Horn abduction and Bayesian networks. Artif. Intell. **64**, 81–129 (1993)
57. Poole, D.: The independent choice logic for modelling multiple agents under uncertainty. Artif. Intell. **94**(1/2), 7–56 (1997)
58. Poole, D.: First-order probabilistic inference. In: International Joint Conference on Artificial Intelligence (IJCAI), pp. 985–991 (2003)
59. Predoiu, L., Stuckenschmidt, H.: Probabilistic models for the semantic web. In: Ma, Z., Wang, H. (eds.) The Semantic Web for Knowledge and Data Management: Technologies and Practices, pp. 74–105. IGI Global, Hershey (2009)
60. De Raedt, L.: Logical and Relational Learning. Springer, Heidelberg (2008)
61. Rettinger, A., Losch, U., Tresp, V., d'Amato, C., Fanizzi, N.: Mining the semantic web - statistical learning for next generation knowledge bases. Data Min. Knowl. Disc. **24**, 613–662 (2012)
62. Richardson, M., Domingos, P.: Markov logic networks. Mach. Learn. **62**(1–2), 107–136 (2006)

63. Riguzzi, F., Bellodi, E., Lamma, E., Zese, R.: Epistemic and statistical probabilistic ontologies. In: International Workshop on Uncertainty Reasoning for the Semantic Web, pp. 1–12 (2012)
64. Sato, T., Kameya, Y.: Parameter learning of logic programs for symbolic-statistical modeling. J. Artif. Intell. Res. **15**, 391–454 (2001)
65. Schild, K.: A correspondence theory for terminological logics: preliminary report. In: International Joint Conference on Artificial Intelligence, pp. 466–471 (1991)
66. Schmidt-Schauss, M., Smolka, G.: Attributive concept descriptions with complements. Artif. Intell. **48**, 1–26 (1991)
67. Sebastiani, F.: A probabilistic terminological logic for modelling information retrieval. In: Croft, W.B., van Rijsbergen, C.J. (eds.) International ACM Conference on Research and Development in Information Retrieval (SIGIR), Dublin, Ireland, pp. 122–130. Springer, London (1994)
68. Singla, P., Domingos, P.: Lifted first-order belief propagation. In: AAAI, pp. 1094–1099 (2008)
69. Staker, R.: Reasoning in expressive description logics using belief networks. In: International Conference on Information and Knowledge Engineering, pp. 489–495 (2002)
70. Taghipour, N., Fierens, D., Van den Broeck, G., Davis, J., Blockeel, H.: Completeness results for lifted variable elimination. In: International Conference on Artificial Intelligence and Statistics, pp. 572–580 (2013)
71. Thomas, A., Spiegelhalter, D., Gilks, W.: BUGS: a program to perform Bayesian inference using Gibbs sampling. In: Bernardo, J., Berger, J., Dawid, A., Smith, A. (eds.) Bayesian Statistics, vol. 4. Oxford University Press, Oxford (1992)
72. van den Broeck, G.: On the completeness of first-order knowledge compilation for lifted probabilistic inference. In: Neural Processing Information Systems (2011)
73. van den Broeck, G., Choi, A., Darwiche, A.: Lifted relax, compensate and then recover: from approximate to exact lifted probabilistic inference. In: Conference on Uncertainty in Artificial Intelligence (2012)
74. Wellman, M.P., Breese, J.S., Goldman, R.P.: From knowledge bases to decision models. Knowl. Eng. Rev. **7**(1), 35–53 (1992)
75. Yedidia, J.S., Freeman, W.T., Weiss, Y.: Constructing free energy approximations and generalized belief propagation algorithms. IEEE Trans. Inf. Theory **51**, 2282–2312 (2005)
76. Yelland, P.M.: Market analysis using a combination of Bayesian networks and description logics. Technical Report SMLI TR-99-78, Sun Microsystems Laboratories (1999)

Information Integration with Provenance on the Semantic Web via Probabilistic Datalog+/−

Thomas Lukasiewicz(✉), Maria Vanina Martinez,
Livia Predoiu, and Gerardo I. Simari

Department of Computer Science, University of Oxford, Oxford, UK
{thomas.lukasiewicz,vanina.martinez,
livia.predoiu,gerardo.simari}@cs.ox.ac.uk

Abstract. In this paper, we explore the use of guarded Datalog+/−
for information integration based on probabilistic data exchange. The
recently introduced Datalog+/− family of tractable ontology languages
is suitable for representing and reasoning over lightweight ontologies,
such as $\mathcal{EL}$ and the *DL-Lite* family of description logics. We study how
Datalog+/− can be used as a mapping language in the context of information integration. We also provide a complexity analysis for deciding
the existence of (deterministic and probabilistic (universal)) solutions
in the context of data exchange. In particular, we show that tractability is preserved for simple probabilistic representations, such as tuple-
independent ones.

1 Introduction

Information integration is widely considered a key (and costly) challenge of our
knowledge society [2,21]. The challenge is complex and includes many interre-
lated sub-challenges, like identifying which data sources to use when answering
a query, creating a common representation for the heterogeneous data sources
identified as relevant, extracting data from the sources, cleaning the extracted
data, eliminating duplicates by identifying the same objects in different data
sources, and transforming the extracted and cleaned data into a unified format.

In the simpler setting of relational information integration, the sources are
databases where information is structured with (different) relational database
schemas; the sources correspond to so-called *source* or *local* schemas. The unified format is the *target* or *global* schema. Heterogeneity is less challenging, as
it is restricted to different kinds of schemas. Unrestricted heterogeneity involves
databases hidden behind applications, document repositories, and other kinds of
unstructured information. In the Semantic Web, with description logics (DLs)
being the underlying knowledge representation formalism, information is structured via ontological schemas. Hence, ontologies replace databases − source
databases are replaced by *source* (or *local*) ontologies, and the target database
is replaced by a *target* (or *global*) ontology. In the following, we use guarded

© Springer International Publishing Switzerland 2014
F. Bobillo et al. (Eds.): URSW 2011-2013, LNAI 8816, pp. 41–62, 2014.
DOI: 10.1007/978-3-319-13413-0_3

Datalog+/– as the language for representing the sources, the target, and the mappings – one of the advantages of this language is that it is capable of representing ontological knowledge, but it keeps the notation used in databases for greater readability.

Information integration is usually achieved via mappings between logical formalizations of data sources (cf. [15, 24, 33] for an overview). There are mainly three ways in which information from different sources can be integrated:

- **Data exchange:** Data structured under a source schema S (or different source schemas $S_1, \ldots, S_k$) is transformed into data structured under a different target schema T, and materialized (merged and acquired) there through the mapping.

- **Data integration:** Heterogeneous data in different sources $S_1, \ldots, S_k$ is queried via a virtual global schema T, i.e., no actual exchange of data is needed.

- **Peer-to-peer data integration:** There is no global schema given. All peers $S_1, \ldots, S_k$ are autonomous and independent from each other, and each peer can hold data and be queried. The peers can be viewed as nodes in a network that are linked to other nodes by means of so-called peer-to-peer (P2P) mappings. That is, each source can also be a target for another source.

In this paper, we investigate probabilistic data exchange, which has been proposed for integrating probabilistic databases with either deterministic or probabilistic mappings [13, 14]. We use guarded Datalog+/– [6] as the underlying relational information integration language, which is an ontology language extending plain Datalog by negative constraints and the possibility of rules with existential quantification and equality in rule heads, while restricting the rule syntax by so-called guards in rule bodies to gain decidability and tractability. Essentially, it extends Datalog to negative constraints, tuple-generating dependencies (TGDs), and equality-generating dependencies (EGDs), but suitably restricted to gain decidability and data tractability. In this way, it is possible to capture the *DL-Lite* family of DLs and also the DL $\mathcal{EL}$. As such, guarded Datalog+/– is a very expressive and convenient language for ontology-based database access, which makes it particularly attractive for data exchange on the Semantic Web. For simplicity, from now on, we often refer to "guarded Datalog+/–" simply as "Datalog+/–". Though general data exchange and probabilistic data exchange frameworks often use standard or weakly acyclic sets of TGDs and EGDs for the target, in this work, we adopt linear and guarded TGDs and non-conflicting keys, as proposed in Datalog+/–.

We also sketch how provenance information can be added to Datalog+/– as a mapping language, to be able to track the origin of a mapping for trust assessment and debugging. Capturing the provenance of mappings allows to resolve inconsistencies of mappings by considering the history of their creation. It also helps to detect whether and how to perform mapping updates if the information sources have changed or evolved. Finally, it allows to capture mapping cycles, debug mappings, and to perform meta-reasoning with mappings and the knowledge bases themselves.

This paper extends [27], where we proposed the use of probabilistic Datalog+/− as a language for information integration. Here, we study a theoretical framework for probabilistic data exchange based on Datalog+/− and provide complexity results for deciding the existence of a solution in linear and guarded Datalog+/− for both the deterministic variant and its probabilistic extension. Another difference to [27] is that here, we study data exchange in the presence of a probabilistic source and consider a very general probabilistic model involving unrestricted probability distributions (in [27], we considered a probabilistic extension of Datalog+/− with Markov logic networks).

The rest of this paper is organized as follows. In Sect. 2, we recall the basics of guarded Datalog+/−. Sections 3 and 4 define our approach to deterministic data exchange on top of Datalog+/− and illustrate the type of ontology mappings that can be expressed in this framework, respectively. In Sects. 5 and 6, we generalize to probabilistic data exchange on top of Datalog+/− and the types of probabilistic ontology mappings that it can express, respectively. Section 7 provides complexity results, and Sect. 8 deals with provenance in our approach. In Sects. 9 and 10, we discuss related work, summarize the main results, and give an outlook on future research.

2   Guarded Datalog+/−

We now describe guarded Datalog+/− [6], which here includes negative constraints and (separable) equality-generating dependencies (EGDs). We first describe some preliminaries on databases and queries, and then tuple-generating dependencies (TGDs) and the concept of chase. We finally recall negative constraints and (separable) EGDs, which are other important ingredients of guarded Datalog+/− ontologies.

Databases and Queries. The elementary ingredients are constants, nulls, and variables, which serve as arguments in atomic formulas in databases, queries, and dependencies: (i) a fixed countably infinite universe of *(data) constants* Δ (which constitute the "normal" domain of a database), (ii) a fixed countably infinite set of *(labeled) nulls* Δ_N (used as "fresh" Skolem terms, which are placeholders for unknown values, and can thus be seen as variables), and (iii) a fixed countably infinite set of variables $\mathcal{V}$ (used in queries and dependencies). Different constants represent different values (*unique name assumption*), while different nulls may represent the same value. We assume a lexicographic order on $\Delta \cup \Delta_N$, with every symbol in Δ_N following all symbols in Δ. We denote by $\mathbf{X}$ sequences of variables $X_1, \ldots, X_k$ with $k \geq 0$.

We next define atomic formulas, which occur in databases, queries, and dependencies, and which are constructed from relation names and terms, as usual. We assume a *relational schema* $\mathcal{R}$, which is a finite set of *relation names* (or *predicate symbols*, or simply *predicates*). A *position* $P[i]$ identifies the i-th argument of a predicate P. A *term* t is a data constant, null, or variable. An *atomic formula* (or *atom*) $\mathbf{a}$ has the form $P(t_1, \ldots, t_n)$, where P is an n-ary predicate, and $t_1, \ldots, t_n$ are terms. We denote by $pred(\mathbf{a})$ and $dom(\mathbf{a})$ its predicate and the set of all its arguments, respectively. The latter two notations are naturally extended to sets of atoms and conjunctions of atoms. A conjunction of atoms is often identified with the set of all its atoms.

We are now ready to define the notion of a database relative to a relational schema, as well as conjunctive and Boolean conjunctive queries to databases. A *database (instance)* D for a relational schema $\mathcal{R}$ is a (possibly infinite) set of atoms with predicates from $\mathcal{R}$ and arguments from Δ. Such D is *ground* iff it contains only atoms with arguments from Δ. A *conjunctive query (CQ)* over $\mathcal{R}$ has the form $Q(\mathbf{X}) = \exists \mathbf{Y}\, \Phi(\mathbf{X}, \mathbf{Y}, \mathbf{C})$, where $\Phi(\mathbf{X}, \mathbf{Y}, \mathbf{C})$ is a conjunction of atoms with the variables $\mathbf{X}$ and $\mathbf{Y}$, and possibly constants $\mathbf{C}$, but without nulls. Note that $\Phi(\mathbf{X}, \mathbf{Y})$ may also contain equalities but no inequalities. A *Boolean CQ (BCQ)* over $\mathcal{R}$ is a CQ of the form $Q()$. We often write a BCQ as the set of all its atoms, having constants and variables as arguments, and omitting the quantifiers. Answers to CQs and BCQs are defined via *homomorphisms*, which are mappings $\mu\colon \Delta \cup \Delta_N \cup \mathcal{V} \to \Delta \cup \Delta_N \cup \mathcal{V}$ such that (i) $c \in \Delta$ implies $\mu(c) = c$, (ii) $c \in \Delta_N$ implies $\mu(c) \in \Delta \cup \Delta_N$, and (iii) μ is naturally extended to atoms, sets of atoms, and conjunctions of atoms. The set of all *answers* to a CQ $Q(\mathbf{X}) = \exists \mathbf{Y}\, \Phi(\mathbf{X}, \mathbf{Y})$ over a database D, denoted $Q(D)$, is the set of all tuples $\mathbf{t}$ over Δ for which there exists a homomorphism $\mu\colon \mathbf{X} \cup \mathbf{Y} \to \Delta \cup \Delta_N$ such that $\mu(\Phi(\mathbf{X}, \mathbf{Y})) \subseteq D$ and $\mu(\mathbf{X}) = \mathbf{t}$. The *answer* to a BCQ $Q()$ over a database D is *Yes*, denoted $D \models Q$, iff $Q(D) \neq \emptyset$.

Tuple-Generating Dependencies (TGDs). Tuple-generating dependencies (TGDs) describe constraints on databases in the form of generalized Datalog rules with existentially quantified conjunctions of atoms in rule heads; their syntax and semantics are as follows. Given a relational schema $\mathcal{R}$, a *tuple-generating dependency (TGD)* σ is a first-order formula of the form $\forall \mathbf{X} \forall \mathbf{Y}\, \Phi(\mathbf{X}, \mathbf{Y}) \to \exists \mathbf{Z}\, \Psi(\mathbf{X}, \mathbf{Z})$, where $\Phi(\mathbf{X}, \mathbf{Y})$ and $\Psi(\mathbf{X}, \mathbf{Z})$ are conjunctions of atoms over $\mathcal{R}$ called the *body* and the *head* of σ, denoted $body(\sigma)$ and $head(\sigma)$, respectively. A TGD is *guarded* iff it contains an atom in its body that involves all variables appearing in the body. The leftmost such atom is the *guard atom* (or *guard*) of σ. The non-guard atoms in the body of σ are the *side atoms* of σ. We usually omit the universal quantifiers in TGDs. Such σ is satisfied in a database D for $\mathcal{R}$ iff, whenever there exists a homomorphism h that maps the atoms of $\Phi(\mathbf{X}, \mathbf{Y})$ to atoms of D, there exists an extension h' of h that maps the atoms of $\Psi(\mathbf{X}, \mathbf{Z})$ to atoms of D. All sets of TGDs are finite here.

Query answering under TGDs, i.e., the evaluation of CQs and BCQs on databases under a set of TGDs is defined as follows. For a database D for $\mathcal{R}$, and a set of TGDs Σ on $\mathcal{R}$, the set of *models* of D and Σ, denoted $mods(D, \Sigma)$, is the set of all (possibly infinite) databases B such that (i) $D \subseteq B$ (ii) every $\sigma \in \Sigma$ is satisfied in B. The set of *answers* for a CQ Q to D and Σ, denoted $ans(Q, D, \Sigma)$, is the set of all tuples $\mathbf{a}$ such that $\mathbf{a} \in Q(B)$ for all $B \in mods(D, \Sigma)$. The *answer* for a BCQ Q to D and Σ is *Yes*, denoted $D \cup \Sigma \models Q$, iff $ans(Q, D, \Sigma) \neq \emptyset$. We recall that query answering under TGDs is equivalent to query answering under TGDs with only single atoms in their heads. We thus often assume w.l.o.g. that every TGD has a single atom in its head.

The Chase. The *chase* was introduced to enable checking implication of dependencies [29] and later also for checking query containment [22]. It is a procedure for repairing a database relative to a set of dependencies, so that the result of the chase satisfies the dependencies. By "chase", we refer both to the chase procedure and to its output. The TGD chase works on a database through so-called TGD *chase rules* (an extended chase with also equality-generating dependencies is discussed below). The TGD chase rule comes in two flavors: *restricted* and *oblivious*, where the restricted one applies TGDs only when they are not satisfied (to repair them), while the oblivious one always applies TGDs (if they produce a new result). We focus on the oblivious one here; the *(oblivious) TGD chase rule* defined below is the building block of the chase.

TGD CHASE RULE. Consider a database D for a relational schema $\mathcal{R}$, and a TGD σ on $\mathcal{R}$ of the form $\Phi(\mathbf{X}, \mathbf{Y}) \to \exists \mathbf{Z}\, \Psi(\mathbf{X}, \mathbf{Z})$. Then, σ is *applicable* to D if there exists a homomorphism h that maps the atoms of $\Phi(\mathbf{X}, \mathbf{Y})$ to atoms of D. Let σ be applicable to D, and h_1 be a homomorphism that extends h as follows: for each $X_i \in \mathbf{X}$, $h_1(X_i) = h(X_i)$; for each $Z_j \in \mathbf{Z}$, $h_1(Z_j) = z_j$, where z_j is a "fresh" null, i.e., $z_j \in \Delta_N$, z_j does not occur in D, and z_j lexicographically follows all other nulls already introduced. The *application of* σ on D adds to D the atom $h_1(\Psi(\mathbf{X}, \mathbf{Z}))$ if not already in D.

The chase algorithm for a database D and a set of TGDs Σ consists of an exhaustive application of the TGD chase rule in a breadth-first (level-saturating) fashion, which leads as result to a (possibly infinite) chase for D and Σ. Formally, the *chase of level up to* 0 of D relative to Σ, denoted $chase^0(D, \Sigma)$, is defined as D, assigning to every atom in D the *(derivation) level* 0. For every $k \geq 1$, the *chase of level up to* k of D relative to Σ, denoted $chase^k(D, \Sigma)$, is constructed as follows: let $I_1, \ldots, I_n$ be all possible images of bodies of TGDs in Σ relative to some homomorphism such that (i) $I_1, \ldots, I_n \subseteq chase^{k-1}(D, \Sigma)$ and (ii) the highest level of an atom in every I_i is $k - 1$; then, perform every corresponding TGD application on $chase^{k-1}(D, \Sigma)$, choosing the applied TGDs and homomorphisms in a (fixed) linear and lexicographic order, respectively, and assigning to every new atom the *(derivation) level* k. The *chase of* D relative to Σ, denoted $chase(D, \Sigma)$, is then defined as the limit of $chase^k(D, \Sigma)$ for $k \to \infty$.

The (possibly infinite) chase relative to TGDs is a *universal model*, i.e., there exists a homomorphism from $chase(D, \Sigma)$ onto every $B \in mods(D, \Sigma)$ [5,10]. This result implies that BCQs Q over D and Σ can be evaluated on the chase for D and Σ, i.e., $D \cup \Sigma \models Q$ is equivalent to $chase(D, \Sigma) \models Q$. In the case of guarded TGDs Σ, such BCQs Q can be evaluated on an initial fragment of $chase(D, \Sigma) \models Q$ of constant depth $k \cdot |Q|$, and thus be done in polynomial time in the data complexity.

Note that sets of guarded TGDs (with single-atom heads) are theories in the guarded fragment of first-order logic [1]. Note also that guardedness is a truly fundamental class ensuring decidability as adding a single unguarded Datalog rule to a guarded Datalog+/− program may destroy decidability as shown in [5].

Negative Constraints. Another crucial ingredient of Datalog+/− for ontological modeling are *negative constraints (NCs, or simply constraints)*, which are first-order formulas of the form $\forall \mathbf{X}\ \Phi(\mathbf{X}) \rightarrow \bot$, where $\Phi(\mathbf{X})$ is a conjunction of atoms (not necessarily guarded). We usually omit the universal quantifiers, and we implicitly assume that all sets of constraints are finite here. Adding negative constraints to answering BCQs Q over databases and guarded TGDs is computationally easy, as for each constraint $\forall \mathbf{X}\Phi(\mathbf{X}) \rightarrow \bot$, we only have to check that the BCQ $\Phi(\mathbf{X})$ evaluates to false; if one of these checks fails, then the answer to the original BCQ Q is true, otherwise the negative constraints can be simply ignored when answering the original BCQ Q.

Equality-Generating Dependencies (EGDs). A further important ingredient of Datalog+/− for modeling ontologies are *equality-generating dependencies* (or *EGDs*) σ, which are first-order formulas $\forall \mathbf{X}\ \Phi(\mathbf{X}) \rightarrow X_i = X_j$, where $\Phi(\mathbf{X})$, called the *body* of σ, denoted $body(\sigma)$, is a (not necessarily guarded) conjunction of atoms, and X_i and X_j are variables from $\mathbf{X}$. We call $X_i = X_j$ the *head* of σ, denoted $head(\sigma)$. Such σ is satisfied in a database D for $\mathcal{R}$ iff, whenever there exists a homomorphism h such that $h(\Phi(\mathbf{X}, \mathbf{Y})) \subseteq D$, it holds that $h(X_i) = h(X_j)$. We usually omit the universal quantifiers in EGDs, and all sets of EGDs are finite here.

An EGD σ on $\mathcal{R}$ of the form $\Phi(\mathbf{X}) \rightarrow X_i = X_j$ is *applicable* to a database D for $\mathcal{R}$ iff there exists a homomorphism $\eta \colon \Phi(\mathbf{X}) \rightarrow D$ such that $\eta(X_i)$ and $\eta(X_j)$ are different and not both constants. If $\eta(X_i)$ and $\eta(X_j)$ are different constants in Δ, then there is a *hard violation* of σ (and, as we will see below, the *chase* fails). Otherwise, the result of the application of σ to D is the database $h(D)$ obtained from D by replacing every occurrence of a non-constant element $e \in \{\eta(X_i), \eta(X_j)\}$ in D by the other element e' (if e and e' are both nulls, then e precedes e' in the lexicographic order). The *chase* of a database D, in the presence of two sets Σ_T and Σ_E of TGDs and EGDs, respectively, denoted $chase(D, \Sigma_T \cup \Sigma_E)$, is computed by iteratively applying (1) a single TGD once, according to the standard order and (2) the EGDs, as long as they are applicable (i.e., until a fixpoint is reached). To assure that adding EGDs to answering BCQs Q over databases and guarded TGDs along with negative constraints does not increase the complexity of query answering, all EGDs are assumed to be *separable* [6] (one such class of separable EGDs are non-conflicting keys [6]). Intuitively, separability holds whenever: (i) if there is a hard violation of an EGD in the chase, then there is also one on the database w.r.t. the set of EGDs alone (i.e., without considering the TGDs); and (ii) if there is no chase failure, then the answers to a BCQ w.r.t. the entire set of dependencies equals those w.r.t. the TGDs alone (i.e., without the EGDs).

Guarded Datalog+/− Ontologies. We define (guarded) Datalog+/− ontologies as follows. A *(guarded) Datalog+/− ontology* consists of a database D, a (finite) set of guarded TGDs Σ_T, a (finite) set of negative constraints Σ_C, and a (finite) set of EGDs Σ_E that are separable from Σ_T.

3   Deterministic Data Exchange

In this section, we recall the classical logical framework of data exchange and data integration [14,15] in the context of both deterministic and probabilistic databases, and tailor the framework to suit data exchange and data integration between Datalog+/− ontologies. The syntax of a schema mapping in Datalog+/− is defined as follows.

Definition 1 (Schema Mapping). A *schema mapping* $\mathcal{M} = (\mathbf{S}, \mathbf{T}, \Sigma)$ consists of a source schema $\mathbf{S} = \{S_1, \ldots, S_n\}$, a target schema $\mathbf{T} = \{T_1, \ldots, T_m\}$ disjoint from $\mathbf{S}$, and a set $\Sigma = \Sigma_{st} \cup \Sigma_t$ of TGDs, negative constraints, and non-conflicting keys, where Σ_{st} are *source-to-target* TGDs, negative constraints, and non-conflicting keys over $\mathbf{S} \cup \mathbf{T}$, and Σ_t are *target* TGDs, negative constraints, and non-conflicting keys over $\mathbf{T}$.

The semantics of schema mappings is defined by relating source and target databases in a semantically meaningful and consistent way. More specifically, in the case of data exchange between deterministic databases, a target database J over Δ is considered to be a *solution* of the source database I over Δ for the data exchange problem specified via the schema mapping $\mathcal{M} = (\mathbf{S}, \mathbf{T}, \Sigma)$ iff $I \cup J \models \Sigma$.

Definition 2 (Solution). A target database J over Δ is a *solution* for a source database I over Δ relative to a schema mapping $\mathcal{M} = (\mathbf{S}, \mathbf{T}, \Sigma)$ iff $I \cup J \models \Sigma$. We denote by $Sol_{\mathcal{M}}$ the set of all pairs (I, J) of source databases I and target databases J with $I \cup J \models \Sigma$.

There are many possible solutions J to a source database I relative to $\mathcal{M}$ in $Sol_{\mathcal{M}}$. Among all such solutions, the preferred solutions are the ones that carry only the necessary information for data exchange; i.e., all the constants of the source database that can be transferred via the mapping are included in the target database. Such solutions are called *universal* solutions. Similar to universal models in the context of the chase derivation of Datalog+/− (see Sect. 2), a universal solution can be homomorphically mapped to all other solutions leaving the constants unchanged.

Definition 3 (Universal Solution). A target database J over Δ is a *universal solution* for a source database I over Δ relative to a schema mapping $\mathcal{M} = (\mathbf{S}, \mathbf{T}, \Sigma)$ iff (i) J is a solution, and (ii) for each solution J' for I relative to $\mathcal{M}$, there is a homomorphism $h\colon J \to J'$. We denote by $USol_{\mathcal{M}}$ $(\subseteq Sol_{\mathcal{M}})$ the set of all pairs (I, J) of source databases I and target databases J such that J is a universal solution for I relative to $\mathcal{M}$.

For defining the data exchange problem over probabilistic databases, we first need to define probabilistic databases, as they serve as source and target databases in the probabilistic data exchange setting.

Definition 4 (Probabilistic Database). A *probabilistic database over* $\mathcal{R}$ is a probability space $Pr = (\mathcal{I}, \mu)$ such that $\mathcal{I}$ is the set of all (possibly infinitely many) standard databases over $\mathcal{R}$, and $\mu \colon \mathcal{I} \to [0,1]$ is a function that satisfies $\sum_{I \in \mathcal{I}} \mu(I) = 1$.

In this paper, we adopt a compact encoding of probabilistic databases by annotating database atoms with Boolean combinations of elementary events, where every annotation describes when the atom is true and is associated with a probability. We first define annotations and annotated atoms.

Definition 5 (Annotations and Annotated Atoms). Let $e_1, \ldots, e_n$ be $n \geq 1$ *elementary events*. A *world* w is a conjunction $\ell_1 \wedge \cdots \wedge \ell_n$, where each ℓ_i, $i \in \{1, \ldots, n\}$, is either the elementary event e_i or its negation $\neg e_i$. An *annotation* λ is any Boolean combination of elementary events (i.e., all elementary events are annotations, and if λ_1 and λ_2 are annotations, then also $\neg \lambda_1$ and $\lambda_1 \wedge \lambda_2$); as usual, $\lambda_1 \vee \lambda_2$ abbreviates $\neg(\neg \lambda_1 \wedge \neg \lambda_2)$. An *annotated atom* has the form $a \colon \lambda$, where a is an atom, and λ is an annotation.

Based on this definition, we can now define the compact encoding of probabilistic databases that we will be using.

Definition 6 (Compact Encoding of Probabilistic Databases). A set D of annotated atoms over Δ along with a probability $\mu(w) \in [0,1]$ for every world w *compactly encodes a probabilistic database* by defining:

1. the probability of every annotation λ as the sum of the probabilities of all worlds in which λ is true, and
2. the probability of every database $\{a_1, \ldots, a_m\}$ such that $\{a_1 \colon \lambda_1, \ldots, a_m \colon \lambda_m\} \subseteq D$ for some annotations $\lambda_1, \ldots, \lambda_m$ as the probability of $\lambda_1 \wedge \cdots \wedge \lambda_m$ (and the probability of every other database as 0).

While the syntax of a deterministic schema mapping over probabilistic databases does not change, its semantics needs to be adapted. Exchanging data between probabilistic databases means that a properly defined joint probability space Pr over the solution relation $Sol_{\mathcal{M}}$ and the universal solution relation $USol_{\mathcal{M}}$ must exist. Note that $Sol_{\mathcal{M}}$ and $USol_{\mathcal{M}}$ exist in probabilistic data exchange as well, because joint events (I, J) that can be constructed for pairs of probabilistic source instances $Pr_s = (\mathcal{I}, \mu_s)$ and probabilistic target instances $Pr_t = (\mathcal{J}, \mu_t)$ need also to satisfy the condition $I \cup J \models \Sigma$ to be considered as semantic components of probabilistic solutions.

As stated more formally below, a properly defined joint probability distribution Pr over the solution relation $Sol_{\mathcal{M}}$ or the universal solution relation $USol_{\mathcal{M}}$ requires to match each of the given marginal distributions Pr_s and Pr_t. These constraints over the marginal distributions of Pr are called $Sol_{\mathcal{M}}$-match and $USol_{\mathcal{M}}$-match [14], as they are defined over $Sol_{\mathcal{M}}$ and $USol_{\mathcal{M}}$, respectively.

Definition 7 (Probabilistic (Universal) Solution). A probabilistic target database $Pr_t = (\mathcal{J}, \mu_t)$ is a *probabilistic solution* (resp., *probabilistic universal solution*) for a probabilistic source database $Pr_s = (\mathcal{I}, \mu_s)$ relative to a schema mapping $\mathcal{M} = (\mathbf{S}, \mathbf{T}, \Sigma)$ iff there exists a probabilistic space $Pr = (\mathcal{I} \times \mathcal{J}, \mu)$ that satisfies the following two conditions:

1. The left and right marginals of Pr are Pr_s and Pr_t, respectively. That is,
   (a) $\sum_{J \in \mathcal{J}} (\mu(I, J)) = \mu_s(I)$ for all $I \in \mathcal{I}$ and
   (b) $\sum_{I \in \mathcal{I}} (\mu(I, J)) = \mu_t(J)$ for all $J \in \mathcal{J}$;
2. $\mu(I, J) = 0$ for all $(I, J) \notin Sol_\mathcal{M}$ (resp., $(I, J) \notin USol_\mathcal{M}$).

For mapping $\mathcal{M} = (\mathbf{S}, \mathbf{T}, \Sigma)$ and query $Q(\mathbf{X}) = \exists \mathbf{Y}\, \Phi(\mathbf{X}, \mathbf{Y}, \mathbf{C})$ (as introduced in Sect. 2), query answering within the data exchange setting of a source database I is defined as deriving the certain answers, i.e., the tuples consisting of constants that belong to $Q(J)$ for all solutions J for I relative to $\mathcal{M}$. In the probabilistic generalization, each probabilistic target database defines a probability with which a tuple of constants belongs to $Q(J)$ (which is the sum of the probabilities of all standard target databases in which the query evaluates to true), and the probability that this tuple belong to the answer of the query is the infimum of all such probabilities. In the following definition, we also generalize queries to unions of conjunctive queries (UCQs).

Definition 8 (UCQs). A *union of conjunctive queries* (or *UCQ*) has the form $Q(\mathbf{X}) = \bigvee_{i=1}^{k} \exists \mathbf{Y}_i\, \Phi_i(\mathbf{X}, \mathbf{Y}_i, \mathbf{C}_i)$, where each $\exists \mathbf{Y}_i\, \Phi_i(\mathbf{X}, \mathbf{Y}_i, \mathbf{C}_i)$ with $i \in \{1, \ldots, k\}$ is a CQ with exactly the variables $\mathbf{X}$ and $\mathbf{Y}_i$, and the constants $\mathbf{C}_i$. Given a schema mapping $\mathcal{M} = (\mathbf{S}, \mathbf{T}, \Sigma)$, a probabilistic source database $Pr_s = (\mathcal{I}, \mu_s)$, a UCQ $Q(\mathbf{X}) = \bigvee_{i=1}^{k} \exists \mathbf{Y}_i\, \Phi_i(\mathbf{X}, \mathbf{Y}_i, \mathbf{C}_i)$, and a tuple $\mathbf{A}$ (being a ground instance of $\mathbf{X}$ in Q) over Δ, the *confidence* of $\mathbf{A}$ relative to Q, denoted $conf_Q(\mathbf{A})$, in Pr_s relative to $\mathcal{M}$ is the infimum of $Pr_t(Q(\mathbf{A}))$ subject to all probabilistic solutions Pr_t for Pr_s relative to $\mathcal{M}$. Here, $Pr_t(Q(\mathbf{A}))$ for $Pr_t = (\mathcal{J}, \mu_t)$ is the sum of all $\mu_t(J)$ such that $Q(\mathbf{A})$ evaluates to true in the database $J \in \mathcal{J}$ (i.e., some BCQ $\exists \mathbf{Y}_i\, \Phi_i(\mathbf{A}, \mathbf{Y}_i, \mathbf{C}_i)$ with $i \in \{1, \ldots, k\}$ evaluates to true in J).

The following are the main computational tasks that we consider in this paper.

Existence of a solution (resp., universal solution): Given a schema mapping $\mathcal{M}$ and a probabilistic source database Pr_s, decide whether there exists a probabilistic (resp., probabilistic universal) solution for Pr_s relative to $\mathcal{M}$.

Materialization of a solution (resp., universal solution): Given a schema mapping $\mathcal{M}$ and a probabilistic source database Pr_s, compute a probabilistic solution (resp., probabilistic universal solution) for Pr_s relative to $\mathcal{M}$, if it exists.

Answering UCQs: Given a schema mapping $\mathcal{M}$, a probabilistic source database Pr_s, a UCQ $Q(\mathbf{X})$, and a tuple $\mathbf{A}$ over Δ, compute $conf_Q(\mathbf{A})$ in Pr_s relative to $\mathcal{M}$.

4   Ontology Mappings with Datalog+/−

Mapping languages are formal knowledge representation languages that are chosen according to specific criteria. The two most important criteria are the *expressive power* needed for specifying desired data interoperability tasks on the one hand and the *tractability* of dealing with that language, i.e., query answering, checking for solutions, materializing solutions, etc., on the other. It is well known that there is a tradeoff between expressivity and tractability [25] – the latter is often attained via algorithmic properties that imply certain structural properties, such as the existence of universal solutions after performing a bounded number of computations.

In the following, we examine Datalog+/– as a mapping language. As a language lying in the intersection of the DL and the logic programming paradigms, Datalog+/– allows to integrate the information available in ontologies and, hence, nicely ties together the results on data exchange and integration in databases and the work on ontology mediation in the Semantic Web.

When integrating ontologies with Datalog+/– via *source-to-target TGDs* (for short, we often refer to them as s-t TGDs), such TGDs correspond to *GLAV (global-local-as-view) dependencies* and are used as mappings. In their most general form, TGDs are (as mentioned above) first-order formulas $\forall \mathbf{X}\, \phi(\mathbf{X}) \rightarrow \exists \mathbf{Y} \psi(\mathbf{X}, \mathbf{Y})$ with $\mathbf{X}$ and $\mathbf{Y}$ being tuples of variables, $\phi(\mathbf{X})$ and $\psi(\mathbf{X}, \mathbf{Y})$ being a conjunction of atomic formulas.

The following two types of dependencies are important special cases of source-to-target TGDs: LAV (local as view) and GAV (global as view):

- A **LAV (local as view)** dependency is a source-to-target TGD with a single atom in the body, i.e., of the form $\forall \mathbf{X}\, A_S(\mathbf{X}) \rightarrow \exists \mathbf{Y} \psi(\mathbf{X}, \mathbf{Y})$, where A_S is an atom over the source schema, and $\psi(\mathbf{X}, \mathbf{Y})$ is a conjunction of atoms over the target schema.
- A **GAV (global as view)** dependency is a source-to-target TGD with a single atom in the head, i.e., of the form $\forall \mathbf{X}\, \phi(\mathbf{X}) \rightarrow A_T(\mathbf{X}')$, where $\phi(\mathbf{X})$ is a conjunction of atoms over the source schema, and $A_T(\mathbf{X}')$ is an atom over the target schema with $\mathbf{X}' \subseteq \mathbf{X}$.

The following mappings that are mentioned in [34] as "essential" can also be represented in Datalog+/– (all examples below stem from a consideration of the OAEI benchmark set; more specifically, ontologies 101 and 301–303):

- **Copy (Nicknaming):** Copy a source relation (or concept or role) (of arbitrary arity n) into a target relation (or concept or role) (of the same arity n like the source relation (or concept or role)) and rename it. Note that this kind of mapping is a LAV and a GAV mapping at the same time. For example:

$$\forall x, y \quad S : location(x, y) \rightarrow \quad T : address(x, y).$$

- **Projection (Column Deletion):** Create a target relation (or concept or role) by deleting one or more columns of a source relation (or concept or role) (of arbitrary arity $n \geq 2$). Note that this kind of mapping is a LAV and GAV mapping at the same time. For instance:

$$\forall x, y \quad S : author(x, y) \rightarrow \quad T : person(x).$$

- **Augmentation (Column Addition):** Create a target relation (or concept or role) (of arbitrary arity $n \geq 2$) by adding one or more columns to the source relation (or concept or role). Note that this is a LAV dependency. A simple example follows:

$$\forall x \quad S : editor(x) \rightarrow \exists z \quad T : hasEditor(z, x).$$

- **Decomposition:** Decompose a source relation (or concept or role) (of arbitrary arity n) into two or more target relations (or concepts or roles). Note that this is a LAV dependency. For instance, we can have:

$$\forall x, y \quad S : publisher(x, y) \rightarrow \quad T : organization(x), \quad T : proceedings(y).$$

Only one mapping construct mentioned in [34] as essential – the join – cannot be represented by guarded Datalog+/−. As each TGD has to be guarded, there must be an atom in the body that contains all non-existentially quantified variables and, hence, a join like $\forall x, y \quad S : book(y), \ S : person(x) \rightarrow T : author(x, y)$ cannot be represented in guarded Datalog+/−. This, however, can also be considered as a benefit, as joins usually need more computing resources, because they require a large number of operations. Note that the join is introduced for query answering by means of conjunctive queries that we are using to query the target database. This is equivalent to the join in databases.

In ontology mediation, a *mapping* or *alignment* is based on correspondences between so-called matchable entities of two ontologies. The following definition is based on [12]; let S and T be two ontologies (the source and the target) that are to be mapped onto each other, and let q be a function that defines the sets of matchable entities $q(S)$ and $q(T)$. Then, a correspondence between S and T is a triple $\langle e_1, e_2, r \rangle$ with $e_1 \in q(S)$, $e_2 \in q(T)$, and r being an alignment relation between the two matchable elements (note that equivalence and implication are examples of such an alignment relations if the chosen mapping language supports them). A *mapping* or *alignment* between S and T is then a set of correspondences $C = \cup_{i,j,k}\{\langle e_i, e_j, r_k \rangle\}$ between S and T. This is a very general definition, which allows to describe many types of mappings.

Definition 9 (Ontology Mapping [12]). Let S be a source ontology, and T be a target ontology. Let q be a function that defines the sets of matchable entities $q(S)$ and $q(T)$. Then, a *correspondence* between S and T is a triple $\langle e_1, e_2, r \rangle$, where $e_1 \in q(S)$, $e_2 \in q(T)$, and r is an alignment relation between the two matchable elements e_1 and e_2. An *ontology mapping* between S and T is a set of correspondences $C = \cup_{i,j,k}\{\langle e_i, e_j, r_k \rangle\}$ between S and T.

Semantic Web and ontology mapping languages usually contain a subset of the aforementioned mapping expressions, plus additional mapping expressions in the form of constraints, which usually are used to specify class disjointness (see, e.g., [31,33]). However, note that both the data exchange and the ontology mediation communities have also proposed mapping languages that are more expressive than source-to-target TGDs, consisting of full general Datalog expressions and also containing existentially quantified variables in the head – e.g., second-order mappings as described in the requirements of [32] or second-order TGDs [16]. Of course, such mapping languages have less desirable tractability properties. In [31], a probabilistic mapping language is presented that is based on Markov logic networks, built by mappings of basic DL axioms onto predicates with the desired semantics. A closer look reveals that the deterministic mapping constructs that are used are renaming, decomposition, and class disjointness constraints, as well as their combinations. Such disjointness constraints can be modeled with Datalog+/−, using negative constraints (NCs), such as:

– **Disjointness of ontology entities with the same arity:** A source relation (or concept or role) with arity n is disjoint to another relation (or concept or role) with the same arity n. The NC below corresponds to class disjointness that specifies that persons cannot be addresses:

$$\forall x \quad S : Person(x), \quad T : Address(x) \to \bot.$$

– **Disjointness of ontology entities with different arity:** A source relation (or concept or role) with arity $n \geq 2$ is disjoint with another relation (or concept or role) with the arity $n > m \geq 1$. The example below specifies that persons cannot be bought and, hence, do not have prices.

$$\forall x, y \quad S : Person(x), \quad T : hasPrice(x, y) \to \bot.$$

EGDs are also part of some mapping languages, especially in the database area, and can be represented by Datalog+/− as long as they are separable from the TGDs. Such kinds of dependencies allow to create mappings like the following one specifying that publishers of the same book or journal in both, the source and target schema (or ontology), have to be the same:

$$\forall x, y, z \quad S : publisher(x, y), \quad T : publishes(y, z) \to x = z.$$

5   Probabilistic Data Exchange

Probabilistic data exchange extends classical data exchange by the demand of two database instances I and J not only meeting the deterministic constraints of solutions, but also the probabilities specified by a probability distribution over a set of deterministic schema mappings, which is expressed in the following definition of probabilistic schema mappings.

Definition 10 (Probabilistic Schema Mapping). A *probabilistic schema mapping* is a tuple of the form $\mathcal{M} = (\mathbf{S}, \mathbf{T}, \Sigma, \mu)$, consisting of a source schema $\mathbf{S} = \{S_1, \ldots, S_n\}$, a target schema $\mathbf{T} = \{T_1, \ldots, T_m\}$ disjoint from $\mathbf{S}$, a set $\Sigma = \Sigma_{st} \cup \Sigma_t$ of TGDs, negative constraints, and non-conflicting keys, where Σ_{st} are *source-to-target* TGDs, negative constraints, and non-conflicting keys over $\mathbf{S} \cup \mathbf{T}$, and Σ_t are *target* TGDs, negative constraints, and non-conflicting keys over $\mathbf{T}$, and a function $\mu \colon 2^{\Sigma} \to [0, 1]$ such that $\sum_{\Sigma' \subseteq \Sigma} \mu(\Sigma') = 1$.

Probabilistic schema mappings are compactly encoded in the same way as probabilistic databases by annotating TGDs, negative constraints, and non-conflicting keys with Boolean combinations of elementary events. If such mappings are given along with probabilistic databases, then both are compactly encoded via two (not necessarily disjoint) sets of elementary events, assuming that the probabilities of common elementary events are the same in both compact encodings. Note that in probabilistic Datalog+/− [18], which we used for the probabilistic mappings in [27], both sets of elementary events coincide, the annotations are conjunctions of elementary events, and the probability of every world is defined via a Markov logic network.

The next definition lifts the notion of probabilistic (universal) solution from probabilistic source databases under deterministic schema mappings to probabilistic source databases under probabilistic schema mappings.

Definition 11 (Probabilistic (Universal) Solution). A probabilistic target database $Pr_t = (\mathcal{J}, \mu_t)$ is a *probabilistic solution* (resp., *probabilistic universal solution*) for a probabilistic source database $Pr_s = (\mathcal{I}, \mu_s)$ relative to a probabilistic schema mapping $\mathcal{M} = (\mathbf{S}, \mathbf{T}, \Sigma, \mu_m)$ iff there exists a probabilistic space $Pr = (\mathcal{I} \times \mathcal{J} \times 2^\Sigma, \mu)$ that satisfies the following two conditions:

1. The three marginals of μ are μ_s, μ_t, and μ_m, such that:
   (a) $\sum_{J \in \mathcal{J}, \, \Sigma' \subseteq \Sigma} \mu(I, J, \Sigma') = \mu_s(I)$ for all $I \in \mathcal{I}$,
   (b) $\sum_{I \in \mathcal{I}, \, \Sigma' \subseteq \Sigma} \mu(I, J, \Sigma') = \mu_t(J)$ for all $J \in \mathcal{J}$, and
   (c) $\sum_{I \in \mathcal{I}, \, J \in \mathcal{J}} \mu(I, J, \Sigma') = \mu_m(\Sigma')$ for all $\Sigma' \subseteq \Sigma$;
2. $\mu(I, J, \Sigma') = 0$ for all $(I, J) \notin Sol_{(\mathbf{S}, \mathbf{T}, \Sigma')}$ (resp., $(I, J) \notin USol_{(\mathbf{S}, \mathbf{T}, \Sigma')}$).

Using the above probabilistic and probabilistic universal solutions for probabilistic source databases under probabilistic schema mappings, the semantics of UCQs can easily be lifted from deterministic to probabilistic schema mappings as follows.

Definition 12 (UCQs). A *union of conjunctive queries* (or *UCQ*) has the form $Q(\mathbf{X}) = \bigvee_{i=1}^{k} \exists \mathbf{Y}_i \, \Phi_i(\mathbf{X}, \mathbf{Y}_i, \mathbf{C}_i)$, where each $\exists \mathbf{Y}_i \, \Phi_i(\mathbf{X}, \mathbf{Y}_i, \mathbf{C}_i)$ with $i \in \{1, \ldots, k\}$ is a CQ with exactly the variables $\mathbf{X}$ and $\mathbf{Y}_i$, and the constants $\mathbf{C}_i$. Given a probabilistic schema mapping $\mathcal{M} = (\mathbf{S}, \mathbf{T}, \Sigma, \mu_m)$, a probabilistic source database $Pr_s = (\mathcal{I}, \mu_s)$, a UCQ $Q(\mathbf{X}) = \bigvee_{i=1}^{k} \exists \mathbf{Y}_i \, \Phi_i(\mathbf{X}, \mathbf{Y}_i, \mathbf{C}_i)$, and a tuple $\mathbf{A}$ over Δ, the *confidence* of $\mathbf{A}$ relative to Q, denoted $conf_Q(\mathbf{A})$, in Pr_s relative to $\mathcal{M}$ is the infimum of $Pr_t(Q(\mathbf{A}))$ subject to all probabilistic solutions Pr_t for Pr_s relative to $\mathcal{M}$. Here, $Pr_t(Q(\mathbf{A}))$ for $Pr_t = (\mathcal{J}, \mu_t)$ is the sum of all $\mu_t(J)$ such that $Q(\mathbf{A})$ evaluates to true in the database $J \in \mathcal{J}$ (i.e., some BCQ $\exists \mathbf{Y}_i \, \Phi_i(\mathbf{A}, \mathbf{Y}_i, \mathbf{C}_i)$ with $i \in \{1, \ldots, k\}$ evaluates to true in J).

Similarly, the main computational tasks of this paper can easily be generalized from deterministic to probabilistic schema mappings as follows.

Existence of a solution (resp., universal solution): Given a probabilistic schema mapping $\mathcal{M}$ and a probabilistic source database Pr_s, decide whether there exists a probabilistic (resp., probabilistic universal) solution for Pr_s relative to $\mathcal{M}$.

Materialization of a solution (resp., universal solution): Given a probabilistic schema mapping $\mathcal{M}$ and a probabilistic source database Pr_s, compute a probabilistic solution (resp., probabilistic universal) solution for Pr_s relative to $\mathcal{M}$, if it exists.

Answering UCQs: Given a probabilistic schema mapping $\mathcal{M}$, a probabilistic source database Pr_s, a UCQ $Q(\mathbf{X})$, and a tuple $\mathbf{A}$ over Δ, compute $conf_Q(\mathbf{A})$ in Pr_s relative to $\mathcal{M}$.

6   Ontology Mappings with Probabilistic Datalog+/−

In general, probabilistic mappings $\mathcal{M} = (\mathbf{S}, \mathbf{T}, \Sigma, \mu)$ in probabilistic Datalog+/− have a similar expressivity as deterministic mappings $\mathcal{M} = (\mathbf{S}, \mathbf{T}, \Sigma)$ in Datalog+/−; that is, they can encode LAV and GAV mappings and also the mappings mentioned in Sect. 4 as being essential like *Copy (Nicknaming), Projection (Column Deletion), Augmentation (Column Addition), Decomposition,* and *Disjointness constraints* of entities with same or different arity (see Sect. 4 for details).

In addition, in probabilistic Datalog+/−, the above-mentioned kinds of deterministic mappings $\mathcal{M} = (\mathbf{S}, \mathbf{T}, \Sigma)$ are extended with uncertainty by defining a probability space $\Omega(2^\Sigma, \mu_m)$ over the dependencies Σ in Datalog+/− such that each mapping m_i holds with a probability $\mu_m(m_i)$. The mappings together with the probability space defined over them is represented by the compact encoding defined in Sect. 3 for probabilistic databases and used in Sect. 5 for probabilistic mappings as well. This compact encoding of probabilistic databases and mappings considers conditions or events under which the mappings are true or not. These events are not part of the databases, but can represent databases or other relevant events. As such, the two tasks of database and mapping dependencies modeling are separated from the task of modeling the uncertainty around the axioms of the ontology. Note that the set of events used to encode the probabilistic mappings can overlap with the set of events used to encode the probabilistic databases, and their probability can depend completely or in part on the same events.

As an example, consider a tuple-independent (see Sect. 7) database and a mapping consisting of a single rule. Without loss of generality, let Σ_{st} consist of only one mapping dependency – e.g., the first one mentioned in Sect. 4 (Copy/Nicknaming):

$$\forall x, y \quad S : location(x, y) \; \rightarrow \; T : address(x, y).$$

The probabilistic version of this mapping for a probabilistic source database with only two possible probabilistic tuples *(location(CS Department, Wolfson Building Oxford), 0.8), (location(Maths Department, Andrew Wiles Building Oxford), 0.9)*, which are independent, has probability 0.98 of being true. In this case, clearly, the probability of the mapping is dependent on the probabilistic source database.

Another, more expressive, possibility to encode the probabilistic events under which the mappings hold is to use Markov logic networks (MLNs) as done in [18] – in [27], we suggested the use of the probabilistic extension of Datalog+/− presented in [18] for mappings. With annotations referring to MLNs, a probabilistic mapping has the form $\mathcal{M} = (\mathbf{S}, \mathbf{T}, \Sigma, M)$, where M is the MLN encoding the probabilistic worlds in which the dependencies can either hold or not hold.

As described in [18], the TGDs, negative constraints, and non-conflicting keys are annotated with probabilistic scenarios λ that correspond to the worlds that they are valid in. The complex probabilistic dependencies that the annotations are involved in are represented by the MLN. In the probabilistic extension of Datalog+/− in [18], annotations cannot refer to elements of the databases or the mappings; hence, again, there is a modeling advantage in separating the two tasks of database and mapping dependencies modeling when modeling the uncertainty around the databases and dependencies.

Note that due to the disconnected representation between the probabilistic dependencies and the ontology, we can encode a part of mapping formulas as predicates encoding a specific semantics like disjointness, renaming, or decomposition, in a similar way as in [31]. With these predicates, an MLN can be created, and the actual mappings can be enriched by ground predicates that add the probabilistic interpretation. However, another more interesting encoding involves using a second ontology describing additional features of the generation of the mappings, and in this way eventually even meta reasoning about the mapping generation is possible. A rather general example of such an additional MLN describing the generation of a mapping is shown in Fig. 1. Here, the MLN describes the generation of a mapping via the matcher that it generates and a set of (possibly dependent) applicability conditions, as well as additional conditions that influence the probability of the mapping besides the result of the matcher.

With such kind of an MLN describing the dependency of different kinds of conditions (also dependencies between matchers are conceivable to combine the results of several different matchers), probabilistic reasoning over data integration

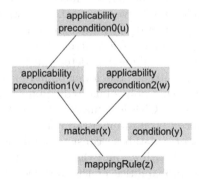

Fig. 1. Example of a Markov logic network describing the generation of mappings by means of applicability conditions and an additional condition that influences the probability of the mapping besides the result of the matcher.

settings can be done in much more precise settings. Hence, such an annotation is clearly much more expressive than the one corresponding to the two cases considered in Sect. 7.

7 Computational Complexity

In this section, we explore the computational complexity of deciding the existence of (universal) solutions for deterministic and probabilistic data exchange problems in our framework with mappings encoded in different variants of Datalog+/−. Due to space limitations, we only sketch the proofs – full proofs are given in the extended paper.

We assume that all annotations are in disjunctive normal form (DNF), i.e., disjunctions of conjunctions of literals, and we consider the following two cases: (i) that elementary events and their negations are pairwise probabilistically independent (i.e., the probability of worlds $\ell_1 \wedge \cdots \wedge \ell_n$ of elementary events ($\ell_i = e_i$) and their negations ($\ell_i = \neg e_i$) is defined as $\Pi_{i=1}^n \nu(\ell_i)$, where $\nu(\ell_i) = \mu(e_i)$ and $\nu(\ell_i) = 1 - \mu(e_i)$, respectively), called *elementary-event-independence*; and (ii) that all annotations are also elementary and that all worlds have a positive probability, called *tuple-independence*.

7.1 Complexity of Deterministic Data Exchange

The following result shows that deciding the existence of probabilistic (or probabilistic universal) solutions for deterministic data exchange problems relative to probabilistic source databases is co-NP-complete (resp., in PTIME) in the elementary-event-independent (resp., tuple-independent) case in the data complexity. The upper bound for the elementary-event-independent case follows from the fact that, relative to a probabilistic source database, a probabilistic universal solution exists iff a probabilistic solution exists, which is in turn equivalent to the existence of a deterministic solution relative to the deterministic database for every world. To decide the contrary, we thus only have to guess a world and check that there is no solution relative to its database, which is in co-NP. In the tuple-independent case, the tractability follows from the fact that a probabilistic solution exists relative to a probabilistic source database iff a deterministic solution exists relative to the maximal possible deterministic source database, which can be decided in polynomial time. The lower bound for the elementary-event-independent case follows from a polynomial reduction from the co-NP-complete problem of deciding whether a CNF formula $\phi = c_1 \wedge \cdots \wedge c_n$, where every c_i is a disjunction of literals over m propositional variables $x_1, \ldots, x_m$, is unsatisfiable. In this reduction, the source database consists of the atoms $E_S(i-1, i)$ for all $i \in \{1, \ldots, n\}$, annotated with c_i, and the atom $L_S(0)$, annotated with the true event $\top$, while the probabilities of the variables x_i are defined as 0.5. The deterministic mapping in Datalog+/− is defined as $E_S(X, Y) \rightarrow E_T(X, Y)$, $L_S(X) \rightarrow L_T(X)$, $L_T(n) \rightarrow \bot$, and $L_T(X) \wedge E_T(X, Y) \rightarrow L_T(Y)$. Then, the above probabilistic database and deterministic mapping have a probabilistic solution iff ϕ is unsatisfiable. Observe that by this construction, hardness for co-NP also holds for the special case where the constructed mapping is formulated without existentially quantified variables.

Theorem 1. *Given a schema mapping* $\mathcal{M} = (\boldsymbol{S}, \boldsymbol{T}, \Sigma_{st} \cup \Sigma_t)$, *where* Σ_{st} *and* Σ_t *are guarded TGDs, negative constraints, and non-conflicting keys over* $\boldsymbol{S} \cup \boldsymbol{T}$ *and* $\boldsymbol{T}$, *respectively, and a probabilistic source database* Pr_s, *deciding whether there exists a probabilistic (or probabilistic universal) solution for* Pr_s *relative to* $\mathcal{M}$ *is* co-NP*-complete (resp., in* PTIME*) in the elementary-event-independent (resp., tuple-independent) case in the data complexity.*

The next result shows that deciding the existence of probabilistic (or probabilistic universal) solutions for deterministic data exchange problems relative to probabilistic source databases is also in PTIME in the elementary-event-independent case in the data complexity, if we restrict the TGDs in the mapping to linear rather than guarded TGDs. The main idea behind this result is that, in the linear case, all the target inconsistencies can be traced back to a polynomial number of potential source inconsistencies, which must be associated with annotations that are either inconsistent or have probability zero, which can be checked in polynomial time in the data complexity.

Theorem 2. *Given a schema mapping* $\mathcal{M} = (\boldsymbol{S}, \boldsymbol{T}, \Sigma_{st} \cup \Sigma_t)$, *where* Σ_{st} *and* Σ_t *are linear TGDs, negative constraints, and non-conflicting keys over* $\boldsymbol{S} \cup \boldsymbol{T}$ *and* $\boldsymbol{T}$, *respectively, and a probabilistic source database* Pr_s, *deciding whether there exists a probabilistic (or probabilistic universal) solution for* Pr_s *relative to* $\mathcal{M}$ *is in* PTIME *in the elementary-event-independent case in the data complexity.*

7.2   Complexity of Probabilistic Data Exchange

The next result shows that deciding the existence of probabilistic (or probabilistic universal) solutions for probabilistic data exchange problems with probabilistic source databases is also co-NP-complete (resp., in PTIME) in the elementary-event-independent (resp., tuple-independent) case in the data complexity. The upper bound for the elementary-event-independent case follows from the fact that relative to a probabilistic source database, a probabilistic universal solution exists iff a probabilistic solution exists, which is in turn equivalent to the existence of a deterministic solution relative to the deterministic database and deterministic mapping for every world. In the tuple-independent case, the tractability follows from the fact that a probabilistic solution exists relative to a probabilistic source database iff a deterministic solution exists relative to the maximal possible deterministic source database and mapping, which can be decided in polynomial time. The co-NP-hardness in the probabilistic case follows from the co-NP-hardness in the less general deterministic case.

Theorem 3. *Given a probabilistic schema mapping* $\mathcal{M} = (\boldsymbol{S}, \boldsymbol{T}, \Sigma_{st} \cup \Sigma_t, \mu)$, *where* Σ_{st} *and* Σ_t *are guarded TGDs, negative constraints, and non-conflicting* *keys over $S \cup T$ and T, respectively, and a probabilistic source database Pr_s, deciding whether there exists a probabilistic (or probabilistic universal) solution for Pr_s relative to $\mathcal{M}$ is co-NP-complete (resp., in PTIME) in the elementary-event-independent (resp., tuple-independent) case in the data complexity.*

Like in the deterministic case, deciding the existence of probabilistic (or probabilistic universal) solutions for probabilistic data exchange problems relative to probabilistic source databases is also in PTIME in the elementary-event-independent case in the data complexity, if we restrict the TGDs in the mapping to linear rather than guarded TGDs. This is because we essentially have to consider one deterministic data exchange problem for every subset of the mapping, which is fixed in the data complexity case.

Theorem 4. *Given a probabilistic schema mapping $\mathcal{M} = (S, T, \Sigma_{st} \cup \Sigma_t, \mu)$, where Σ_{st} and Σ_t are linear TGDs, negative constraints, and non-conflicting keys over $S \cup T$ and T, respectively, and a probabilistic source database Pr_s, deciding whether there exists a probabilistic (or probabilistic universal) solution for Pr_s relative to $\mathcal{M}$ is in PTIME in the elementary-event-independent case in the data complexity.*

8 Provenance

With the compact encoding of probabilistic databases via annotated atoms with events representing the worlds in which they are valid, we are essentially using a data provenance formalism for tracing the probabilistic worlds that an atom belongs to.

Provenance information adds value to data by explaining how it was obtained, thus allowing to validate its correctness, truthfulness, trustworthiness, and compliance. In information integration, when pieces of data from distributed databases or ontologies are integrated, provenance information allows to check the trustworthiness and correctness of the results of queries and debug them as well as trace the errors back to where they were created. Hence, an information integration framework should be equipped with some form of provenance information.

Provenance research distinguishes between *workflow* and *data* provenance. The former is about capturing the processes (i.e., flows and transformations) that a piece of data has gone through before arriving at its current destination or its current version [4]. Some of these processes cannot be accessed and remain a black box in workflow provenance. Contrary to workflow provenance, data provenance is much more fine-grained and focuses on the lineage of data for a query – i.e., the whereabouts of its derivation in a database itself. While we are mainly interested in data provenance, workflow provenance certainly is also of much relevance for data exchange. Hence, the recent W3C recommendation PROV [30], which is a language for specifying workflow provenance in the Web, is relevant to our work as well. However, as we have a fine-grained logical formalization of the data exchange process, we focus on data provenance.

In data provenance, there is a principal distinction made among where-, why- and how-provenance [8]. How-provenance [19] is the most expressive one and the most appropriate for annotating mappings and tracing back the origin of query results. How-provenance can be modeled via an algebraic structure, called a *semiring*, and it is possible to construct different kinds of semirings, depending on what kind of information is to be captured, and which operations on that information are to be allowed. Besides formalizing different kinds of provenance annotations with certain kinds of semirings (called K-relations) based on the positive relational algebra, [19] provides a formalization of plain Datalog without negation with K-relations, used within the collaborative data sharing system ORCHESTRA [20] also for modeling TGDs without existential quantifiers. To capture applications of mappings in ORCHESTRA, [23] proposes to use a so-called $\mathcal{M}$-semiring, which allows to annotate the mappings with names $m_1, \ldots, m_k$ (unary functions), one for each mapping. This can be combined with the formalization of negation-free Datalog (with a procedural semantics based on the least fixpoint operator to construct the model) with positive K-relations as presented in [19].

Clearly, such a formalization for the probabilistic Datalog+/− information integration framework in this paper allows to capture provenance and annotate the mappings with an id such that the integration paths can be traced back to their origin. In this way, routes that can be used to debug mappings like in [9] can be captured. In addition, as shown in [19], when the mappings are the only probabilistic or uncertain elements, the probabilities can also be computed more efficiently, as the captured provenance also carries the information where the probabilities are propagated from. In addition, cycles can be detected, and the trustworthiness of query results can also be estimated, as it can be detected where the data that is involved in the query result has been integrated from. For this purpose, the trustworthiness of data sets and possibly also peers who provide access to data sets need to be assessed beforehand.

A similar approach to the aforementioned ORCHESTRA system for the application of the chase within probabilistic Datalog+/− with a semiring formalization can be constructed and is currently under development. In probabilistic data integration with Datalog+/−, lineage is restricted by the guards, which help to direct the chase towards the answer of a query through the annotated guarded chase forest. In [18,26], a similar kind of tuple annotation as proposed here was used in combination with the chase to speed up the reasoning process.

9   Related Work

Probabilistic data exchange with different combinations of source-to-target TGDs with and without existential quantified variables in the head, target equality constraints, and weakly-acyclic target TGDs with and without existential quantified variables in the head have been studied in [14]. In contrast, Datalog+/− allows to deal with ontologies on the Semantic Web and as such allows to integrate information residing in ontologies. One work on integrating information in ontologies is [3], which tackles the problem of knowledge base data exchange; however, they assume that ontologies in *DL-Lite* are used to exchange data – guarded Datalog+/– strictly subsumes *DL-Lite* and goes well beyond its expressive power.

Other articles that are closely related to our work and that of [14] are [11,17]. There, the source data is deterministic and the mappings are probabilistic. By-table and by-tuple solutions are defined; while the former correspond to a restriction of general probabilistic mappings, by-tuple solutions correspond to mappings that translate only single facts and hence correspond to inclusion dependencies.

In the Semantic Web, the integration of information in ontologies has also been addressed in [31]. There, predicates are defined encoding specific semantics like disjointness, renaming, or decomposition, and an MLN is built from them. Very much related is also our prior work in [7,28]. There, ontologies are mapped with Bayesian description logic programs, which correspond to a subset of probabilistic Datalog+/– and tightly coupled description logics programs with negation under different semantics like the answer set semantics and the well-founded semantics.

Provenance for information integration is used in ORCHESTRA [20] for integrating XML data. In contrast, we exchange data between databases, and we also deal with uncertain and incomplete databases.

10   Summary and Outlook

In this paper, we have studied probabilistic data exchange via probabilistic Datalog+/– and used annotations encoding probabilistic data provenance of atoms. We have also considered using provenance for tracking the origin of integrated information and using provenance for tracking the origin of mappings themselves, including the conditions for their applicability for trust assessment and debugging.

By means of Datalog+/– [6], which can represent *DL-Lite* and $\mathcal{EL}$, we use a tractable language with dependencies that allows to nicely tie together the theoretical results on information integration in databases and the work on ontology mediation in the Semantic Web. The separation between the ontology and the probabilistic dependencies allows us to either model the mappings with specific newly-invented predicates like disjointness, renaming, or decomposition, etc. or – more interestingly – with a probabilistic meta ontology describing the matching process.

Our work shows how classical and probabilistic (guarded and linear) Datalog+/– can be used to model information integration settings, and sketches a deterministic mapping language based on Datalog+/– and a probabilistic generalization based on the rather loosely coupled probabilistic extension of Datalog+/– with worlds represented by propositional events. We also justify why data provenance needs to be captured and represented within such a probabilistic information integration framework, and propose to use an adaptation of K-relations as proposed by [19]. Such an extension with provenance allows to track how results of queries to the framework have been created, and also debug mappings, since errors can be traced back to their origin.

An interesting topic for future research is to develop the proposed framework for provenance capture and, among others, investigate how to use the chase for reasoning with probabilistic Datalog+/− within a semiring-framework.

Acknowledgments. This work was supported by the EU (FP7/2007-2013) Marie-Curie Intra-European Fellowship "PRODIMA", the Engineering and Physical Sciences Research Council (EPSRC) grant EP/J008346/1 "PrOQAW: Probabilistic Ontological Query Answering on the Web", the European Research Council (FP7/2007-2013/ERC) grant 246858 ("DIADEM"), and by a Yahoo! Research Fellowship.

References

1. Andréka, H., Németi, I., van Benthem, J.: Modal languages and bounded fragments of predicate logic. J. Philos. Logic **27**(3), 217–274 (1998)
2. Bernstein, P.A., Haas, L.M.: Information integration in the enterprise. Commun. ACM **9**(51), 72–79 (2008)
3. Botoeva, E.: Description Logic Knowledge Base Exchange. Ph.D. thesis, Free University of Bozen-Bolzano (2014)
4. Buneman, P.: The providence of provenance. In: Gottlob, G., Grasso, G., Olteanu, D., Schallhart, C. (eds.) BNCOD 2013. LNCS, vol. 7968, pp. 7–12. Springer, Heidelberg (2013)
5. Calì, A., Gottlob, G., Kifer, M.: Taming the infinite chase: Query answering under expressive integrity constraints. In: Proceedings of KR, pp. 70–80. AAAI Press (2008)
6. Calì, A., Gottlob, G., Lukasiewicz, T.: A general datalog-based framework for tractable query answering over ontologies. J. Web Sem. **14**, 57–83 (2012)
7. Calì, A., Lukasiewicz, T., Predoiu, L., Stuckenschmidt, H.: Rule-based approaches for representing probabilistic ontology mappings. In: da Costa, P.C.G., d'Amato, C., Fanizzi, N., Laskey, K.B., Laskey, K.J., Lukasiewicz, T., Nickles, M., Pool, M. (eds.) URSW 2005 - 2007. LNCS (LNAI), vol. 5327, pp. 66–87. Springer, Heidelberg (2008)
8. Cheney, J., Chiticariu, L., Tan, W.C.: Provenance in databases: why, how and where. Found. Trends Databases **1**(4), 379–474 (2009)
9. Chiticariu, L., Tan, W.C.: Debugging schema mappings with routes. In: Proceedings of VLDB, pp. 79–90. ACM Press (2006)
10. Deutsch, A., Nash, A., Remmel, J.: The chase revisited. In: Proceedings of PODS, pp. 149–158. ACM Press (2008)
11. Dong, X.L., Halevy, A., Yu, C.: Data integration with uncertainty. VLDB J. **18**, 469–500 (2009)
12. Euzenat, J., Shvaiko, P.: Ontology Matching. Springer, Heidelberg (2007)
13. Fagin, R., Kimelfeld, B., Kolaitis, P.G.: Probabilistic data exchange. In: Proceedings of ICDT, pp. 76–88. ACM Press (2010)
14. Fagin, R., Kimelfeld, B., Kolaitis, P.G.: Probabilistic data exchange. J. ACM **58**, 15 (2011)
15. Fagin, R., Kolaitis, P.G., Miller, R.J., Popa, L.: Data exchange: semantics and query answering. Theor. Comput. Sci. **336**(1), 89–124 (2005)

16. Fagin, R., Kolaitis, P.G., Popa, L., Tan, W.C.: Composing schema mappings: second-order dependencies to the rescue. ACM T. Database Syst. **30**, 994–1055 (2005)
17. Gal, A., Martinez, M.V., Simari, G.I., Subrahmanian, V.S.: Aggregate query answering under uncertain schema mappings. In: Proceedings of ICDE, pp. 940–951. IEEE Computer Society (2009)
18. Gottlob, G., Lukasiewicz, T., Martinez, M.V., Simari, G.I.: Query answering under probabilistic uncertainty in Datalog+/- ontologies. Ann. Math. Artif. Intell. **69**(1), 37–72 (2013)
19. Green, T.J., Karvounarakis, G., Tannen, V.: Provenance semirings. In: Proceedings of PODS, pp. 31–40. ACM Press (2007)
20. Green, T.J., Karvounarakis, G., Taylor, N.E., Biton, O., Ives, Z.G., Tannen, V.: ORCHESTRA: facilitating collaborative data sharing. In: Proceedings of SIGMOD, pp. 1131–1133. ACM Press (2007)
21. Haas, L.: Beauty and the beast: the theory and practice of information integration. In: Schwentick, T., Suciu, D. (eds.) ICDT 2007. LNCS, vol. 4353, pp. 28–43. Springer, Heidelberg (2006)
22. Johnson, D.S., Klug, A.: Testing containment of conjunctive queries under functional and inclusion dependencies. J. Comput. Syst. Sci. **28**, 167–189 (1984)
23. Karvounarakis, G.: Provenance in collaborative data sharing. Ph.D. thesis, University of Pennsylvania (2009)
24. Lenzerini, M.: Data integration: A theoretical perspective. In: Proceedings of PODS, pp. 233–246. ACM Press (2002)
25. Levesque, H.J., Brachman, R.J.: Expressiveness and tractability in knowledge representation and reasoning. Comput. Intell. **3**, 78–93 (1987)
26. Lukasiewicz, T., Martinez, M.V., Simari, G.I.: Consistent answers in probabilistic Datalog+/- ontologies. In: Krötzsch, M., Straccia, U. (eds.) RR 2012. LNCS, vol. 7497, pp. 156–171. Springer, Heidelberg (2012)
27. Lukasiewicz, T., Predoiu, L.: Information integration with provenance on the Semantic Web via probabilistic Datalog+/-. In: Proceedings of URSW. CEUR Workshop Proceedings (2013)
28. Lukasiewicz, T., Predoiu, L., Stuckenschmidt, H.: Tightly integrated probabilistic description logic programs for representing ontology mappings. Ann. Math. Artif. Intell. **63**(3/4), 385–425 (2011)
29. Maier, D., Mendelzon, A.O., Sagiv, Y.: Testing implications of data dependencies. ACM Trans. Database Syst. **4**(4), 455–469 (1979)
30. Moreau, L., Groth, P.: Provenance: An Introduction to PROV. Morgan and Claypool, San Rafael (2013)
31. Niepert, M., Noessner, J., Meilicke, C., Stuckenschmidt, H.: Probabilistic-logical web data integration. In: Polleres, A., d'Amato, C., Arenas, M., Handschuh, S., Kroner, P., Ossowski, S., Patel-Schneider, P. (eds.) Reasoning Web 2011. LNCS, vol. 6848, pp. 504–533. Springer, Heidelberg (2011)
32. Scharffe, F., de Bruijn, J.: A language to specify mappings between ontologies. In: Proceedings of SITIS, pp. 267–271. Dicolor Press (2005)
33. Serafini, L., Stuckenschmidt, H., Wache, H.: A formal investigation of mapping language for terminological knowledge. In: Proceedings of IJCAI, pp. 576–581. Professional Book Center (2005)
34. ten Cate, B., Kolaitis, P.G.: Structural characterizations of schema-mapping languages. Commun. ACM **53**, 101–110 (2010)

Learning Probabilistic Description Logics

Fabrizio Riguzzi[1]([✉]), Elena Bellodi[2], Evelina Lamma[2],
Riccardo Zese[2], and Giuseppe Cota[2]

[1] Dipartimento di Matematica e Informatica, University of Ferrara,
Via Saragat 1, 44122 Ferrara, Italy
fabrizio.riguzzi@unife.it
[2] Dipartimento di Ingegneria, University of Ferrara,
Via Saragat 1, 44122 Ferrara, Italy
{elena.bellodi,evelina.lamma,riccardo.zese}@unife.it,
giuseppe.cota@student.unife.it

Abstract. We consider the problem of learning both the structure and
the parameters of Probabilistic Description Logics under the DISPONTE
semantics. DISPONTE is based on the distribution semantics for Probabilistic Logic Programming and assigns a probability to assertional and
terminological axioms. The system EDGE, given a DISPONTE knowledge base (KB) and sets of positive and negative examples in the form
of concept assertions, returns the value of the probabilities associated
with axioms. We present the system LEAP that learns both the structure and the parameters of DISPONTE KBs explointing EDGE. LEAP
is based on the system CELOE for ontology engineering and exploits its
search strategy in the space of possible axioms. LEAP uses the axioms
returned by CELOE to build a KB so that the likelihood of the examples
is maximized. We present experiments showing the potential of EDGE
and LEAP.

1 Introduction

Recently, the problem of representing uncertainty in Description Logics (DLs)
has received an increasing attention due to the ubiquity of uncertain information
in real world domains. Various authors have studied the use of probabilistic
DLs and many proposals have been presented for allowing DLs to represent
uncertainty [10,13,16,17,28].

In addition, some works have started to appear about learning the probabilities or the whole structure of probabilistic ontologies. These arise, on one hand,
from the fact that specifying the values of the probabilities is a difficult task for
humans and data is usually available that could be leveraged for tuning them,
and, on the other hand, from the fact that in some domains there exist poor-
structured knowledge bases which could be improved [13,14]. A knowledge base
with a refined structure and instance data coherent with it allows more powerful
reasoning, better consistency checking and improved querying possibilities.

In [2,18,19,23] we proposed an approach for the integration of probabilistic
information in DLs called DISPONTE for "DIstribution Semantics for

© Springer International Publishing Switzerland 2014
F. Bobillo et al. (Eds.): URSW 2011-2013, LNAI 8816, pp. 63–78, 2014.
DOI: 10.1007/978-3-319-13413-0_4

Probabilistic ONTologiEs". DISPONTE applies the distribution semantics for probabilistic logic programming [25] to DLs.

In this paper we present an approach for learning the structure of probabilistic DLs following the DISPONTE semantics. The approach is based on the algorithm EDGE for "Em over bDds for description loGics paramEter learning" [21,22] that starts from examples of instances and non-instances of concepts and learns the parameters of a probabilistic theory. EDGE builds Binary Decision Diagrams (BDDs) for representing the explanations of the examples from the theory. The parameters are then tuned using an EM algorithm [8] in which the required expectations are computed directly on the BDDs in an efficient way.

The algorithm for learning the structure is called LEAP for "LEArning Probabilistic description logics" and combines the learning system CELOE with EDGE. The former provides a method to build new (equivalence and subsumption) axioms that can be added to the KB, the latter is used to learn the parameters of these probabilistic axioms.

We provide a performance evaluation of both EDGE and LEAP. For EDGE, we extend the evaluation of [21] with a new dataset and that of [22] by including a cross-validation result. For LEAP, we present a comparison between a theory before and after applying LEAP. The experiments with EDGE show that it achieves statistically significant greater areas under the Precision Recall and the Receiver Operating Characteristics curves (AUCPR and AUCROC) with respect to a theory where the probabilities are obtained from an Association Rule learner. The experiments with LEAP show that it improves the AUCPR and AUCROC of the theory with the difference being statistically significant for AUCROC.

The paper is organized as follows. Section 2 introduces Description Logics and the DISPONTE semantics. Section 3 describes EDGE. In Sect. 4 we introduce LEAP. Section 5 discusses related works and Sect. 6 shows the results of experiments for both systems. Section 7 concludes the paper.

2   Description Logics and the DISPONTE Semantics

Description Logics (DLs) are knowledge representation formalisms that are particularly useful for representing ontologies. Their syntax is usually based on concepts and roles. A concept corresponds to a set of individuals of the domain while a role corresponds to a set of couples of individuals of the domain. In the following we consider and describe $\mathcal{ALC}$ [26].

Let $\mathbf{A}$, $\mathbf{R}$ and $\mathbf{I}$ be sets of *atomic concepts*, *roles* and *individuals*, respectively. *Concepts* are defined by induction as follows. Each $A \in \mathbf{A}$ is a concept and $\bot$ and $\top$ are concepts. If C, $C1$ and $C2$ are concepts and $R \in \mathbf{R}$, then $(C_1 \sqcap C_2)$, $(C_1 \sqcup C_2)$ and $\neg C$ are concepts, as well as $\exists R.C$ and $\forall R.C$. A *TBox* $\mathcal{T}$ is a finite set of *concept inclusion axioms* $C \sqsubseteq D$, where C and D are concepts; we use $C \equiv D$ to abbreviate $C \sqsubseteq D$ and $D \sqsubseteq C$. An *ABox* $\mathcal{A}$ is a finite set of *concept membership axioms* $a : C$, *role membership axioms* $(a, b) : R$, *equality axioms* $a = b$ and *inequality axioms* $a \neq b$. A *knowledge base* (KB) $\mathcal{K} = (\mathcal{T}, \mathcal{A})$ consists of a TBox $\mathcal{T}$ and an ABox $\mathcal{A}$.

A knowledge base $\mathcal{K}$ is usually assigned a semantics in terms of set-theoretic interpretations and models of the form $\mathcal{I} = (\Delta^{\mathcal{I}}, \cdot^{\mathcal{I}})$, where $\Delta^{\mathcal{I}}$ is a non-empty *domain* and $\cdot^{\mathcal{I}}$ is the *interpretation function* that assigns an element in $\Delta^{\mathcal{I}}$ to each $a \in \mathbf{I}$, a subset of $\Delta^{\mathcal{I}}$ to each $C \in \mathbf{A}$ and a subset of $\Delta^{\mathcal{I}} \times \Delta^{\mathcal{I}}$ to each $R \in \mathbf{R}$. The mapping $\cdot^{\mathcal{I}}$ is extended to all concepts (where $R^{\mathcal{I}}(x) = \{y | (x, y) \in R^{\mathcal{I}}\}$) as:

$$\top^{\mathcal{I}} = \Delta^{\mathcal{I}}$$
$$(\neg C)^{\mathcal{I}} = \Delta^{\mathcal{I}} \setminus C^{\mathcal{I}}$$
$$(C_1 \sqcup C_2)^{\mathcal{I}} = C_1^{\mathcal{I}} \cup C_2^{\mathcal{I}}$$
$$(\exists R.C)^{\mathcal{I}} = \{x \in \Delta^{\mathcal{I}} | R^{\mathcal{I}}(x) \cap C^{\mathcal{I}} \neq \emptyset\}$$
$$\bot^{\mathcal{I}} = \emptyset$$
$$(C_1 \sqcap C_2)^{\mathcal{I}} = C_1^{\mathcal{I}} \cap C_2^{\mathcal{I}}$$
$$(\forall R.C)^{\mathcal{I}} = \{x \in \Delta^{\mathcal{I}} | R^{\mathcal{I}}(x) \subseteq C^{\mathcal{I}}\}$$

A query over a knowledge base is an axiom for which we want to test the entailment from the knowledge base. The entailment test may be reduced to checking the unsatisfiability of a concept in the knowledge base, i.e., the emptiness of the concept.

DISPONTE applies the distribution semantics [25] to probabilistic ontologies. In DISPONTE a *probabilistic knowledge base* $\mathcal{K}$ is a set of certain and probabilistic axioms: *certain axioms* take the form of regular DL axioms, *probabilistic axioms* take the form $p :: E$, where p is a real number in $[0, 1]$ and E is a DL axiom. The probability p can be interpreted as an *epistemic* probability, i.e., as the degree of our belief in axiom E. A DISPONTE KB defines a distribution over DL KBs called *worlds*. Each world w is obtained by including every certain axiom. For each probabilistic axiom, we decide whether or not to include it in w. A world therefore is a non probabilistic KB that can be assigned a semantics in the usual way. By multiplying the probability of the choices made to obtain a world we can assign a probability to it. The probability of a query is then the sum of the probabilities of the worlds where the query holds true.

The system BUNDLE [18–20, 23, 24] computes the probability of a query w.r.t. ontologies that follow the DISPONTE semantics by first computing all the explanations for the query and then building a Binary Decision Diagram (BDD) that represents them. A *set of explanations* for a query Q is a set of sets of pairs (E_i, k) where E_i is the ith probabilistic axiom and $k \in \{0, 1\}$ indicates whether E_i is chosen to be included in a world ($k = 1$) or not ($k = 0$). Given the set of explanations K for a query Q, we can define the Disjunctive Normal Form (DNF) Boolean formula f_K as $f_K(\mathbf{X}) = \bigvee_{\kappa \in K} \bigwedge_{(E_i, 1)} X_i \bigwedge_{(E_i, 0)} \overline{X_i}$. The variables $\mathbf{X} = \{X_i | (E_i, k) \in \kappa, \kappa \in K\}$ are independent Boolean random variables and the probability that $f_K(\mathbf{X})$ takes on value 1 is equal to the probability of Q. A BDD for a function of Boolean variables is a rooted graph that has one level for each Boolean variable. A node n has two children: one corresponding to the 1 value of the variable associated with the level of n, indicated with $child_1(n)$, and one corresponding to the 0 value of the variable, indicated with $child_0(n)$. When drawing BDDs, the 0-branch - the one going to $child_0(n)$ - is distinguished from the 1-branch by drawing it with a dashed line. The leaves store either 0 or 1.

Explanations are found by using the Pellet reasoner [27] and are then translated into a BDD that allows to compute the probability of Q with a dynamic programming algorithm in polynomial time in the size of the diagram [7].

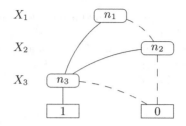

Fig. 1. BDD for Example 1.

The system TRILL [29] implements the BUNDLE's inference algorithm in Prolog and compute the probability of a query w.r.t. KBs that follow the DISPONTE semantics.

Example 1. Let us consider the following knowledge base, inspired by the ontology `people+pets` proposed in [15]:

$$\exists hasAnimal.Pet \sqsubseteq NatureLover$$
$$(kevin, fluffy) : hasAnimal$$
$$(kevin, tom) : hasAnimal$$

(E_1) $0.4 :: fluffy : Cat$

(E_2) $0.3 :: tom : Cat$

(E_3) $0.6 :: Cat \sqsubseteq Pet$

Individuals that own an animal which is a pet are nature lovers and *kevin* owns the animals *fluffy* and *tom*. We believe in the fact that *fluffy* and *tom* are cats and that cats are pets with the specified probability. This KB has eight worlds and the query axiom $Q = kevin : NatureLover$ is true in three of them, corresponding to the following choices: $\{(E_1, 1), (E_2, 0), (E_3, 1)\}$, $\{(E_1, 0), (E_2, 1), (E_3, 1)\}, \{(E_1, 1), (E_2, 1), (E_3, 1)\}$. The probability is therefore $P(Q) = 0.4 \cdot 0.7 \cdot 0.6 + 0.6 \cdot 0.3 \cdot 0.6 + 0.4 \cdot 0.3 \cdot 0.6 = 0.348$. If we associate the random variables X_1 to the axiom E_1, X_2 to E_2 and X_3 to E_3, the BDD representing the set of explanations is shown in Fig. 1.

3   Parameter Learning of Probabilistic DLs

EDGE [21,22] performs parameter learning of probabilistic ontologies under the DISPONTE semantics and is inspired by the algorithm EMBLEM [3,4], which was developed for learning the parameters of probabilistic logic programs under the distribution semantics. The parameters correspond to the epistemic probabilities previously introduced and are tuned using an Expectation Maximization (EM) algorithm [8], an iterative method to estimate some unknown parameters Θ of a model: in particular, it finds maximum likelihood or maximum a posteriori (MAP) estimates of Θ. EM alternates between performing an Expectation (E) step, where the missing data are estimated given the observed data and current estimate of the model parameters, and a Maximization (M) step, which computes the parameters maximizing the likelihood of the data given the sufficient statistics on the data computed in the E step.

EDGE takes as input a DL KB and a number of examples that represent the queries. Tipically, the queries are concept assertions and are divided into *positive* examples (set E^+) - representing true information, for which we would like to get high probability - and *negative* examples (set E^-) - representing false information, for which we would like to get low probability. EDGE first computes, for each query Q, the BDD encoding its explanations using the reasoner BUNDLE. A limit on the maximum number of explanations to be found ($NumE$) or a time limit for the search for explanations (TLE) can be possibly set for BUNDLE. For negative examples, EDGE computes the explanations of the query, builds the BDD and then negates it. For example, if the negative example is $a : C$, EDGE executes the query $a : C$, finds the BDD and then negates it. Given the knowledge base of Example 1 and the positive example $kevin : NatureLover$, we obtain the BDD in Fig. 1.

EDGE main procedure consists of the EM cycle in which the steps of Expectation and Maximization are repeated until the log-likelihood (LL) of the examples reaches a local maximum, as shown in Algorithm 1. At each iteration the LL of the example increases, i.e., the probability of positive examples increases and that of negative examples decreases. The EM algorithm is guaranteed to find a local maximum, which however may not be the global maximum. Procedure EXPECTATION returns the LL of the data that is used in the stopping criterion: EDGE stops when the difference between the LL of the current iteration and the one of the previous iteration (LL_0) drops below a threshold ϵ or when this difference is below a fraction δ of the LL.

Algorithm 1. Function EDGE.

```
 1: function EDGE(K, E+, E−, ε, δ, NumE, TLE)
 2:     Build BDDs      ▷ BUNDLE builds all the BDDs according to the limits NumE and TLE
 3:     LL = −inf
 4:     repeat
 5:         LL0 = LL
 6:         LL = EXPECTATION(BDDs)
 7:         MAXIMIZATION
 8:     until LL − LL0 < ε ∨ LL − LL0 < −LL · δ
 9:     return (LL, K)
10: end function
```

Procedure EXPECTATION (shown in Algorithm 2) takes as input a list of BDDs, one for each example Q, and computes the expectations $\mathbf{E}[c_{i0}|Q]$ and $\mathbf{E}[c_{i1}|Q]$ for all axioms E_i directly over the BDDs. Let c_{ix} be the number of times a Boolean random variable X_i takes on value x for $x \in \{0, 1\}$:

$$\mathbf{E}[c_{ix}|Q] = P(X_i = x|Q).$$

Then it sums up the contributions of all examples: $\mathbf{E}[c_{ix}] = \sum_Q \mathbf{E}[c_{ix}|Q]$. Finally, $P(X_i = x|Q)$ is given by $\frac{P(X_i=x,Q)}{P(Q)}$. In this procedure we use $\eta^x(i)$

Algorithm 2. Function Expectation.

```
 1: function EXPECTATION(BDDs)
 2:     LL = 0
 3:     for all i ∈ Axioms do
 4:         E[c_i0] = 0; E[c_i1] = 0
 5:     end for
 6:     for all BDD ∈ BDDs do
 7:         for all i ∈ Axioms do
 8:             η⁰(i) = 0; η¹(i) = 0
 9:         end for
10:         for all variables X do
11:             ς(X) = 0
12:         end for
13:         GETFORWARD(root(BDD))
14:         Prob=GETBACKWARD(root(BDD))
15:         T = 0
16:         for l = 1 to levels(BDD) do
17:             Let X_i be the variable associated with level l
18:             T = T + ς(X_i)
19:             η⁰(i) = η⁰(i) + T × (1 − p_i)
20:             η¹(i) = η¹(i) + T × p_i
21:         end for
22:         for all i ∈ Axioms do
23:             E[c_i0] = E[c_i0] + η⁰(i)/Prob
24:             E[c_i1] = E[c_i1] + η¹(i)/Prob
25:         end for
26:         LL = LL + log(Prob)
27:     end for
28:     return LL
29: end function
```

---

to indicate $P(X_i = x, Q)$. EXPECTATION first calls procedures GETFORWARD and GETBACKWARD that compute the forward and the backward probability of nodes and $\eta^x(i)$ for non-deleted paths only. These are the paths that have not been deleted when building the BDDs. Forward and backward probabilities in each node represent the probability mass of paths from the root to the node and that of the paths from the node to the leaves respectively. The expression

$$P(X_i = x, Q) = \sum_{n \in N(Q), v(n) = X_i} F(n)B(child_x(n))\pi_{ix},$$

with $N(Q)$ the set of BDD nodes for query Q, $v(n)$ the variable associated with node n, $\pi_{i1} = p_i$, $\pi_{i0} = 1 − p_i$, $F(n)$ the forward probability of n, $B(n)$ the backward probability of n, represents the probability mass of each path passing through each node associated with X_i and going down its x-branch. We use the notation $e^x(n)$ to indicate the expression inside the sum. Computing the two types of probability in the nodes requires two traversals of the graph, so its cost is linear in the number of nodes.

Procedure GETFORWARD computes the value of the forward probabilities for every node. It traverses the diagram one level at a time starting from the root level, where $F(root) = 1$, and for each node n computes its contribution to the forward probabilities of its children. Then the forward probabilities of both children are updated. Function GETBACKWARD computes the backward probability of nodes by traversing recursively the tree from the leaves to the root. It returns the backward probability of the root corresponding to the probability of the query $P(Q)$, indicated as $Prob$ at line 14 of Algorithm 2.

When the calls of GETBACKWARD for both children of a node n return, we compute the $e^x(n)$ and $\eta^x(i)$ values for non-deleted paths. An array ς is used to store the contributions of the deleted paths by starting from the root level and accumulating $\varsigma(l)$ for the various levels l. See [22] for more details.

Expectations are updated for all axioms (lines 23–24) and finally the log-likelihood of the current example is added to the overall log likelihood.

Procedure MAXIMIZATION computes the parameters values for the next EM iteration by relative frequency for all axioms E_i:

$$p_i = \frac{\mathbf{E}[c_{i1}]}{\mathbf{E}[c_{i0}] + \mathbf{E}[c_{i1}]}.$$

4   Structure Learning of Probabilistic DLs

LEAP performs structure and parameter learning of probabilistic ontologies under the DISPONTE semantics by exploiting: (1) CELOE [11] for the structure, and (2) EDGE (Sect. 3) for the parameters. We first introduce CELOE before describing LEAP.

4.1   CELOE

CELOE [11] stands for "Class Expression Learning for Ontology Engineering" and is available in the Java open-source framework DL-Learner[1] for OWL and DLs.

Let us consider a knowledge base $\mathcal{K}$ and a class Target whose formal description we want to learn. Target has (inferred or asserted) instances in $\mathcal{K}$. CELOE can take as input a target class, a set of positive and negative examples (i.e. individuals) or a set of positive only examples.

If Target is already described by a class expression C through axioms such as Target $\sqsubseteq C$ or Target $\equiv C$, it is possible to learn a description for Target by refining C or by relearning from scratch, as stated in Definition 1.

Definition 1 (Class Learning Problem). Let an existing named class Target be in a knowledge base $\mathcal{K}$. Let $R_\mathcal{K}(C)$ be a retrieval reasoner operation that returns the set of all instances of C. The class learning problem is to find an expression C such that $R_\mathcal{K}(\text{Target}) = R_\mathcal{K}(C)$.

CELOE finds a set of n class expressions C_i ($1 \leq i \leq n$) sorted according to a heuristic. Such expressions are candidates for adding axioms of the form Target $\equiv C_i$ or Target $\sqsubseteq C_i$.

---

[1] http://dl-learner.org/Projects/DLLearner

On the other hand, if a set of positive and negative examples or a set of only positive examples is given, CELOE can be seen as a learning algorithm that solves a problem of learning from examples, as described in Definition 2.

Definition 2 (Learning from Examples Problem).
   Given:

- *a concept name* Target*;*
- *a knowledge base* $\mathcal{K}$ *not containing* Target*;*
- *a space of possible concepts* $\mathcal{C}$*;*
- *a set of positive examples* E^+ *with elements of the form* $a :$ Target $(a \in \mathbf{I})$*;*
- *a set of negative examples* E^- *with elements of the form* $a :$ Target $(a \in \mathbf{I})$*;*

Find a concept expression $C \in \mathcal{C}$ *such that:*

- Target *does not occur in* C *(acyclic definition);*
- $\forall e^+ \in E^+, \mathcal{K} \cup \{$Target $\equiv C\} \models e^+$*;*
- $\forall e^- \in E^-, \mathcal{K} \cup \{$Target $\equiv C\} \not\models e^-$*.*

If $\mathcal{K}' = \mathcal{K} \cup \{$Target $\equiv C\}$ we say that a concept C *covers* an example $e \in E^+ \cup E^-$ if $\mathcal{K}' \models e$. We distinguish the two cases in which both sets E^+ and E^- of individuals are given or only the set E^+ is given as *Positive and Negative Examples Learning Problem* (*LP*) and *Positive Examples Learning Problem* respectively.

CELOE is a top-down algorithm that starts from the $\top$ concept and uses the $\mathcal{ALCQ}$ refinement operator defined in [12]. Each generated class expression is evaluated using one of five available heuristics, whose resulting value is used to guide the search in the learning process. All these heuristics need a set of examples in order to be computed; in the case the algorithm took as input only the target class to be described, we can consider as positive examples the existing instances (inferred or asserted) of the target class and the remaining instances in the KB as negative examples.

4.2   LEAP

In order to learn an ontology, LEAP first finds good candidate axioms (subsumption axioms) by means of CELOE, then it performs a greedy search in the space of theories.

LEAP main procedure is shown in Algorithm 3: it takes as input the knowledge base $\mathcal{K}$ and the type of learning problem LP_{type}; the maximum number of class expressions $NumC$ and the time limit TLC for CELOE; the values of ϵ and δ, the maximum number of explanations $NumE$ and the time limit TLE for the computation of the BDDs for each example for EDGE. Note that CELOE's default is $NumC = 10$ and $TLC = 10$ s and EDGE's default is $NumE = TLE = \infty$.

In the first phase, a set of class expressions is generated by using CELOE (line 2), then the sets of positive (P_I) and negative (N_I) individuals are extracted according to the following rules:

- if a set of positive and negative individuals has been given as input to CELOE (LP_{type} = *Positive and Negative Examples Learning Problem*), then no extraction is necessary;
- if a set of positive only individuals has been given (LP_{type} = *Positive Examples Learning Problem*), then the set of negative examples will be composed of all the individuals of $\mathcal{K}$ except the positive ones;
- if a target class has been given (LP_{type} = *Class Learning Problem*, cf. Definition 1), then we consider the existing instances (inferred or asserted) of the target class as positive individuals and the remaining instances as negative individuals.

After the extraction, the *assertional* axioms, which represent the examples (i.e. queries) for EDGE, are created (see lines 4–9). Then EDGE is applied to the KB to compute the initial value of the parameters and of the LL.

In the second phase, LEAP performs a greedy search in the space of theories, described in lines 11–19. For each element of the class expressions set, one probabilistic subsumption axiom at a time of the form $p :: CE \sqsubseteq \texttt{Target}$ is added to the ontology $\mathcal{K}$; p is either a random probabilistic value or the accuracy returned by CELOE. After each addition, EDGE is run on the extended theory to compute the log-likelihood of the data LL and the updated parameters (line 14). If LL is better than the current best LL_0, the new axiom is kept in the knowledge base, otherwise the new axiom is discarded (lines 15–18). The final theory, obtained from the union of the initial ontology and the probabilistic subsumption axioms learned, is returned to the user.

LEAP is a client-server Java RMI application. The server side contains a class called EDGERemote, which performs the EDGE algorithm. The client side, instead, runs a modified version of CELOE called ProbCELOE and a class called EDGE that invokes the remote methods of EDGERemote in order to compute the log-likelihood and the parameters. Figure 2 illustrates the communication between the LEAP client and the server.

5 Related Work

GoldMiner [10, 28] is an algorithm that exploits Association Rules (ARs) for building ontologies. GoldMiner extracts information about individuals, named classes and roles using SPARQL queries. From these data, it builds two *transaction tables*: one that stores the classes to which each individual belongs and one that stores the roles to which each couple of individuals belongs. Finally, the APRIORI algorithm [1] is applied to each table in order to find ARs. Implications of the form $A \Rightarrow B$ can be converted to subclass axioms of the form $A \sqsubseteq B$. Moreover, the confidence p of an AR can be interpreted as the probability of the axiom $p :: A \sqsubseteq B$. So GoldMiner can be used to obtain a probabilistic knowledge base.

The structure learner LEAP is inspired to SLIPCOVER, an algorithm proposed for learning probabilistic logic programs based on distribution semantics [5]. LEAP shares with it the search strategy and the use of the log-likelihood of the data as the score of the learnt theories. Like SLIPCOVER, it divides the

Algorithm 3. Function LEAP.

```
1: function LEAP(𝒦, LP_type, NumC, TLC, ε, δ, NumE, TLE)
2:     ClassExpressions = up to NumC or until TLC is reached          ▷ generated by CELOE
3:     (P_I, N_I) = EXTRACTINDIVIDUALS(LP_type)
4:     for all ind ∈ P_I do                                           ▷ P_I: set of positive individuals
5:         Add ind : Target to P_E                                    ▷ P_E: set of positive examples
6:     end for
7:     for all ind ∈ N_I do                                          ▷ N_I: set of negative individuals
8:         Add ind : Target to N_E                                    ▷ N_E: set of negative examples
9:     end for
10:    (LL_0, 𝒦) = EDGE(𝒦, P_E, N_E, ε, δ, NumE, TLE)
11:    for all CE ∈ ClassExpressions do
12:        Axiom = p::CE ⊑ Target
13:        𝒦' = 𝒦 ∪ {Axiom}
14:        (LL, 𝒦') = EDGE(𝒦', P_E, N_E, ε, δ, NumE, TLE)
15:        if LL > LL_0 then
16:            𝒦 = 𝒦'
17:            LL_0 = LL
18:        end if
19:    end for
20:    return 𝒦
21: end function
```

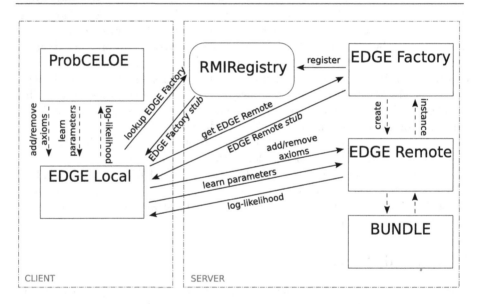

Fig. 2. LEAP as a client-server Java RMI application.

search between learning promising axioms and building in a greedy way a theory whose parameters are optimized by relying on a parameter learning algorithm.

A work that integrates parameter and structure learning for a probabilistic extension of $\mathcal{ALC}$, named CR$\mathcal{ALC}$, is [13]. CR$\mathcal{ALC}$ allows statistical axioms of the form $P(C|D) = \alpha$, meaning that for any element x in $\mathcal{D}$, the probability that it is in C given that is in D is α, and of the form $P(R) = \beta$, meaning that for each couple of elements x and y in $\mathcal{D}$, the probability that x is linked to y by the role R is β. CR$\mathcal{ALC}$ does not allow to express a degree of belief in axioms as DISPONTE. An algorithm is presented in [13] that learns parameters and structure of CR$\mathcal{ALC}$ KBs. It starts from positive and negative examples for a single concept and from the general concept $\top$ in the root of a search tree to be refined. For a set of candidate concept definitions, their probabilistic parameters are learned using an EM algorithm and a score is assigned to the corresponding node. If the best score in the tree is above a threshold, a deterministic concept definition is returned, otherwise a probabilistic inclusion C_i is searched on a weighted spanning tree, where the target concept is added as a parent of each vertex and probabilities are learned as $P(C_i|Parents(C_i))$. We share the top-down procedure for building axioms (CELOE) but we exploit the BDD structures instead of resorting to inference in a graphical model to compute the expected counts for EM.

The paper [14] presents a Statistical Relational Learning system for learning terminological naïve Bayesian classifiers, which estimate the probability that an individual a belongs to a certain target concept given its membership to a set of induced DL (feature) concepts. The classifier consists of a Bayesian Network (BN) modelling the dependency relations between the feature concepts and the target one. The learning process handles three different assumptions that can be made about the lack of knowledge (under OWA) regarding concept-membership, reflecting in the adoption of different scoring functions and search strategies of the optimal network and parameters. Under one of the assumptions - the probability of concept-membership of a depends on the knowledge on a available in $\mathcal{K}$ - the EM method is proposed to train the BN parameters. The classifier can be seen as a learner of probabilistic assertional axioms, while LEAP learns probabilistic terminological axioms. We exploit BDDs instead of BNs, while we share with them the use of EM.

6   Experiments

In order to test the performances of EDGE and of LEAP we performed several experiments. First we have executed two tests on EDGE which are inspired by the ones presented in [21,22]. Then, once shown that EDGE achieves good results, we have done a preliminary test for comparing LEAP with it.

6.1   Parameter Learning

EDGE has been compared with Association Rules (ARs) [21,22] over two real world datasets from the Linked Open Data cloud: education.data.gov.uk[2] and an extract of DBPedia[3]. We extend the experiments from [21] by including the DBPedia dataset and those of [22] by presenting the results of a cross-validation rather than of a single training-test split.

In the experiments, we wanted to simulate the situation in which an expert provides the structure of the ontology together with information on a set of

---

[2] http://education.data.gov.uk/
[3] http://dbpedia.org/About

individuals. The ontologies were obtained with GoldMiner: we extracted 10,000 individuals and 5,545 axioms for education.data.gov.uk and 7,200 individuals and 6,228 axioms for DBPedia and we learned ARs from the resulting transaction tables. The ARs were then converted into subclass axioms.

In order to generate a set of examples (queries) for EDGE, for each extracted individual a we sampled three named classes: A and B were sampled from the named classes to which a explicitly belonged, while C was sampled from the named classes to which a did not explicitly belong but that exhibited at least one explanation for the query $a : C$. The axiom $a : A$ is added to the KB, while $a : B$ is considered as a positive example and $a : C$ as a negative example. We used a 5-fold cross validation to test the system. In the training phase, we ran EDGE on the ontology obtained by GoldMiner where we consider all the axioms as probabilistic. We randomly set the initial values of the parameters. EDGE, for handling 5,000 examples, took about 15,000 s in average for DBPedia, about 3 s per example, and about 173,000 s in average for education.data.gov.uk, about 43 s per example. Most of the runtime was spent finding the explanations and building the BDDs, while the execution of the EM iterations took only about 6 s for DBPedia and about 2 s for education.data.gov.uk. In the testing phase, we computed the probability of the queries using BUNDLE. For a negative example of the form $a : C$ we compute the probability p of $a : C$ and we assign probability $1 - p$ to the example.

We compare the parameters learned by EDGE with ARs' confidence. For each AR corresponding to the subclass axiom $A \sqsubseteq B$, we computed the confidence by running two SPARQL queries over the training KBs, one for finding all the individuals that belong to $A \sqcap B$ and one for those that belong to A. The confidence is then given by the ratio of the number of individuals in $A \sqcap B$ over those in A. We created 330 different SPARQL queries for education.data.gov.uk and 2,243 for DBPedia.

In the testing phase, we computed the probability of the examples in the test set using BUNDLE, according to the theory learned by EDGE and to the theory composed of the ARs with the confidence as probability. We drew the Precision-Recall (PR) and the Receiver Operating Characteristics (ROC) curves and computed the Area Under the Curve (AUCPR and AUCROC) following the methods of [6,9]. Table 1 shows the AUCPR, the AUCROC, the execution times averaged over the five folds and the p-value of a paired two-tailed t-test at the 5 % significance level of the difference in AUCROC and AUCPR. The times are referred to the learning time for EDGE and to the SPARQL queries execution time for ARs. Note that the elapsed time for EDGE depends on the number of executed queries and the number of different explanations involved in each query, while the elapsed time for ARs depends on the number of classes in the KB. EDGE achieves greater areas in a time that is of the same or lower order of magnitude with respect to ARs. For both areas and KBs, the differences are statistically significant at the 5 % level.

Table 1. Areas under the ROC and PR curves with standard deviation, execution times and p-value of a paired two-tailed t-test at the 5 % significance level for EDGE and Association Rules.

Datasets		EDGE	ARs	p-value
education.data.gov.uk	PR	0.9702 ± 0.0289	0.8804 ± 0.0165	0.0051
	ROC	0.9796 ± 0.0166	0.9158 ± 0.0171	0.0093
	Time (s)	173,528	10,490	
DBPedia	PR	0.9784 ± 0.0483	0.5916 ± 0.0999	0.0013
	ROC	0.9902 ± 0.0219	0.4346 ± 0.1319	0.0007
	Time (s)	14,883	578,420	

6.2 Structure Learning

LEAP has been evaluated on the Carcinogenesis[4] KB which contains 22,372 individuals and 74,405 axioms.

We randomly selected 180 individuals, 103 of which representing positive examples for the class *Compound* (P_E), i.e. individuals that belong to the class *Compound*, and 77 representing negative examples (N_E), i.e. individuals that do not belong to the class *Compound*. We assigned a random probability to every axiom of the KB and we applied a 5-fold cross validation.

In the training phase, we first ran EDGE on the original KB for learning the parameters associated with the probabilistic axioms, with $NumE = 3$ and $TLE = \infty$ for the call to BUNDLE (cf. Algorithm 1) in order to limit the runtime. Then, we separately ran LEAP on the original KB for learning probabilistic subsumption axioms for the class *Compound* and the associated parameters, with $LP_{type} = Positive$ and *Negative Examples Learning Problem*. LEAP learned 1 probabilistic subsumption axiom. For CELOE, we set $NumC = 3$ and a timeout TLC for its execution of 120 s: when the timeout expires or 3 class expressions are found, the current set of them is returned to the caller. For EDGE, we set $NumE$ and TLE as before.

In the testing phase, we computed the probability of the examples (queries) in the test set according to the KB learnt by LEAP and the original one, by applying BUNDLE. We drew the PR and ROC curves and computed the AUCPR and AUCROC. Table 2 shows the AUCPR and the AUCROC averaged over the folds together with the standard deviation. Table 3 reports the learning time in seconds, most of which was spent by EDGE for computing the explanations of the examples and building the corresponding BDDs.

The p-value of a paired two-tailed t-test of the difference in AUCPR and AUCROC between the LEAP ontology and the initial one is 0.0603 for AUCPR and 0.0360 for AUCROC, thus showing that LEAP can achieve better areas under both the PR and ROC curves, with the difference in AUCROC being statistically significant at the 5 % significance level.

---

[4] http://dl-learner.org/wiki/Carcinogenesis

Table 2. Results of the experiments in terms of AUCPR and AUCROC averaged over the folds. Standard deviations are also shown.

Original KB		LEAP	
AUCPR	AUCROC	AUCPR	AUCROC
0.534 ± 0.1082	0.4452 ± 0.0510	0.8006 ± 0.2399	0.798 ± 0.2463

Table 3. Learning time in seconds for LEAP, divided into stages. 'Other' refers to the initialization of the systems and the time spent for sending information between ProbCELOE and EDGE.

	Time (s)
ProbCELOE	139
EDGE	1,765
Other	0.206
Total	1,905

7   Conclusions

We have discussed two algorithms for learning the parameters and the structure of probabilistic DLs following the DISPONTE semantics. EDGE applies an EM algorithm for learning the parameters. It exploits the BDDs that are built during inference to efficiently compute the expectations for hidden variables. The experiments over two real world datasets show that EDGE achieves larger areas both under the PR and the ROC curve with respect to an algorithm based on Association Rules in a comparable or smaller time, thus demonstrating that EDGE is a viable alternative to ARs.

LEAP learns the structure by first performing a search in the space of promising axioms, by exploiting CELOE to learn class expressions of target concepts, and then a greedy search in the space of the ontologies. In this second phase the probabilities of the new axioms are computed by EDGE. The experiments over a real world dataset show that LEAP, by learning the target class expressions, achieves larger areas under both the PR and the ROC curve than a single execution of EDGE. Both EDGE and LEAP are available for download from http://sites.unife.it/ml/.

References

1. Agrawal, R., Srikant, R.: Fast algorithms for mining association rules in large databases. In: International Conference on Very Large Data Bases, pp. 487–499. Morgan Kaufmann (1994)
2. Bellodi, E., Lamma, E., Riguzzi, F., Albani, S.: A distribution semantics for probabilistic ontologies. In: Uncertainty Reasoning for the Semantic Web. CEUR Workshop Proceedings, vol. 778, pp. 75–86. Sun SITE Central Europe (2011)

3. Bellodi, E., Riguzzi, F.: Experimentation of an expectation maximization algorithm for probabilistic logic programs. Intelligenza Artificiale **8**(1), 3–18 (2012)
4. Bellodi, E., Riguzzi, F.: Expectation Maximization over binary decision diagrams for probabilistic logic programs. Intell. Data Anal. **17**(2), 343–363 (2013)
5. Bellodi, E., Riguzzi, F.: Structure learning of probabilistic logic programs by searching the clause space. Theory and Practice of Logic Programming FirstView Articles (to appear, 2014)
6. Davis, J., Goadrich, M.: The relationship between precision-recall and ROC curves. In: International Conference on Machine Learning, pp. 233–240. ACM (2006)
7. De Raedt, L., Kimmig, A., Toivonen, H.: ProbLog: A probabilistic Prolog and its application in link discovery. In: International Joint Conference on Artificial Intelligence, vol. 7, pp. 2462–2467 (2007)
8. Dempster, A.P., Laird, N.M., Rubin, D.B.: Maximum likelihood from incomplete data via the EM algorithm. J. Roy. Stat. Soc. Ser. B **39**(1), 1–38 (1977)
9. Fawcett, T.: An introduction to ROC analysis. Pattern Recogn. Lett. **27**(8), 861–874 (2006)
10. Fleischhacker, D., Völker, J.: Inductive learning of disjointness axioms. In: Meersman, R., et al. (eds.) OTM 2011, Part II. LNCS, vol. 7045, pp. 680–697. Springer, Heidelberg (2011)
11. Lehmann, J., Auer, S., Bühmann, L., Tramp, S.: Class expression learning for ontology engineering. J. Web Semant. **9**(1), 71–81 (2011)
12. Lehmann, J., Hitzler, P.: Concept learning in description logics using refinement operators. Mach. Learn. **78**, 203–250 (2010)
13. Ochoa-Luna, J.E., Revoredo, K., Cozman, F.G.: Learning probabilistic description logics: a framework and algorithms. In: Batyrshin, I., Sidorov, G. (eds.) MICAI 2011, Part I. LNCS, vol. 7094, pp. 28–39. Springer, Heidelberg (2011)
14. Minervini, P., d'Amato, C., Fanizzi, N.: Learning probabilistic description logic concepts: Under different assumptions on missing knowledge. In: ACM Symposium on Applied Computing, pp. 378–383. ACM (2012)
15. Patel-Schneider, P.F., Horrocks, I., Bechhofer, S.: Tutorial on OWL (2003)
16. Riguzzi, F., Bellodi, E., Lamma, E.: Probabilistic Datalog+/- under the distribution semantics. In: International Workshop on Description Logics. CEUR Workshop Proceedings, vol. 846. Sun SITE Central Europe (2012)
17. Riguzzi, F., Bellodi, E., Lamma, E.: Probabilistic ontologies in Datalog+/-. In: Italian Conference on Computational Logic. CEUR Workshop Proceedings, vol. 857, pp. 221–235. Sun SITE Central Europe (2012)
18. Riguzzi, F., Bellodi, E., Lamma, E., Zese, R.: Epistemic and statistical probabilistic ontologies. In: Uncertainty Reasoning for the Semantic Web. CEUR Workshop Proceedings, vol. 900, pp. 3–14. Sun SITE Central Europe (2012)
19. Riguzzi, F., Bellodi, E., Lamma, E., Zese, R.: BUNDLE: a reasoner for probabilistic ontologies. In: Faber, W., Lembo, D. (eds.) RR 2013. LNCS, vol. 7994, pp. 183–197. Springer, Heidelberg (2013)
20. Riguzzi, F., Bellodi, E., Lamma, E., Zese, R.: Computing instantiated explanations in OWL DL. In: Baldoni, M., Baroglio, C., Boella, G., Micalizio, R. (eds.) AI*IA 2013. LNCS, vol. 8249, pp. 397–408. Springer, Heidelberg (2013)
21. Riguzzi, F., Bellodi, E., Lamma, E., Zese, R.: Parameter learning for probabilistic ontologies. In: Faber, W., Lembo, D. (eds.) RR 2013. LNCS, vol. 7994, pp. 265–270. Springer, Heidelberg (2013)

22. Riguzzi, F., Bellodi, E., Lamma, E., Zese, R.: Learning the parameters of probabilistic description logics. In: Inductive Logic Programming Late Breaking papers.
CEUR Workshop Proceedings, vol. 1187, pp. 46–51. Sun SITE Central Europe
(2014)
23. Riguzzi, F., Bellodi, E., Lamma, E., Zese, R.: Probabilistic description logics under
the distribution semantics. Semantic Web - Interoperability, Usability, Applicability (to appear, 2014)
24. Riguzzi, F., Lamma, E., Bellodi, E., Zese, R.: Semantics and inference for probabilistic ontologies. In: Popularize Artificial Intelligence Workshop. CEUR Workshop Proceedings, vol. 860, pp. 41–46. Sun SITE Central Europe (2012)
25. Sato, T.: A statistical learning method for logic programs with distribution semantics. In: International Conference on Logic Programming, pp. 715–729. MIT Press
(1995)
26. Schmidt-Schauß, M., Smolka, G.: Attributive concept descriptions with complements. Artif. Intell. **48**(1), 1–26 (1991)
27. Sirin, E., Parsia, B., Cuenca-Grau, B., Kalyanpur, A., Katz, Y.: Pellet: a practical
OWL-DL reasoner. J. Web Semant. **5**(2), 51–53 (2007)
28. Völker, J., Niepert, M.: Statistical schema induction. In: Antoniou, G., Grobelnik,
M., Simperl, E., Parsia, B., Plexousakis, D., De Leenheer, P., Pan, J. (eds.) ESWC
2011, Part I. LNCS, vol. 6643, pp. 124–138. Springer, Heidelberg (2011)
29. Zese, R., Bellodi, E., Lamma, E., Riguzzi, F.: A description logics tableau reasoner in Prolog. In: Italian Conference on Computational Logic. CEUR Workshop
Proceedings, vol. 1068, pp. 33–47. Sun SITE Central Europe (2013)

Semantics and Inference for Probabilistic Description Logics

Riccardo Zese[1], Elena Bellodi[1], Evelina Lamma[1], Fabrizio Riguzzi[2](✉),
and Fabiano Aguiari[1]

[1] Dipartimento di Ingegneria, University of Ferrara, Ferrara, Italy
{riccardo.zese,elena.bellodi,evelina.lamma}@unife.it
[2] Dipartimento di Matematica e Informatica, University of Ferrara,
Via Saragat 1, 44122 Ferrara, Italy
fabrizio.riguzzi@unife.it

Abstract. We present a semantics for Probabilistic Description Logics that is based on the distribution semantics for Probabilistic Logic Programming. The semantics, called DISPONTE, allows to express assertional probabilistic statements. We also present two systems for computing the probability of queries to probabilistic knowledge bases: BUNDLE and TRILL. BUNDLE is based on the Pellet reasoner while TRILL exploits the declarative Prolog language. Both algorithms compute a propositional Boolean formula that represents the set of explanations to the query. BUNDLE builds a formula in Disjunctive Normal Form in which each disjunct corresponds to an explanation while TRILL computes a general Boolean pinpointing formula using the techniques proposed by Baader and Peñaloza. Both algorithms then build a Binary Decision Diagram (BDD) representing the formula and compute the probability from the BDD using a dynamic programming algorithm. We also present experiments comparing the performance of BUNDLE and TRILL.

1 Introduction

The main idea of the Semantic Web is making information available in a form that is understandable and automatically manageable by machines [16]. In order to realize this vision, the W3C has supported the development of a family of knowledge representation formalisms of increasing complexity for defining ontologies, called Web Ontology Language (OWL). In particular, OWL defines the sublanguages OWL-Lite, OWL-DL (based on Description Logics) and OWL-Full. Since the real world often contains uncertain information, it is fundamental to be able to represent and reason with such information. This problem has been investigated by various authors both in the general case of First Order Logic (FOL) [5,14,27] and in the case of restricted logics, such as Description Logics (DLs) and Logic Programming (LP).

In particular, in LP the distribution semantics [39] has emerged as one of the most effective approaches for representing probabilistic information and it

© Springer International Publishing Switzerland 2014
F. Bobillo et al. (Eds.): URSW 2011-2013, LNAI 8816, pp. 79–99, 2014.
DOI: 10.1007/978-3-319-13413-0_5

underlies many probabilistic LP languages such as Probabilistic Horn Abduction [30], PRISM [39,40], Independent Choice Logic [29], Logic Programs with Annotated Disjunctions [47], ProbLog [9] and CP-logic [46].

In [7,34,36,37] we applied the distribution semantics to DLs obtaining DISP-ONTE for "DIstribution Semantics for Probabilistic ONTologiEs" (Spanish for "get ready"), in which we annotate axioms of a theory with a probability and assume that each axiom is independent of the others. A DISPONTE knowledge base (KB for short) defines a probability distribution over regular KBs (worlds) and the probability of a query is obtained from the joint probability of the worlds and the query.

In order to fully support the development of the Semantic Web, efficient DL reasoners, such us Pellet [44], RacerPro [12] and HermiT [43], are used to extract implicit information from the modeled ontologies, and probabilistic DL reasoners, such as PRONTO [21], are used to compute the probability of the inferred information. Most DL reasoners implement a tableau algorithm in a procedural language. However, some tableau expansion rules are non-deterministic, requiring the developers to implement a search strategy in an or-branching search space. Moreover, in some cases we want to compute all explanations for a query, thus requiring the exploration of all the non-deterministic choices of the tableau algorithm.

We present the algorithm BUNDLE for "Binary decision diagrams for Uncertain reasoNing on Description Logic thEories", that performs inference over DISPONTE DLs. BUNDLE exploits an underlying reasoner such as Pellet [44] that returns explanations for queries.

Moreover, we present the system TRILL for "Tableau Reasoner for descrIption Logics in proLog", a tableau reasoner implemented in the declarative Prolog language. Prolog's search strategy is exploited for taking into account the non-determinism of the tableau rules. TRILL uses the Thea2 library [45] for parsing OWL in its various dialects. Thea2 translates OWL files into a Prolog representation in which each axiom is mapped into a fact. TRILL can check the consistency of a concept and the entailment of an axiom from an ontology, and can return the "pinpointing formula" for queries.

Both BUNDLE and TRILL use the inference techniques developed for probabilistic logic programs under the distribution semantics, in particular Binary Decision Diagrams (BDDs), for computing the probability of queries from the set of all explanations and the pinpointing formula respectively. They encode the results of the inference process in a BDD from which the probability can be computed in time linear in the size of the diagram.

In the following, Sect. 2 briefly introduces $\mathcal{ALC}$ and $\mathcal{SHOIN}(\mathbf{D})$ DLs. Section 3 presents the DISPONTE semantics while Sect. 4 defines the problem of answering queries to DLs. Sections 5 and 6 describe BUNDLE and TRILL respectively. Section 7 illustrates related work. Section 8 shows experiments and Sect. 9 concludes the paper.

2   Description Logics

Description Logics (DLs) are knowledge representation formalisms that possess nice computational properties such as decidability and/or low complexity, see [1,2] for excellent introductions. DLs are particularly useful for representing ontologies and have been adopted as the basis of the Semantic Web.

While DLs can be translated into FOL, they are usually represented using a syntax based on concepts and roles. A concept corresponds to a set of individuals of the domain while a role corresponds to a set of pairs of individuals of the domain. We first briefly describe $\mathcal{ALC}$ and then $\mathcal{SHOIN}(\mathbf{D})$, showing the difference with $\mathcal{ALC}$.

Let $\mathbf{A}$, $\mathbf{R}$ and $\mathbf{I}$ be sets of *atomic concepts, roles* and *individuals*, respectively. *Concepts* are defined by induction as follows. Each $C \in \mathbf{A}$ is a concept, $\bot$ and $\top$ are concepts. If C, C_1 and C_2 are concepts and $R \in \mathbf{R}$, then $(C_1 \sqcap C_2)$, $(C_1 \sqcup C_2)$ and $\neg C$ are concepts, as well as $\exists R.C$ and $\forall R.C$. A *TBox* $\mathcal{T}$ is a finite set of *concept inclusion axioms* $C \sqsubseteq D$, where C and D are concepts. We use $C \equiv D$ to abbreviate the conjunction of $C \sqsubseteq D$ and $D \sqsubseteq C$. An *ABox* $\mathcal{A}$ is a finite set of *concept membership axioms* $a : C$, *role membership axioms* $(a, b) : R$, *equality axioms* $a = b$ and *inequality axioms* $a \neq b$, where C is a concept, $R \in \mathbf{R}$ and $a, b \in \mathbf{I}$. A *knowledge base* $\mathcal{K} = (\mathcal{T}, \mathcal{A})$ consists of a TBox $\mathcal{T}$ and an ABox $\mathcal{A}$. A knowledge base $\mathcal{K}$ is usually assigned a semantics in terms of interpretations $\mathcal{I} = (\Delta^{\mathcal{I}}, \cdot^{\mathcal{I}})$, where $\Delta^{\mathcal{I}}$ is a non-empty *domain* and $\cdot^{\mathcal{I}}$ is the *interpretation function* that assigns an element in $\Delta^{\mathcal{I}}$ to each $a \in \mathbf{I}$, a subset of $\Delta^{\mathcal{I}}$ to each $C \in \mathbf{A}$ and a subset of $\Delta^{\mathcal{I}} \times \Delta^{\mathcal{I}}$ to each $R \in \mathbf{R}$.

The mapping $\cdot^{\mathcal{I}}$ is extended to all concepts (where $R^{\mathcal{I}}(x) = \{y | (x, y) \in R^{\mathcal{I}}\}$) as:

$$\top^{\mathcal{I}} = \Delta^{\mathcal{I}}$$
$$\bot^{\mathcal{I}} = \emptyset$$
$$(C_1 \sqcap C_2)^{\mathcal{I}} = C_1^{\mathcal{I}} \cap C_2^{\mathcal{I}}$$
$$(C_1 \sqcup C_2)^{\mathcal{I}} = C_1^{\mathcal{I}} \cup C_2^{\mathcal{I}}$$
$$(\neg C)^{\mathcal{I}} = \Delta^{\mathcal{I}} \setminus C^{\mathcal{I}}$$
$$(\forall R.C)^{\mathcal{I}} = \{x \in \Delta^{\mathcal{I}} | R^{\mathcal{I}}(x) \subseteq C^{\mathcal{I}}\}$$
$$(\exists R.C)^{\mathcal{I}} = \{x \in \Delta^{\mathcal{I}} | R^{\mathcal{I}}(x) \cap C^{\mathcal{I}} \neq \emptyset\}$$

The *satisfaction* of an axiom E in an interpretation $\mathcal{I} = (\Delta^{\mathcal{I}}, \cdot^{\mathcal{I}})$, denoted by $\mathcal{I} \models E$, is defined as follows: (1) $\mathcal{I} \models C \sqsubseteq D$ iff $C^{\mathcal{I}} \subseteq D^{\mathcal{I}}$, (2) $\mathcal{I} \models a : C$ iff $a^{\mathcal{I}} \in C^{\mathcal{I}}$, (3) $\mathcal{I} \models (a, b) : R$ iff $(a^{\mathcal{I}}, b^{\mathcal{I}}) \in R^{\mathcal{I}}$, (4) $\mathcal{I} \models a = b$ iff $a^{\mathcal{I}} = b^{\mathcal{I}}$, (5) $\mathcal{I} \models a \neq b$ iff $a^{\mathcal{I}} \neq b^{\mathcal{I}}$. $\mathcal{I}$ satisfies a set of axioms $\mathcal{E}$, denoted by $\mathcal{I} \models \mathcal{E}$, iff $\mathcal{I} \models E$ for all $E \in \mathcal{E}$. An interpretation $\mathcal{I}$ *satisfies* a knowledge base $\mathcal{K} = (\mathcal{T}, \mathcal{A})$, denoted $\mathcal{I} \models \mathcal{K}$, iff $\mathcal{I}$ satisfies $\mathcal{T}$ and $\mathcal{A}$. In this case we say that $\mathcal{I}$ is a *model* of $\mathcal{K}$.

In following we describe $\mathcal{SHOIN}(\mathbf{D})$ showing what it adds to $\mathcal{ALC}$. A *role* is either an atomic role $R \in \mathbf{R}$ or the inverse R^- of an atomic role $R \in \mathbf{R}$. We use $\mathbf{R}^-$ to denote the set of all inverses of roles in $\mathbf{R}$. An *RBox* $\mathcal{R}$ consists of a finite set of *transitivity axioms* $Trans(R)$, where $R \in \mathbf{R}$, and *role inclusion axioms* $R \sqsubseteq S$, where $R, S \in \mathbf{R} \cup \mathbf{R}^-$.

If $a \in \mathbf{I}$, then $\{a\}$ is a concept called *nominal*, and if C, C_1 and C_2 are concepts and $R \in \mathbf{R} \cup \mathbf{R}^-$, then $\geq nR$ and $\leq nR$ for an integer $n \geq 0$ are also concepts. A $\mathcal{SHOIN}(\mathbf{D})$ KB $\mathcal{K} = (\mathcal{T}, \mathcal{R}, \mathcal{A})$ consists of a TBox $\mathcal{T}$, an RBox $\mathcal{R}$ and an ABox $\mathcal{A}$.

The mapping $\cdot^{\mathcal{I}}$ is extended to all new concepts (where $\#X$ denotes the cardinality of the set X) as:

$$(R^-)^{\mathcal{I}} = \{(y, x) | (x, y) \in R^{\mathcal{I}}\}$$
$$\{a\}^{\mathcal{I}} = \{a^{\mathcal{I}}\}$$
$$(\geq nR)^{\mathcal{I}} = \{x \in \Delta^{\mathcal{I}} | \#R^{\mathcal{I}}(x) \geq n\}$$
$$(\leq nR)^{\mathcal{I}} = \{x \in \Delta^{\mathcal{I}} | \#R^{\mathcal{I}}(x) \leq n\}$$

$\mathcal{SHOIN}(\mathbf{D})$ allows the definition of datatype roles, i.e., roles that map an individual to an element of a datatype such as integers, floats, etc. Then new concept definitions involving datatype roles are added that mirror those involving roles introduced above. We also assume that we have predicates over the datatypes.

The *satisfaction* of an axiom E in an interpretation $\mathcal{I} = (\Delta^{\mathcal{I}}, \cdot^{\mathcal{I}})$, denoted by $\mathcal{I} \models E$, is defined as for $\mathcal{ALC}$, plus the following ones regarding RBox axioms: (6) $\mathcal{I} \models Trans(R)$ iff $R^{\mathcal{I}}$ is transitive, (7) $\mathcal{I} \models R \sqsubseteq S$ iff $R^{\mathcal{I}} \subseteq S^{\mathcal{I}}$. An interpretation $\mathcal{I}$ *satisfies* a knowledge base $\mathcal{K} = (\mathcal{T}, \mathcal{R}, \mathcal{A})$, denoted $\mathcal{I} \models \mathcal{K}$, iff $\mathcal{I}$ satisfies $\mathcal{T}$, $\mathcal{R}$ and $\mathcal{A}$. In this case we say that $\mathcal{I}$ is a *model* of $\mathcal{K}$.

Each DL is *decidable* if the problem of checking the satisfiability of a KB is decidable. In particular, $\mathcal{SHOIN}(\mathbf{D})$ is decidable iff there are no number restrictions on non-simple roles. A role is *non-simple* iff it is transitive or it has transitive subroles.

A query Q over a KB $\mathcal{K}$ is usually an axiom for which we want to test the entailment from the KB, written $\mathcal{K} \models Q$. The entailment test may be reduced to checking the unsatisfiability of a concept in the knowledge base, i.e., the emptiness of the concept. For example, the entailment of the axiom $C \sqsubseteq D$ may be tested by checking the unsatisfiability of the concept $C \sqcap \neg D$.

Example 1. The following KB is inspired by the ontology people+pets [28]:

$$\exists hasAnimal.Pet \sqsubseteq NatureLover$$
$$fluffy : Cat$$
$$tom : Cat$$
$$Cat \sqsubseteq Pet$$
$$(kevin, fluffy) : hasAnimal$$
$$(kevin, tom) : hasAnimal$$

It states that individuals that own an animal which is a pet are nature lovers and that *kevin* owns the animals *fluffy* and *tom*. Moreover, *fluffy* and *tom* are cats and cats are pets. The query $Q = kevin : NatureLover$ is entailed by the KB.

3 The DISPONTE Semantics

DISPONTE [37] applies the distribution semantics [39] of probabilistic logic programming to DLs. A program following this semantics defines a probability distribution over normal logic programs called *worlds*. Then the distribution is extended to queries and the probability of a query is obtained by marginalizing the joint distribution of the query and the programs.

In DISPONTE, a *probabilistic knowledge base* $\mathcal{K}$ contains a set of *probabilistic axioms* which take the form

$$p :: E \tag{1}$$

where p is a real number in $[0, 1]$ and E is a DL axiom.

The idea of DISPONTE is to associate independent Boolean random variables to the probabilistic axioms. To obtain a *world* w we decide whether to include each probabilistic axiom or not in w. A world therefore is a non probabilistic KB that can be assigned a semantics in the usual way. A query is entailed by a world if it is true in every model of the world.

The probability p can be interpreted as an *epistemic probability*, i.e., as the degree of our belief in axiom E. For example, a probabilistic concept membership axiom $p :: a : C$ means that we have degree of belief p in $C(a)$. A probabilistic concept inclusion axiom of the form $p :: C \sqsubseteq D$ represents the fact that we believe in the truth of $C \sqsubseteq D$ with probability p.

Formally, an *atomic choice* is a couple (E_i, k) where E_i is the ith probabilistic axiom and $k \in \{0, 1\}$. k indicates whether E_i is chosen to be included in a world ($k = 1$) or not ($k = 0$). A *composite choice* κ is a consistent set of atomic choices, i.e., $(E_i, k) \in \kappa, (E_i, m) \in \kappa$ implies $k = m$ (only one decision is taken for each axiom). The probability of a composite choice κ is $P(\kappa) = \prod_{(E_i,1)\in\kappa} p_i \prod_{(E_i,0)\in\kappa}(1 - p_i)$, where p_i is the probability associated with axiom E_i. A *selection* σ is a total composite choice, i.e., it contains an atomic choice (E_i, k) for every axiom of the theory. A selection σ identifies a theory w_σ called a *world* in this way: $w_\sigma = \{E_i | (E_i, 1) \in \sigma\}$. Let us indicate with $\mathcal{S}_\mathcal{K}$ the set of all selections and with $\mathcal{W}_\mathcal{K}$ the set of all worlds. The probability of a world w_σ is $P(w_\sigma) = P(\sigma) = \prod_{(E_i,1)\in\sigma} p_i \prod_{(E_i,0)\in\sigma}(1 - p_i)$. $P(w_\sigma)$ is a probability distribution over worlds, i.e., $\sum_{w\in\mathcal{W}_\mathcal{K}} P(w) = 1$.

We can now assign probabilities to queries. Given a world w, the probability of a query Q is defined as $P(Q|w) = 1$ if $w \models Q$ and 0 otherwise. The probability of a query can be defined by marginalizing the joint probability of the query and the worlds:

$$P(Q) = \sum_{w\in\mathcal{W}_\mathcal{K}} P(Q, w) = \sum_{w\in\mathcal{W}_\mathcal{K}} P(Q|w)P(w) = \sum_{w\in\mathcal{W}_\mathcal{K}:w\models Q} P(w) \tag{2}$$

4   Querying KBs

In order to answer queries to DL KBs, a *tableau algorithm* [42] can be used. Such an algorithm decides whether an axiom is entailed or not by a KB by refutation: axiom E is entailed if $\neg E$ has no model in the KB. The algorithm works on *completion graphs* also called *tableaux*: they are ABoxes that can also be seen as graphs, where each node represents an individual a and is labeled with the set of concepts $\mathcal{L}(a)$ it belongs to. Each edge $\langle a, b \rangle$ in the graph is labeled with the set of roles $\mathcal{L}(\langle a, b \rangle)$ to which the couple (a, b) belongs. The algorithm starts from a tableau that contains the ABox of the KB and the negation of the axiom to be proved. For example, if the query is a membership one, $C(a)$, it adds $\neg C$ to the label of a. If we query for the emptyness (unsatisfiability) of a concept C, the algorithm adds a new anonymous node a to the tableau and adds C to the label of a. The axiom $C \sqsubseteq D$ can be proved by showing that $C \sqcap \neg D$ is unsatisfiable. The algorithm repeatedly applies a set of consistency preserving *tableau expansion rules* (see [35] for a list of expansion rules for $\mathcal{SHOIN}(\mathbf{D})$) until a clash (i.e., a contradiction) is detected or a clash-free graph is found to which no more rules are applicable.

Some of the rules used by the tableau algorithm are non-deterministic, i.e., they generate a finite set of tableaux. Thus the algorithm keeps a set of tableaux T. If a non-deterministic rule is applied to a graph G in T, then G is replaced by the resulting set of graphs.

An *event* during the execution of the algorithm can be [18]: (1) $Add(C, a)$, the addition of a concept C to $\mathcal{L}(a)$; (2) $Add(R, \langle a, b \rangle)$, the addition of a role R to $\mathcal{L}(\langle a, b \rangle)$; (3) $Merge(a, b)$, the merging of the nodes a, b; (4) $\neq(a, b)$, the addition of the inequality $a \neq b$ to the relation $\neq$; (5) $Report(g)$, the detection of a clash g. We use $\mathcal{E}$ to denote the set of events recorded during the execution of the algorithm. A clash is either:

– a couple (C, a) where C and $\neg C$ are present in the label of node a, i.e. $\{C, \neg C\} \subseteq \mathcal{L}(a)$;
– a couple $(Merge(a, b), \neq(a, b))$, where the events $Merge(a, b)$ and $\neq(a, b)$ belong to $\mathcal{E}$.

Each time a clash is detected in a completion graph G, the algorithm stops applying rules to G. Once every completion graph in T contains a clash or no more expansion rules can be applied to it, the algorithm terminates. If all the completion graphs in the final set T contain a clash, the algorithm returns *unsatisfiable* as no model can be found. Otherwise, any one clash-free completion graph in T represents a possible model for $C(a)$ and the algorithm returns *satisfiable*.

In order to perform probabilistic inference, we need not only to answer queries but also to compute explanations for queries. In fact, computing the probability of a query by generating the worlds of the KB would be impractical as there is an exponential number of them. By computing explanations, we find compact representations of the set of worlds where the query is true, as shown below.

4.1   Finding Explanations

The problem of finding explanations for a query has been investigated by various authors [13,18–20,41]. It was called *axiom pinpointing* in [41] and considered as a non-standard reasoning service useful for tracing derivations and debugging ontologies. In particular, Schlobach and Cornet [41] define *minimal axiom sets* or *MinAs* for short.

Definition 1 (MinA). *Let $\mathcal{K}$ be a knowledge base and Q an axiom that follows from it, i.e., $\mathcal{K} \models Q$. We call a set $M \subseteq \mathcal{K}$ a* minimal axiom set *or* MinA *for Q in $\mathcal{K}$ if $M \models Q$ and it is minimal w.r.t. set inclusion.*

We also *explanation* a MinA. The problem of enumerating all MinAs is called MIN-A-ENUM in [41]. ALL-MINAS$(Q, \mathcal{K})$ is the set of all MinAs for query Q in the knowledge base $\mathcal{K}$.

We report here the techniques used by Pellet [44] to compute explanations for queries. Pellet first finds a single MinA by using a modified version of the tableau algorithm and then finds the others with a black box method: axioms are iteratively removed from the KB and new MinAs are computed until all possible MinAs have been found. The modified tableau algorithm is shown in Algorithm 1.

Algorithm 1. Tableau algorithm.

```
 1: function TABLEAU(C, K)
 2:     Input: C (the concept to be tested for unsatisfiability)
 3:     Input: K (the knowledge base)
 4:     Output: S (a set of axioms) or null
 5:     Let G₀ be an initial completion graph from K containing an anonymous individual a and
        C ∈ L(a)
 6:     T ← {G₀}
 7:     repeat
 8:         Select a rule r applicable to a clash-free graph G from T
 9:         T ← T \ {G}
10:         Let G = {G'₁, ..., G'ₙ} be the result of applying r to G
11:         T ← T ∪ G
12:     until All graphs in T have a clash or no rule is applicable
13:     if All graphs in T have a clash then
14:         S ← ∅
15:         for all G ∈ T do
16:             let s_G the result of τ for the clash of G
17:             S ← S ∪ s_G
18:         end for
19:         S ← S \ {C(a)}
20:         return S
21:     else
22:         return null
23:     end if
24: end function
```

In this algorithm, each expansion rule updates as well a *tracing function* τ, which associates sets of axioms with events in the derivation. For example, $\tau(Add(C, a))$ $(\tau(Add(R, \langle a, b \rangle)))$ is the set of axioms needed to explain the event $Add(C, a)$ $(Add(R, \langle a, b \rangle))$. For the sake of brevity, we define τ for couples (concept, individual) and (role, couple of individuals) as $\tau(C, a) = \tau(Add(C, a))$ and $\tau(R, \langle a, b \rangle) = \tau(Add(R, \langle a, b \rangle))$ respectively. The function τ is initialized as the empty set for all the elements of its domain except for $\tau(C, a)$ and $\tau(R, \langle a, b \rangle)$ to which the values $\{a : C\}$ and $\{(a, b) : R\}$ are assigned if $a : C$ and $(a, b) : R$ are in the ABox respectively. The expansion rules add axioms to values of τ.

If $g_1, \ldots, g_n$ are the clashes, one for each tableau of the final set, the output of the algorithm TABLEAU is $S = \bigcup_{i \in \{1, \ldots, n\}} \tau(g_i) \setminus \{C(a)\}$ where a is the anonymous individual initially assigned to C.

TABLEAU returns a single MinA. To solve MIN-A-ENUM, Pellet uses the *hitting set algorithm* [31]. The algorithm, described in detail in [18], starts from a MinA S and initializes a labeled tree called *Hitting Set Tree* (HST) with S as the label of its root v. Then it selects an arbitrary axiom E in S, it removes it from $\mathcal{K}$, generating a new knowledge base $\mathcal{K}' = \mathcal{K} - \{E\}$, and tests the unsatisfiability of C w.r.t. $\mathcal{K}'$. If C is still unsatisfiable, we obtain a new explanation. The algorithm adds a new node w and a new edge $\langle v, w \rangle$ to the tree, then it assigns this new explanation to the label of w and the axiom E to the label of the edge. The algorithm repeats this process until the unsatisfiability test returns negative: in that case the algorithm labels the new node with OK, makes it a leaf, backtracks to a previous node, selects a different axiom to be removed from the KB and repeats these operations until the HST is fully built. The algorithm also eliminates extraneous unsatisfiability tests based on previous results: once a path leading to a node labeled OK is found, any superset of that path is guaranteed to be a path leading to a node where C is satisfiable, and thus no additional unsatisfiability test is needed for that path, as indicated by a X in the node label. When the HST is fully built, all leaves of the tree are labeled with OK or X. The distinct non leaf nodes of the tree collectively represent the set ALL-MINAS$(C, \mathcal{K})$.

In [3,4], Baader and Peñaloza presented the problem of finding a *pinpointing formula* instead of ALL-MINAS$(Q, \mathcal{K})$ for queries. The pinpointing formula is a monotone Boolean formula in which each Boolean variable corresponds to an axiom of the KB. This formula is built using the variables and the conjunction and disjunction connectives. It compactly encodes the set of all MinAs. Let assume that each axiom E of a KB $\mathcal{K}$ is associated with a propositional variable, indicated with $var(E)$. The set of all propositional variables is indicated with $var(\mathcal{K})$. A valuation ν of a monotone Boolean formula is the set of propositional variables that are true. For a valuation $\nu \subseteq var(\mathcal{K})$, let $\mathcal{K}_\nu := \{t \in \mathcal{K} | var(t) \in \nu\}$.

Definition 2 (Pinpointing formula). *Given a query Q and a KB $\mathcal{K}$, a monotone Boolean formula ϕ over $var(\mathcal{K})$ is called a* pinpointing formula *for Q if for every valuation $\nu \subseteq var(\mathcal{K})$ it holds that $\mathcal{K}_\nu \models Q$ iff ν satisfies ϕ.*

In Lemma 2.4 of [4], the authors proved that we can obtain all MinAs from a pinpointing formula by transforming the formula into DNF and removing disjuncts implying other disjuncts. The example below illustrates axiom pinpointing and the pinpointing formula.

Example 2 (Pinpointing formula). Consider the KB of Example 1. We associate Boolean variables with axioms as follows: $F_1 = \exists hasAnimal.Pet \sqsubseteq NatureLover$, $F_2 = (kevin, fluffy) : hasAnimal$, $F_3 = (kevin, tom) : hasAnimal$, $F_4 = fluffy : Cat$, $F_5 = tom : Cat$ and $F_6 = Cat \sqsubseteq Pet$. Let $Q = kevin : NatureLover$ be the query, then ALL-MINAS$(Q, \mathcal{K}) = \{\{F_2, F_4, F_6, F_1\}, \{F_3, F_5, F_6, F_1\}\}$, while the pinpointing formula is $((F_2 \wedge F_4) \vee (F_3 \wedge F_5)) \wedge F_6 \wedge F_1$.

A tableau algorithm can be modified to find the pinpointing formula. See [4] for the details.

4.2   Probabilistic Inference

We do not have to generate all worlds where a query is true in order to compute its probability, finding a pinpointing formula is enough.

From a pinpointing formula ϕ for Q we can compute the probability $P(\phi)$ of ϕ being true from the probability of the Boolean variables that appear in ϕ assuming all the variables are independent. $P(\phi)$ is the sum of the probabilities of the valuations that make the formula true. The probability of a valuation is given by

$$P(\nu) = \prod_{var(E_i)\in\nu} p_i \prod_{var(E_i)\in var(\mathcal{K})\setminus\nu} (1 - p_i)$$

where p_i is the probability associated with axiom E_i. Computing $P(\phi)$ is equivalent to performing *weighted model counting* [38]: each variable $var(E_i)$ has a weight p_i when set to true and a weight $1 - p_i$ when set to false, the weight of a truth assignment is the product of the weights of its literals and the weighted model count of a formula is the sum of the weights of its satisfying assignments.

Theorem 1. *If ϕ is a pinpointing formula for the query Q from a KB $\mathcal{K}$, then $P(Q) = P(\phi)$.*

Proof. Every valuation $\nu \subseteq var(\mathcal{K})$ that satisfies ϕ uniquely corresponds to a world where Q is true. Thus the sum of the probability of the valuations that satisfy ϕ is equal to the sum of the probabilities of the worlds where Q is true.

The pinpointing formula can be obtained directly from the inference algorithm or can be built starting from the set of all explanations $K = \text{ALL-MINAS}(Q, \mathcal{K})$ in this way

$$\phi_K = \bigvee_{\kappa\in K} \bigwedge_{(E_i,1)\in\kappa} var(E_i).$$

It is easy to see that every valuation that makes ϕ_K true uniquely corresponds to a world where Q is true. ϕ_K is in Disjunctive Normal Form (DNF).

Weighted model counting is a #P-complete problem [11]. A practical approach for solving it involves *knowledge compilation* [8]: we translate the formula to a target language that allows weighted model counting in polynomial time. In this case the complexity is confined in the compilation process.

5   BUNDLE

BUNDLE is based on Pellet [44] and extends it in order to allow the computation of the probability of queries from a probabilistic knowledge base that follows the DISPONTE semantics. BUNDLE can answer concept and role membership queries, subsumption queries, and can find explanations both for the unsatifiability of one or all concepts contained in the KB and for the inconsistency of a KB.

Irrespective of which representation of the explanations we choose, a DNF or a general pinpointing formula, we can apply knowledge compilation and transform it into a Binary Decision Diagram (BDD), from which we can compute the probability (perform weighted model counting) of the query with a dynamic programming algorithm that is linear in the size of the BDD.

A BDD for a function of Boolean variables is a rooted graph that has one level for each Boolean variable. A node n in a BDD has two children: one corresponding to the 1 value of the variable associated with the level of n, indicated with $child_1(n)$, and one corresponding to the 0 value of the variable, indicated with $child_0(n)$. When drawing BDDs, the 0-branch - the one going to $child_0(n)$ - is distinguished from the 1-branch by drawing it with a dashed line. The leaves store either 0 or 1. Figure 1 shows a BDD for the function $f(\mathbf{X}) = (X_1 \wedge X_3) \vee (X_2 \wedge X_3)$, where the variables $\mathbf{X} = \{X_1, X_2, X_3\}$ are independent Boolean random variables whose probability of being true is p_i for the variable X_i.

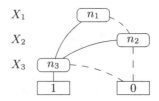

Fig. 1. BDD representing the function $f(\mathbf{X}) = (X_1 \wedge X_3) \vee (X_2 \wedge X_3)$.

A BDD performs a Shannon expansion of the Boolean formula $f(\mathbf{X})$, so that, if X is the variable associated with the root level of a BDD, the formula $f(\mathbf{X})$ can be represented as $f(\mathbf{X}) = X \wedge f^X(\mathbf{X}) \vee \overline{X} \wedge f^{\overline{X}}(\mathbf{X})$ where $f^X(\mathbf{X})$ ($f^{\overline{X}}(\mathbf{X})$) is the formula obtained by $f(\mathbf{X})$ by setting X to 1 (0). Now the two disjuncts are pairwise exclusive and the probability of $f(\mathbf{X})$ can be computed as $P(f(\mathbf{X})) = P(X)P(f^X(\mathbf{X})) + (1 - P(X))P(f^{\overline{X}}(\mathbf{X}))$. Algorithm 2 shows function PROB that implements the dynamic programming algorithm for computing the probability of a formula encoded as a BDD. The function should also store the value of already visited nodes in a table so that, if a node is visited again, its probability can be retrieved from the table. For the sake of simplicity the algorithm does not show this optimization but it is fundamental to achieve linear cost in the number of nodes, as without it the cost of function PROB would be proportional to 2^n where n is the number of Boolean variables.

The main BUNDLE function, shown in Algorithm 3, first builds a data structure $PMap$ that associates each DL axiom E_i with its probability p_i. In the OWL files the probabilistic information is specified using the annotation system allowed by the OWL language. Then BUNDLE uses Pellet's EXPHST$(C, \mathcal{K})$ function that computes all the MinAs for the unsatisfiability of a concept C using the Hitting Set Tree algorithm. BUNDLE exploits the version of this function in which we can specify the maximum number of explanations to be found.

Algorithm 2. Computation of the probability of a formula encoded as a BDD.

```
1: function PROB(node)
2:     Input: a BDD node
3:     Output: the probability of the Boolean function associated with the node
4:     if node is a terminal then
5:         return value(node)                                      ▷ value(node) is 0 or 1
6:     else
7:         let X be v(node)                    ▷ v(node) is the variable associated with node
8:         P₁ ←PROB(child₁(node))
9:         P₀ ←PROB(child₀(node))
10:        return P(X) · P₁ + (1 − P(X)) · P₀
11:    end if
12: end function
```

Algorithm 3. Function BUNDLE: computation of the probability of unsatisfiability of C given $\mathcal{K}$.

```
1: function BUNDLE(𝒦, C, maxEx, maxTime)
2:     Input: 𝒦 (the knowledge base)
3:     Input: C (the concept to be tested for unsatisfiability)
4:     Input: maxEx (the maximum number of explanations to be found)
5:     Input: maxTime (time limit for the search for explanations)
6:     Output: the probability of the unsatisfiability of C w.r.t. 𝒦
7:     Build Map PMap from DL axioms to sets of couples (axiom, probability)
8:     MinAs ←EXPHST(C, 𝒦, maxEx)                                          ▷ Call to Pellet
9:     Initialize VarAx to empty           ▷ VarAx is an array of couples (Axiom, Prob)
10:    BDD ←BDDZERO
11:    for all MinA ∈ MinAs do
12:        BDDE ←BDDONE
13:        for all Ax ∈ MinA do
14:            p ← PMap(Ax)
15:            Scan VarAx looking for Ax
16:            if !found then
17:                Add to VarAx a new cell containing (Ax, p)
18:            end if
19:            Let i be the position of (Ax, p) in VarAx
20:            BDDA ← BDDGETITHVAR(i)
21:            BDDE ←BDDAND(BDDE, BDDA)
22:        end for
23:        BDD ←BDDOR(BDD, BDDE)
24:    end for
25:    return PROB(BDD)                     ▷ VarAx is used to compute P(X) in PROB
26: end function
```

Two data structures are initialized: $VarAx$ is an array that maintains the association between Boolean random variables (whose index is the array index) and couples (axiom, probability), and BDD stores a BDD. BDD is initialized to the zero Boolean function.

Then BUNDLE performs two nested loops that build a BDD representing the pinpointing formula in DNF. To manipulate BDDs we used JavaBDD[1] that is an interface to a number of underlying BDD manipulation packages. As the underlying package we used CUDD.

In the outer loop, BUNDLE combines BDDs for different explanations. In the inner loop, BUNDLE generates the BDD for a single explanation.

In the outer loop, $BDDE$ is initialized to the "one" Boolean function. In the inner loop, the axioms of each MinA are considered one by one. The value p

---

[1] http://javabdd.sourceforge.net/

associated with the axiom is extracted from $PMap$. The axiom is searched for in $VarAx$ to see if it was already assigned a random variable. If not, a cell is added to $VarAx$ to store the couple. At this point we know the couple position i in $VarAx$ and so the index of its Boolean variable X_i. We obtain a BDD representing $X_i = 1$ with BDDGETITHVAR and we conjoin it with $BDDE$. After the two cycles, function PROB of Algorithm 2 is called over BDD and its result is returned to the user.

6   TRILL

TRILL implements a tableau algorithm that computes the pinpointing formula representing the set of MinAs. After generating the pinpointing formula, TRILL converts it into a BDD and computes the probability of the query. TRILL can answer concept and role membership queries, subsumption queries, and can find explanations both for the unsatifiability of a concept contained in the KB and for the inconsistency of the entire KB. TRILL was implemented in Prolog, so the management of the non-determinism of the rules is delegated to this language.

We use the Thea2 library [45] for converting OWL DL KBS into Prolog. Thea2 performs a direct translation of the OWL axioms into Prolog facts. For example, a simple subclass axiom between two named classes $Cat \sqsubseteq Pet$ is written using the subClassOf/2 predicate as subClassOf('Cat','Pet'). For more complex axioms, Thea2 exploits the list construct of Prolog, so the axiom $NatureLover \equiv PetOwner \sqcup GardenOwner$ becomes equivalentClasses(['NatureLover', unionOf(['PetOwner', 'GardenOwner'])]). When a probabilistic KB is given as input, for each probabilistic axiom of the form $Prob ::$ $Axiom$ a fact p(Axiom,Prob) is asserted in the Prolog KB.

In order to represent the tableau, TRILL uses a couple $Tableau = (A, T)$, where A is a list containing information about nominal individuals and class assertions with the corresponding value of the pinpointing formula, while T is a triple (G, RBN, RBR) in which G is a directed graph that contains the structure of the tableau, RBN is a red-black tree (a key-value dictionary) in which a key is a couple of individuals and its value is the set of the labels of the edge between the two individuals, and RBR is a red-black tree in which a key is a role and its value is the set of couples of individuals that are linked by the role. This representation allows to quickly find the information needed during the execution of the tableau algorithm. For managing the *blocking* system we use a predicate for each blocking state: nominal/2, blockable/2, blocked/2, indirectly_blocked/2 and safe/3. Each predicate takes as arguments the individual Ind and the tableau (A, T); safe/3 takes as input also the role R. For each individual Ind in the ABox we add the atom $nominal(Ind)$ to A, then every time we have to check the blocking status of an individual we call the corresponding predicate that returns the status by checking the tableau.

Deterministic and non-deterministic tableau expansion rules are treated differently. Non-deterministic rules are implemented by a predicate $rule\_name(Tab, TabList)$ that, given the current tableau Tab, returns the list of tableaux $TabList$

created by the application of the rule on Tab, while deterministic rules are implemented by a predicate $rule\_name(Tab, Tab1)$ that, given the current tableau Tab, returns the tableau $Tab1$ obtained by the application of the rule on Tab.

Expansion rules are applied in order by `apply_all_rules/2`, first the non-deterministic ones and then the deterministic ones. The predicate `apply_nondet_rules(RuleList,Tab,Tab1)` takes as input the list of non-deterministic rules and the current tableau and returns a tableau obtained by the application of one of the rules. `apply_nondet_rules/3` is called as `apply_nondet_rules([or_rule], Tab,Tab1)` and is shown in Fig. 2. If a non-deterministic rule is applicable, the list of tableaux obtained by its application is returned by the predicate corresponding to the applied rule, a cut is performed to avoid backtracking to other rule choices and a tableau from the list is non-deterministically chosen with the `member/2` predicate.

```
apply_all_rules(Tab,Tab2):-
  apply_nondet_rules([or_rule],Tab,Tab1),
  (Tab=Tab1 -> Tab2=Tab1 ; apply_all_rules(Tab1,Tab2)).


apply_nondet_rules([],Tab,Tab1):-
  apply_det_rules([and_rule,unfold_rule,add_exists_rule,
    forall_rule,exists_rule],Tab,Tab1).

apply_nondet_rules([H|T],Tab,Tab1):-
  C=..[H,Tab,L],
  call(C),!
  member(Tab1,L),
  Tab \= Tab1.

apply_nondet_rules([_|T],Tab,Tab1):-
  apply_nondet_rules(T,Tab,Tab1).
```

Fig. 2. Definition of the non-deterministic expansion rules by means of the predicates `apply_all_rules/2` and `apply_nondet_rules/3`.

If no non-deterministic rule is applicable, deterministic rules are tried sequentially with the predicate `apply_det_rules/3`, shown in Fig. 3, that is called as `apply_det_rules(RuleList,Tab,Tab1)`. It takes as input the list of deterministic rules and the current tableau and returns a tableau obtained with the application of one of the rules.

After the application of a deterministic rule, a cut avoids backtracking to other possible choices for the deterministic rules. If no rule is applicable, the input tableau is returned and rule application stops, otherwise a new round of rule application is performed.

Once the pinpointing formula is built, TRILL builds the corresponding BDD by using the `build_bdd/2` predicate, shown in Fig. 4, that takes as input a

```
apply_det_rules([],Tab,Tab).

apply_det_rules([H|T],Tab,Tab1):-
  C=..[H,Tab,Tab1],
  call(C),!.

apply_det_rules([_|T],Tab,Tab1):-
  apply_det_rules(T,Tab,Tab1).
```

Fig. 3. Definition of the deterministic expansion rules by means of the predicate apply_det_rules/3.

```
build_bdd(and(A),B):-!,
  one(B0),
  bdd_and(A,B0,B).

build_bdd(or(A),B):-!,
  zero(B0),
  bdd_or(A,B0,B).

build_bdd(A,B):-
  p(A,Prob),!,
  ProbN is 1-Prob,
  get_var_n([X],[],[Prob,ProbN],VX),
  equality(VX,0,B).

build_bdd(A,B):-
  one(B).

bdd_and([],B,B).

bdd_and([H|T],B0,B):-
  build_bdd(H,B1),
  and(B0,B1,B2),
  bdd_and(T,B2,B).

bdd_or([],B,B).

bdd_or([H|T],B0,B):-
  build_bdd(H,B1),
  or(B0,B1,B2),
  bdd_or(T,B2,B).
```

Fig. 4. Code of the predicates build_bdd_rules/2.

pinpointing formula and returns the correspondig BDD. It scans the pinpointing formula and, for each variable, it searches for the probabilistic axiom corresponding to the variable with the query p(Axiom,Prob). If the query succeeds, it creates the corresponding BDD and combines it with the BDD representing the pinpointing formula. Finally, it computes the probability of the query from the BDD so built using the predicate compute_prob/2. The predicates one/1 and zero/1 return BDDs representing the Boolean constants 1 and 0; and/3 and or/3 execute Boolean operations between BDDs. get_var_n/4 returns the random variable associated with axiom X and list of probabilities [Prob,ProbN], where ProbN = 1 − Prob. equality/3 returns the BDD B associated with the expression VX=val where VX is a random variable and val is 0 or 1. The predicate p/2 is used for specifying the association between axioms and probability, i.e. p(subClassOf('A','B'),0.9) asserts the axiom $A \sqsubseteq B$ is associated with a probability of 0.9. The predicates compute_prob/2, one/1, zero/1, and/3, or/3, get_var_n/4 and equality/3 are imported from a Prolog library of the cplint suite [33].

7   Related Work

While there are many works that propose approaches for combining probability and DLs, there are relatively fewer inference algorithms. One of these is PRONTO [21] that, similarly to BUNDLE, is based on Pellet. PRONTO performs inference on P-$\mathcal{SHIQ}$(**D**) [25] KBs instead of DISPONTE. In these KBs the probabilistic part contains *conditional constraints* of the form $(D|C)[l, u]$ that informally mean "generally, if an object belongs to C, then it belongs to D with a probability in the interval $[l, u]$". P-$\mathcal{SHIQ}$(**D**) uses probabilistic lexicographic entailment from probabilistic default reasoning and allows both terminological and assertional probabilistic knowledge about instances of concepts and roles. P-$\mathcal{SHIQ}$(**D**) is based on Nilsson's probabilistic logic [27] that defines probabilistic interpretations instead of a single probability distribution over theories.

Differently from BUNDLE and PRONTO, reasoners written in Prolog can exploit Prolog's backtracking facilities for performing the search. This has been observed in various work. Beckert and Posegga [6] proposed a tableau reasoner in Prolog for First Order Logic (FOL) based on free-variable semantic tableaux. However, the reasoner is not tailored to DLs.

Hustadt, Motik and Sattler [17] presented the KAON2 algorithm that exploits basic superposition, a refutational theorem proving method for FOL with equality, and a new inference rule, called decomposition, to reduce a $\mathcal{SHIQ}$ KB into a disjunctive datalog program, while DLog [24] is an ABox reasoning algorithm for the $\mathcal{SHIQ}$ language that allows to store the content of the ABox externally in a database and to answer instance check and instance retrieval queries by transforming the KB into a Prolog program.

Meissner [26] presented the implementation of a Prolog reasoner for the DL $\mathcal{ALCN}$. This work was the basis of the work of Herchenröder [15], that considered $\mathcal{ALC}$ and improved the work of Meissner by implementing heuristic search techniques to reduce the running time. Faizi [10] added to [15] the possibility of returning information about the steps executed during the inference process for queries but still handled only $\mathcal{ALC}$.

A different approach is the one of Ricca et al. [32] that presented OntoDLV, a system for reasoning on a logic-based ontology representation language called OntoDLP. This is an extension of (disjunctive) ASP and can interoperate with OWL. OntoDLV rewrites the OWL KB into the OntoDLP language, can retrieve information directly from external OWL Ontologies and answers queries by using ASP.

TRILL differs from the previous works for the target description logics ($\mathcal{ALC}$) and for the fact that those reasoners do not return explanations for the given queries. Moreover, TRILL differs in particular from DLog for the possibility of answering general queries instead of instance check and instance retrieval only.

8 Experiments

In this section, we evaluate the performance of TRILL and BUNDLE. We first compare BUNDLE with the publicly available version of PRONTO on four probabilistic ontologies. The experiments have been performed on Linux machines with a 3.10 GHz Intel Xeon E5-2687W with 2 GB memory allotted to Java.

The first ontology is BRCA[2] that models breast cancer risk assessment. It contains a certain part and a probabilistic part. The tests were defined following [22]: we randomly sampled axioms from the probabilistic part of this ontology which are then added to the certain part. So each sample was a probabilistic KB with the full certain part of the BRCA ontology and a subset of the probabilistic constraints. We varied the number of these constraints from 9 to 15, and, for each number, we generated 100 different consistent ontologies. In order to generate a query, an individual a is added to the ontology. a is randomly assigned to each class that appears in the sampled conditional constraints with probability 0.6. If the class is composite, as for example *PostmenopausalWomanTakingTestosterone*, a is assigned to the component classes rather than to the composite one. In the example above, a will be added to *PostmenopausalWoman* and *WomanTakingTestosterone*. The ontologies are then translated into DISPONTE by replacing the constraint $(D|C)[l, u]$ with the axiom $u :: C \sqsubseteq D$.

For each ontology we perform the query $a : C$ where the class C is randomly selected among those that represent women under increased and lifetime risk such as *WomanUnderLifetimeBRCRisk* and *WomanUnderStronglyIncreasedBR-CRisk*. We then applied both BUNDLE and PRONTO to each generated test and we measured the execution time and the memory used. Figure 5(a) shows the execution time averaged over the 100 KBs as a function of the number of probabilistic axioms and, similarly, Fig. 5(b) shows the average amount of memory used. As one can see, execution times are similar for small KBs, but the difference between the two reasoners rapidly increases for larger knowledge bases. The memory usage for BUNDLE is always less than 53 % with respect of PRONTO.

The other three ontologies are an extract from the Cell[3] ontology that represents cell types of the prokaryotic, fungal, and eukaryotic organisms, an extract

---

[2] http://sites.google.com/a/unife.it/ml/bundle/brca
[3] http://cellontology.org/

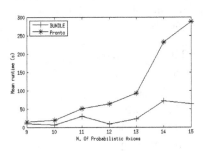
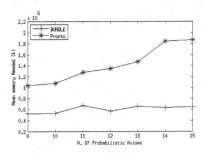

(a) Average execution times (s).   (b) Average memory used (Kb).

Fig. 5. Comparison between BUNDLE and PRONTO on the BRCA KB.

Table 1. Average execution time for the queries to the Cell, Teleost and NCI KBs. The first column reports the size of the non-probabilistic TBox of each KB.

Dataset	TBox axioms		Probabilistic TBox Size				
			0	250	500	750	1000
Cell	1263	time(s)	0.76	2.84	3.88	3.94	4.53
Teleost	3406	time(s)	2.11	8.87	31.80	33.82	36.33
NCI	5423	time(s)	3.02	11.37	11.37	16.37	24.90

from the NCI Thesaurus[4] that describes human anatomy and an extract from the Teleost_anatomy[5] ontology (Teleost for short) that is a multi-species anatomy ontology for teleost fishes. For each of these KBs we considered the versions of increasing size used in [23]: the authors added 250, 500, 750 and 1000 new probabilistic conditional constraints to the extract of the publicly available non-probabilistic version of each ontology. We converted these KBs into DISPONTE in the same way presented for the BRCA ontology and we created a set of 100 different random subclass queries for each KB, such as $CL\_0000802 \sqsubseteq CL\_0000800$ for the Cell KB, $NCI\_C32042 \sqsubseteq NCI\_C32890$ for the NCI Thesaurus and $TAO\_0001102 \sqsubseteq TAO\_0000139$ for the Teleost KB. For generating the queries we built the hierarchy of each KB and we randomly selected two classes connected in the hierarchy for each query, so that it had at least one explanation.

In Table 1 we report, for each version of the datasets, the average execution time for BUNDLE to perform inference. In addition, for each KB we report its number of non-probabilistic TBox axioms. With these datasets, PRONTO always terminated with out-of-memory error.

---

[4] http://ncit.nci.nih.gov/
[5] http://phenoscape.org/wiki/Teleost_Anatomy_Ontology

Table 2. Results of the experiments on BRCA, DBPedia, Biopax and Vicodi KBs in terms of average times for computing the probability of queries. The first column reports the size of the non-probabilistic TBox of each KB.

Dataset	TBox axioms	TRILL time(s)	BUNDLE time(s)
BRCA	322	5.55	6.96
DBPedia	535	16.70	3.79
Biopax level 3	826	0.11	1.85
Vicodi	220	0.19	1.12

As can be seen, BUNDLE needs lower amount of memory and is faster than the publicly available version of PRONTO. BUNDLE can answer most queries in a few seconds and manage larger KBs with respect to PRONTO.

Finally, we tested TRILL performance when computing probability of queries by comparing it to BUNDLE. The experiments have been performed on a Linux machine with a 2.33 GHz Intel Dual Core E6550 with 2 GB memory allotted to Java. We consider four different knowledge bases of various complexity: BRCA already used for the comparison with PRONTO, an extract of the DBPedia[6] ontology obtained from Wikipedia, Biopax level 3[7] that models metabolic pathways and Vicodi[8] that contains information about European history. For the tests, we used the DBPedia and the Biopax KBs without ABox while for BRCA and Vicodi we used a small ABox containing 1 individual for the first one and 19 individuals for the second one. We added 50 probabilistic axioms to each KB. For BRCA we used the probabilistic axioms already created for the previous test, while for the other KBs we created the probabilistic axioms by randomly selecting certain axioms from them and associating a random probability. For each dataset we randomly created 100 different queries. In particular, for the DBPedia and Biopax we created 100 subclass-of queries while for the other KBs we created 80 subclass-of and 20 instance-of queries. Some examples of queries are $Village \sqsubseteq PopulatedPlace$ for DBPedia, $TransportWithBiochemicalReaction \sqsubseteq Entity$ for Biopax and $Creator(Anthony\text{-}van\text{-}Dyck\text{-}is\text{-}Painter\text{-}in\text{-}Flanders)$ for Vicodi KB. The queries generated for the BRCA KB are similar with those used in the test of BUNDLE. For generating the subclass-of queries, we randomly selected two classes that are connected in the class hierarchy, while for the instance-of queries we randomly selected an individual a and a class to which a belongs by following the class hierarchy, starting from the class to which a explicitly belongs, so that each query had at least one explanation. Table 2 shows, for each ontology, the number of non-probabilistic axioms and the average time in seconds that TRILL and BUNDLE took for answering the queries.

---

[6] http://dbpedia.org/

[7] http://www.biopax.org/

[8] http://www.vicodi.org/

These preliminary tests show that TRILL is sometimes able to outperform BUNDLE, thanks to the fact that the translation of the set of explanations into a DNF formula is not required. However, on DBPedia, its longer running time may be due to the lack of all the optimizations that BUNDLE inherits from Pellet. This represents evidence that a Prolog implementation of a Semantic Web tableau reasoner is feasible and that may lead to a practical system.

9   Conclusions

In this paper we presented the DISPONTE semantics for probabilistic DLs that is inspired by the distribution semantics of probabilistic logic programming. We also presented the systems BUNDLE and TRILL for reasoning on DISPONTE KBs and their implementations. Both systems are tested on real world datasets. The experiments show that BUNDLE uses less memory and is faster than the publicly available version of the probabilistic reasoner PRONTO and is able to manage larger KBs. Moreover, the results for TRILL show that Prolog is a viable language for implementing DL reasoning algorithms and that its performance is comparable with that of a state-of-the-art reasoner. Both TRILL and BUNDLE are able to deal with ontologies of significant complexity.

References

1. Baader, F., Calvanese, D., McGuinness, D.L., Nardi, D., Patel-Schneider, P.F. (eds.): The Description Logic Handbook: Theory, Implementation, and Applications. Cambridge University Press, Cambridge (2003)
2. Baader, F., Horrocks, I., Sattler, U.: Description logics. In: Handbook of Knowledge Representation, chap. 3, pp. 135–179. Elsevier, Amsterdam (2008)
3. Baader, F., Peñaloza, R.: Automata-based axiom pinpointing. J. Autom. Reasoning 45(2), 91–129 (2010)
4. Baader, F., Peñaloza, R.: Axiom pinpointing in general tableaux. J. Log. Comput. 20(1), 5–34 (2010)
5. Bacchus, F.: Representing and Reasoning with Probabilistic Knowledge - A Logical Approach to Probabilities. MIT Press, Cambridge (1990)
6. Beckert, B., Posegga, J.: leantap: Lean tableau-based deduction. J. Autom. Reasoning 15(3), 339–358 (1995)
7. Bellodi, E., Lamma, E., Riguzzi, F., Albani, S.: A distribution semantics for probabilistic ontologies. In: International Workshop on Uncertainty Reasoning for the Semantic Web. CEUR Workshop Proceedings, vol. 778. Sun SITE Central Europe (2011)
8. Darwiche, A., Marquis, P.: A knowledge compilation map. J. Artif. Intell. Res. 17, 229–264 (2002)
9. De Raedt, L., Kimmig, A., Toivonen, H.: ProbLog: a probabilistic Prolog and its application in link discovery. In: International Joint Conference on Artificial Intelligence, pp. 2462–2467 (2007)
10. Faizi, I.: A description logic prover in prolog, Bachelor's thesis, Informatics Mathematical Modelling, Technical University of Denmark (2011)

11. Gomes, C.P., Sabharwal, A., Selman, B.: Model counting. In: Biere, A. (ed.) Handbook of Satisfiability. IOS Press, Amsterdam (2008)
12. Haarslev, V., Hidde, K., Möller, R., Wessel, M.: The racerpro knowledge representation and reasoning system. Semant. Web **3**(3), 267–277 (2012)
13. Halaschek-Wiener, C., Kalyanpur, A., Parsia, B.: Extending tableau tracing for ABox updates. Technical report, University of Maryland (2006)
14. Halpern, J.Y.: An analysis of first-order logics of probability. Artif. Intell. **46**(3), 311–350 (1990)
15. Herchenröder, T.: Lightweight semantic web oriented reasoning in prolog: tableaux inference for description logics. Master's thesis, School of Informatics, University of Edinburgh (2006)
16. Hitzler, P., Krötzsch, M., Rudolph, S.: Foundations of Semantic Web Technologies. CRC Press, Boca Raton (2009)
17. Hustadt, U., Motik, B., Sattler, U.: Deciding expressive description logics in the framework of resolution. Inf. Comput. **206**(5), 579–601 (2008)
18. Kalyanpur, A.: Debugging and repair of OWL ontologies. Ph.D. thesis, The Graduate School of the University of Maryland (2006)
19. Kalyanpur, A., Parsia, B., Horridge, M., Sirin, E.: Finding all justifications of OWL DL entailments. In: Aberer, K., et al. (eds.) ISWC/ASWC 2007. LNCS, vol. 4825, pp. 267–280. Springer, Heidelberg (2007)
20. Kalyanpur, A., Parsia, B., Sirin, E., Hendler, J.A.: Debugging unsatisfiable classes in OWL ontologies. J. Web Sem. **3**(4), 268–293 (2005)
21. Klinov, P.: Pronto: a non-monotonic probabilistic description logic reasoner. In: Bechhofer, S., Hauswirth, M., Hoffmann, J., Koubarakis, M. (eds.) ESWC 2008. LNCS, vol. 5021, pp. 822–826. Springer, Heidelberg (2008)
22. Klinov, P., Parsia, B.: Optimization and evaluation of reasoning in probabilistic description logic: towards a systematic approach. In: Sheth, A.P., Staab, S., Dean, M., Paolucci, M., Maynard, D., Finin, T., Thirunarayan, K. (eds.) ISWC 2008. LNCS, vol. 5318, pp. 213–228. Springer, Heidelberg (2008)
23. Klinov, P., Parsia, B.: A hybrid method for probabilistic satisfiability. In: Bjørner, N., Sofronie-Stokkermans, V. (eds.) CADE 2011. LNCS, vol. 6803, pp. 354–368. Springer, Heidelberg (2011)
24. Lukácsy, G., Szeredi, P.: Efficient description logic reasoning in prolog: the dlog system. TPLP **9**(3), 343–414 (2009)
25. Lukasiewicz, T.: Expressive probabilistic description logics. Artif. Int. **172**(6–7), 852–883 (2008)
26. Meissner, A.: An automated deduction system for description logic with alcn language. Studia z Automatyki i Informatyki **28–29**, 91–110 (2004)
27. Nilsson, N.J.: Probabilistic logic. Artif. Intell. **28**(1), 71–87 (1986)
28. Patel-Schneider, P.F., Horrocks, I., Bechhofer, S.: Tutorial on OWL (2003)
29. Poole, D.: The Independent Choice Logic for modelling multiple agents under uncertainty. Artif. Intell. **94**(1–2), 7–56 (1997)
30. Poole, D.: Probabilistic horn abduction and Bayesian networks. Artif. Intell. **64**(1), 81–129 (1993)
31. Reiter, R.: A theory of diagnosis from first principles. Artif. Intell. **32**(1), 57–95 (1987)
32. Ricca, F., Gallucci, L., Schindlauer, R., Dell'Armi, T., Grasso, G., Leone, N.: Ontodlv: an asp-based system for enterprise ontologies. J. Log. Comput. **19**(4), 643–670 (2009)
33. Riguzzi, F.: Extended semantics and inference for the Independent Choice Logic. Log. J. IGPL **17**(6), 589–629 (2009)

34. Riguzzi, F., Bellodi, E., Lamma, E.: Probabilistic Datalog+/- under the distribution semantics. In: Kazakov, Y., Lembo, D., Wolter, F. (eds.) International Workshop on Description Logics (2012)
35. Riguzzi, F., Bellodi, E., Lamma, E., Zese, R.: Computing instantiated explanations in OWL DL. In: Baldoni, M., Baroglio, C., Boella, G., Micalizio, R. (eds.) AI*IA 2013. LNCS, vol. 8249, pp. 397–408. Springer, Heidelberg (2013)
36. Riguzzi, F., Bellodi, E., Lamma, E., Zese, R.: Probabilistic description logics under the distribution semantics. Semant. Web J. (to appear, 2014)
37. Riguzzi, F., Lamma, E., Bellodi, E., Zese, R.: Epistemic and statistical probabilistic ontologies. In: Uncertainty Reasoning for the Semantic Web. CEUR Workshop Proceedings, vol. 900, pp. 3–14. Sun SITE Central Europe (2012)
38. Sang, T., Beame, P., Kautz, H.A.: Performing bayesian inference by weighted model counting. In: Proceedings of AAAI, pp. 475–482. AAAI Press/The MIT Press, Palo Alto, Pittsburgh, 9–13 July 2005
39. Sato, T.: A statistical learning method for logic programs with distribution semantics. In: International Conference on Logic Programming, pp. 715–729. MIT Press (1995)
40. Sato, T., Kameya, Y.: Parameter learning of logic programs for symbolic-statistical modeling. J. Artif. Intell. Res. 15, 391–454 (2001)
41. Schlobach, S., Cornet, R.: Non-standard reasoning services for the debugging of description logic terminologies. In: International Joint Conference on Artificial Intelligence, pp. 355–362. Morgan Kaufmann (2003)
42. Schmidt-Schauß, M., Smolka, G.: Attributive concept descriptions with complements. Artif. Intell. 48(1), 1–26 (1991)
43. Shearer, R., Motik, B., Horrocks, I.: Hermit: A highly-efficient owl reasoner. In: OWLED (2008)
44. Sirin, E., Parsia, B., Cuenca-Grau, B., Kalyanpur, A., Katz, Y.: Pellet: a practical OWL-DL reasoner. J. Web Sem. 5(2), 51–53 (2007)
45. Vassiliadis, V., Wielemaker, J., Mungall, C.: Processing owl2 ontologies using thea: an application of logic programming. In: International Workshop on OWL: Experiences and Directions. CEUR Workshop Proceedings, vol. 529. CEUR-WS.org (2009)
46. Vennekens, J., Denecker, M., Bruynooghe, M.: CP-logic: a language of causal probabilistic events and its relation to logic programming. Theory Pract. Log. Program. 9(3), 245–308 (2009)
47. Vennekens, J., Verbaeten, S., Bruynooghe, M.: Logic programs with annotated disjunctions. In: Demoen, B., Lifschitz, V. (eds.) ICLP 2004. LNCS, vol. 3132, pp. 431–445. Springer, Heidelberg (2004)

A Metaontology for Annotating Ontology Entities with Vagueness Descriptions

Panos Alexopoulos[1]([✉]), Silvio Peroni[2,3], Boris Villazón-Terrazas[1],
Jeff Z. Pan[4], and José Manuel Gómez-Pérez[1]

[1] iSOCO, Madrid, Spain
{palexopoulos,bvillazon,jmgomez}@isoco.com
[2] Department of Computer Science and Engineering,
University of Bologna, Bologna, Italy
[3] STLab-ISTC, CNR, Rome, Italy
silvio.peroni@unibo.it
[4] Department of Computing Science, University of Aberdeen, Aberdeen, UK
jeff.z.pan@abdn.ac.uk

Abstract. The emergence in the last years of initiatives like the Linked
Open Data (LOD) has led to a significant increase in the amount of structured semantic data on the Web. Central role to this development has
been played by ontologies, as these enable the representation of real world
domains in an explicit and formal way and, thus, the production of commonly understood and shareable semantic data. Nevertheless, the shareability and wider reuse of such data can be hampered by the existence
of vagueness within it, as this makes the data's meaning less explicit.
With that in mind, in this paper we present and evaluate the Vagueness
Ontology, a metaontology that enables the explicit identification and
description of vague entities and their vagueness-related characteristics
in ontologies. The rationale is that such descriptions, when accompanying vague ontologies, may narrow the possible interpretations that the
latter's vague elements may assume by its users.

1 Introduction

Ontologies are formal shareable conceptualisations of domains, describing the
meaning of domain aspects in a common, machine-processable form by means
of concepts and their interrelations [10]. As such, their role in the Semantic
Web is very important as they enable the production and sharing of structured
data that can be commonly understood among human and software agents. To
achieve this common understanding, one needs to ensure that the meaning of
ontology elements is explicit and shareable. In other words, all their users have
a clear, unambiguous and consensual understanding of what each ontological
element actually represents. That's in fact the reason why, towards this goal,
a number of relevant techniques and best practices have been proposed by the
literature, such as for example the use of argumentation processes [18,34] for
consensus building on the structure and the content of an ontology. Despite

© Springer International Publishing Switzerland 2014
F. Bobillo et al. (Eds.): URSW 2011-2013, LNAI 8816, pp. 100–121, 2014.
DOI: 10.1007/978-3-319-13413-0_6

these practices, however, a phenomenon that still affects, in a negative way, shareability and reusability of ontologies and semantic data is **vagueness**.

Vagueness is a common human knowledge and language phenomenon, typically manifested by terms and concepts like *High, Expert, Bad, Near* etc., and related to our inability to precisely determine the extensions of such concepts in certain domains and contexts. That is because vague concepts have typically blurred boundaries which do not allow for a sharp distinction between the entities that fall within their extension and those that do not [16,30]. For example, some people are borderline tall: not clearly *"tall"* and not clearly *"not tall"*.

The potential and actual existence of vague terminology in ontologies and semantic datasets has already been identified by the community [2,6,21,33,35]. A characteristic group of such elements are categorisation relations where entities are assigned to categories with no clear applicability criteria. An example is the relation *"hasFilmGenre"*, found in LinkedMDB[1] and DBpedia[2], that relates films with the genres they belong to. As most genres have no clear applicability criteria there will be films for which it is difficult to decide whether or not they belong to a given genre. A similar argument can be made for the DBpedia relations *"dbpedia-owl:ideology"* and *"dbpedia-owl:movement"*. Another group of vague elements comprises specialisations of concepts according to some vague property of them. Examples include *"Famous Person"* and *"Big Building"*, in the Cyc Ontology[3], and *"Competitor"*, found in the Business Role Ontology[4].

The important thing to notice in these examples is the lack of any further definitions that may clarify the intended meaning of the vague entities. For example, the definition of the concept *"Famous Person"* does not include the dimensions of fame according to which someone is judged as famous or not. This may lead to problematic situations.

More specifically, vague ontological definitions can cause **disagreements** among the people who develop, maintain or use it. Such a situation arose in a real life scenario where we faced significant difficulties in defining concepts like *"Critical System Process"* or *"Strategic Market Participant"* while trying to develop an electricity market ontology. When we asked our domain experts to provide exemplary instances of critical processes, there was dispute among them about whether certain processes qualified. Not only did different domain experts have different criteria of process criticality, but neither could anyone really decide which of those criteria were sufficient for the classification. In other words, the problem was the vagueness of the predicate *"critical"*.

While disagreements may be overcome by consensus, they are inevitable as more users alter, extend, or use ontologies. Imagine an enterprise ontology where the concept *"Strategic Client"* was initially created and populated by the company's executive board, their implicit membership criterion being the amount of revenue the clients generate for the company. Imagine also the new R&D Direc-

---

[1] Available at http://linkedmdb.org.
[2] Available at http://dbpedia.org.
[3] Available at http://www.cyc.com/platform/opencyc.
[4] Available at http://www.ip-super.org.

tor querying the instances of this concept while crafting an R&D strategy. If their own applicability criteria for the term *"Strategic"* do not coincide with the board's, using the returned list of clients might lead to poor decisions. Generalising these examples, some typical use-case scenarios where vagueness may be cause problems include:

1. **Structuring Data with a Vague Ontology:** When domain experts are asked to define instances of vague concepts and relations, then disagreements may occur on whether particular entities constitute instances of them.
2. **Utilising Vague Facts in Ontology-Based Systems:** When knowledge-based systems reason with vague facts, their output might not be optimal for those users who disagree with these facts.
3. **Integrating Vague Semantic Information:** When semantic data from several sources need to be merged then the merging of particular vague elements can lead to data that will not be valid for all its users.
4. **Evaluating Vague Semantic Datasets for Reuse:** When data practitioners need to decide whether a particular dataset is suitable for their needs, the existence of vague elements can make this decision harder. It can be quite difficult for them to assess *a priori* whether the data related to these elements are valid for their application context.

To reduce the negative effects of vagueness, we have put forward the notion of **vagueness-aware ontologies** [2], informally defined as *"ontologies whose vague elements are accompanied by comprehensive metainformation that describes the nature and characteristics of their vagueness"*. A simple example of such metainformation is whether an ontology entity (e.g., a class) is vague or not; this is important as many ontology users may not immediately realise this. A more sophisticated example, as we will explain in subsequent sections, is the particular type of the entity's vagueness or the applicability context of its definition. In all cases, our premise is that having such metainformation, explicitly represented and published along with (vague) ontologies, can improve the latter?s comprehensibility and shareability, by narrowing the possible interpretations that its vague elements may assume by human and software agents.

The focus of this paper is how vagueness-related metainformation may best be represented and applied to actual ontologies. For that, we describe here the **Vagueness Ontology (VO)**, an OWL metaontology that defines the necessary concepts, relations and attributes for creating explicit descriptions of vague ontology entities and (certain of) their characteristics. VO is meant to be used by both producers and consumers of ontologies; the former will utilise it to **annotate** the vague part of their produced ontologies with relevant vagueness metainformation while the latter will **query** this metainformation and use it to make a better use of the vague ontologies.

The motivation behind the development of VO is that, in our view, the vagueness-related metainformation should not be merely part of the ontology's informal documentation, neither its representation can be facilitated by simply using OWL's standard annotation properties such as `rdfs:comment`. The latter is because, as we will show in subsequent sections, one or more `rdfs:comment` values in an ontology entity cannot capture the more complex relations that exist between certain vagueness aspects.

The structure of the rest of the paper is as follows. In the next section we present related work while in Sect. 3 we provide a detailed description of the Vagueness Ontology, including the requirements it is designed to cover, the conceptual elements (classes, relations etc.) it comprises and usage examples. In Sect. 4 we present the results of a user-driven evaluation of the Vagueness Ontology, focusing on comprehensibility and usability aspects. Finally, in Sect. 5 we cover some important discussion points regarding the benefits and current limitations of our approach, while in Sect. 6 we summarise our work and outline its future directions.

2   Related Work

The practice of using ontologies for annotating various types of resources with metainformation has been exemplified by many works, including the *NLP Interchange Format (NIF)* [14], the *Extremely Annotational RDF Markup (EARMARK)* [4], and *Annotea* [17] for textual resources, as well as the more generic *Open Annotation Data Model (OADM)* [28] and *Provenance Ontology (PROV-O)* [19]. There are also several existing efforts for annotating ontologies. For general purpose ontology metadata we have Ontology Metadata Vocabulary (OMV) [13], Vocabulary of a Friend (VOAF)[5]. For metadata regarding ontology design and evolution there are the OWL 2 change ontology [24] and the Change and Annotations Ontology (CHAO) [23] as well as the C-ODO OWL metamodel for collaborative ontology design [12]. Finally, LexOMV [22] and Lemon [8] define metadata about multilinguality.

While the above vocabularies cover a large range of possible metainformation for ontologies, there is not yet, to the best of our knowledge, any specialised vocabulary for vagueness. The latter has so far been treated in the Semantic Web community mainly via fuzzy description logics, fuzzy ontologies and fuzzy query services [6,25,31], whose focus, however, is on enabling the definition and automated processing of fuzzy degrees of vague ontology entities and not so much on clarifying their intended interpretation (e.g. the concept membership criteria of a given vague concept). Thus, for example, a fuzzy ontology may contain the statement *"John is expert at ontologies to a degree of 0.8"* but there is no information on how the notion of expertise should be interpreted in the given domain or context. Therefore, as it will become clear in the rest of the paper, our approach is complementary to fuzzy ontology related works and it may be used to enhance the comprehensibility of fuzzy degrees.

3   The Vagueness Ontology

The Vagueness Ontology[6] has been developed following the *SAMOD*[7] (*Simplified Agile Methodology for Ontology Development*) methodology and its relevant

---

[5] Available at http://lov.okfn.org/vocab/voaf/v2.1/index.html.

[6] Available at http://www.essepuntato.it/2013/10/vagueness.

[7] Available at http://www.essepuntato.it/samod.

documentation is available online[8]. In this section, we focus on describing the requirements the ontology has been designed to satisfy and the main elements it consists of.

3.1 Vagueness Ontology Requirements

In an ontology, vagueness may primarily appear in the definitions of classes, object and datatype properties, and datatypes. A class is vague if, in the given domain, context or application scenario, it admits borderline cases, namely if there are (or could be) individuals for which it is indeterminate whether they instantiate the class. Typical vague classes are attributions, namely classes that reflect qualitative states of entities (e.g., *"TallPerson"*, *"ExperiencedResearcher"*, etc.). Similarly, an object property (relation) is vague if there are (or could be) pairs of individuals for which it is indeterminate whether they stand in the relation (e.g., *"hasGenre"*, *"hasIdeology"*, etc.). The same applies for datatype properties and pairs of individuals and literal values. Finally, a vague datatype consists of a set of vague terms. An example is the datatype *"RestaurantPriceRange"* when this comprises the terms *"cheap"*, *"moderate"* and *"expensive"*.

The Vagueness Ontology should enable the annotation of an ontological entity (class, relation or datatype) with a description of the nature and characteristic of its vagueness. In particular, the first thing such a description should explicitly state is whether the entity is actually vague or not. For example, the ontology class *"StrategicClient"* defined as *"A client that has a high value for the company"* is (and should be annotated as) vague while the definition of *"AmericanCompany"* as "A company that has legal status in the Unites States" is not. Moreover, it can often be the case that a seemingly vague element can have a non-vague definition (e.g. *"TallPerson"* when defined as "A person whose height is at least 180 cm"). Then this element is not vague in the given ontology and that is something that needs to be explicitly stated.

The second important vagueness characteristic to be explicitly represented is its type. Vagueness can be described according to at least two complementary types: quantitative (or degree) vagueness and qualitative (or combinatory) vagueness [16]. A predicate has degree-vagueness if the existence of borderline cases stems from the lack of precise boundaries for the predicate along one or more dimensions (e.g. *"bald"* lacks sharp boundaries along the dimension of hair quantity while *"red"* can be vague for both brightness and saturation). A predicate has combinatory vagueness if there are a variety of conditions pertaining to the predicate, but it is not possible to make any crisp identification of those combinations which are sufficient for application. A classical example of this type is *"religion"* as there are certain features that all religions share (e.g. beliefs in supernatural beings, ritual acts) yet it is not clear which are able to classify something as a religion. Based on this typology, we suggest that for a given vague entity it is important to represent and share the following explicitly:

---

[8] Available at http://www.essepuntato.it/2013/10/vagueness/documentation.

- **The type of the entity's vagueness:** Knowing whether an entity has quantitative or qualitative vagueness is important as elements with an intended (but not explicitly stated) quantitative vagueness can be considered by others as having qualitative vagueness and vice versa. Assume, for example, that a company's CEO does not make explicit that for a client to be classified as strategic, the amount of its R&D budget should be the only factor to be considered. Then, even though according to the CEO the vague class *"StrategicClient"* has quantitative vagueness in the dimension of the R&D budget amount, it will be hard for other company members to share the same view as this term has typically qualitative vagueness.
- **The dimensions of the term's quantitative vagueness:** When the entity has quantitative vagueness it is important to state explicitly its intended dimensions. E.g., if a CEO does not make explicit that for a client to be classified as strategic, its R&D budget should be the only pertinent factor, it will be rare for other company members to share the same view as the vagueness of the term *"strategic"* is multi-dimensional.

Furthermore, vagueness is **subjective** and **context dependent**. The first has to do with the same vague entity being interpreted differently by different users. For example, two company executives might have different criteria for the entity *"StrategicClient"*, the one the amount of revenue this client has generated and the other the market in which it operates. Similarly, context dependence has to do with the same vague entity being interpreted or applied differently in different contexts even by the same user; hiring a researcher in industry is different to hiring one in academia when it comes to judging his/her expertise and experience.

Therefore we additionally suggest that one should explicitly represent the **creator** of a vagueness annotation of a certain entity as well as the **applicability context** for which the entity is defined or in which it is used in a vague way. In particular, context-dependent can be (i) the description of vagueness of an entity (i.e. the same entity can be vague in one context and non-vague in another) and (ii) the dimensions related to a description of vagueness having quantitative type (i.e. the same entity can be vague in dimension A in one context and in dimension B in another). Please note that here we adopt the context-as-a-box metaphor [5] according to which a context is a "box" that contains knowledge in form of logical statements and whose boundaries are determined by specific contextual attributes (e.g. location, time, purpose etc.). When a vague term is related to a particular context, then this context has the jurisdiction to interpret the term's meaning and assess its validity in given statements [3].

Summarising the above, the Vagueness Ontology should enable users to ask the following competency questions about the entities of an ontology:

- *What entities have been explicitly defined either as vague or non-vague?*
- *What entities that have been defined both as vague and non-vague at the same time and why?*
- *What entities of a specific type (e.g., classes) have been defined either as vague or non-vague?*

- *What entities are characterised by a specific vagueness type?*
- *What entities have been recognised as vague, by whom and according to which vagueness type (if any)?*
- *What entities have quantitative vagueness and in what dimensions?*
- *What entities have quantitative vagueness, in what dimensions and what is the context of their dimensions (if any)?*
- *What entities are vague, in what contexts and according to whom?*

3.2 Ontology Anatomy

An overall view of the Vagueness Ontology (VO) is depicted in Fig. 1 via a Graffoo diagram [11] that describes its main classes and properties. VO uses several entities defined in external ontologies, i.e., the PROV-O [19] (prefix *prov*), OADM [28] (prefix *oa*), and the Situation ontology design pattern[9] (prefix *sit*). To show how to use the various entities of the ontology to describe vagueness/non-vagueness annotations, we introduce the following natural language scenario:

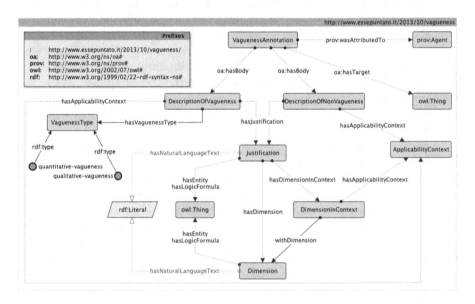

Fig. 1. The Graffoo diagram of the overall structure of the Vagueness Ontology.

The object property *ex:isExpertInResearchArea* is considered vague by John Doe in the context of researcher hiring. Moreover, he describes it as quantitatively vague since, for him, expertise is relevant to the number of her publications and projects; two different dimensions that he thinks relates to the contexts of Academia (i.e., number of relevant publications) and Industry (i.e., number of relevant projects).

---

[9] Available at http://www.ontologydesignpatterns.org/cp/owl/situation.owl.

The main class of the ontology is *VaguenessAnnotation*, which describes any annotation (i.e., *oa:Annotation*) of an ontological entity with information about its vagueness. A vagueness annotation is a particular act done by someone (i.e., an agent, identified by an individual of the class *prov:Agent*) who associates a description of vagueness/non-vagueness (called the body of the annotation, and defined through the property *oa:hasBody*) to the entity in consideration (called the target of the annotation, and defined through the property *oa:hasTarget*). This is formalised as follows[10]:

```
Class: VaguenessAnnotation
  SubClassOf: prov:Entity that prov:wasAttributedTo some prov:Agent,
              oa:Annotation that (oa:hasTarget exactly 1) and
              (oa:hasBody min 1
               (DescriptionOfNonVagueness or DescriptionOfVagueness))
```

Considering the aforementioned example, the annotation made by John Doe can be expressed as follows:

```
@prefix : <http://www.essepuntato.it/2013/10/vagueness/> .
@prefix ex: <http://www.essepuntato.it/resource/> .
ex:annotation a :VaguenessAnnotation ; prov:wasAttributedTo ex:john-doe ;
  oa:hasBody ex:description-of-vagueness ;
  oa:hasTarget ex:isExpertInResearchArea .
ex:isExpertInResearchArea a owl:ObjectProperty .
ex:john-doe a prov:Agent .
```

A vagueness annotation must specify a description of vagueness or non-vagueness for the annotated entity, in the form of an instance of the class *DescriptionOfVagueness* or *DescriptionOfNonVagueness* respectively. Vagueness descriptions must specify a vagueness type (one of the individuals of the class *VaguenessType*, i.e., *quantitative-vagueness* and *qualitative-vagueness*), and must provide at least one justification (i.e., an individual of the class *Justification*) for considering the target ontological entity vague. The individuals of the class *DescriptionOfNonVagueness*, instead, require only the specification of at least one justification. This class is meant to be used for entities that would typically be considered vague but which, for some reason, in the particular ontology are not (e.g. the *"TallPerson"* example in Sect. 3.1). Formalisation here is as follows:

```
Class: DescriptionOfNonVagueness SubClassOf: hasJustification min 1
Class: DescriptionOfVagueness
  SubClassOf: hasJustification min 1 , hasVaguenessType exactly 1
ObjectProperty: hasJustification
  Domain: DescriptionOfNonVagueness or DescriptionOfVagueness
  Range: Justification
ObjectProperty: hasVaguenessType
  Domain: DescriptionOfVagueness Range: VaguenessType
```

Considering again the previous example, the John Doe's description of vagueness can be defined as follows:

```
ex:description-of-vagueness a :DescriptionOfVagueness ;
  :hasJustification ex:justification ;
  :hasVaguenessType :quantitative-vagueness .
```

---

[10] All the entities of the Vagueness Ontology are introduced in Manchester Syntax [15], while the examples of use of the ontology are presented in Turtle [27].

The justifications of descriptions of vagueness/non-vagueness (i.e., individuals of the class *Justification*) aim at explaining the possible reasons behind such descriptions. Vagueness dimensions, in turn, (i.e., individuals of the class *Dimension* referred by the object property *hasDimension* and being always part of a justification) always refer to descriptions of quantitative vagueness and indicate some measurable characteristic of the annotated entity in which it is vague. Both justifications and dimensions can be defined as natural language text (i.e., the data property *hasNaturalLanguageText*), an entity (i.e. the object property *hasEntity*), a more complex logic formula (i.e., the object property *hasLogicFormula*) or any combination of them. The relevant formalisation is as follows:

```
Class: Justification
  SubClassOf: hasNaturalLanguageText some rdfs:Literal or
              hasEntity some owl:Thing or hasLogicFormula some owl:Thing
ObjectProperty: hasDimension
  Domain: Justification that inverse hasJustification only
          (DescriptionOfVagueness that
           (hasVaguenessType value quantitative-vagueness))
  Range: Dimension
Class: Dimension
  SubClassOf: hasNaturalLanguageText some rdfs:Literal or
              hasEntity some owl:Thing or hasLogicFormula some owl:Thing
```

Please note that while the properties *hasEntity* and *hasLogicFormula* share the same range class, i.e., *owl:Thing*, their intended meaning is different. The former property can be used to specify a certain resource (e.g., *dbpedia:H-index*) as (part of) a justification of a certain description. The latter property, instead, is used to link to a resource, which provides a justification, that actually "puns" a particular restriction or constraint on certain entities, e.g., *ex:hasNumberOfPublication some integer[>0]*.

Continuing the previous example, the justification and the related dimensions can be described as follows:

```
ex:justification a :Justification ;
  :hasNaturalLanguageText"It is not possible to define the exact minimum
        number of relevant publications and projects that make a a researcher
        expert in a given area." ;
  :hasDimension ex:dimension-publications , ex:dimension-projects .
ex:dimension-publications a :Dimension ;
  :hasNaturalLanguageText "The number of relevant publications." .
ex:dimension-projects a :Dimension ;
  :hasNaturalLanguageText "The number of relevant projects." .
```

As introduced before, descriptions of vagueness/non-vagueness and related dimensions can be characterised by particular contexts of application (i.e., individuals of the class *ApplicabilityContext*), which means that they can be applied within the boundaries of such particular contexts (i.e., the same entity can be vague in one context and non-vague in another). The contextualisation of descriptions is facilitated by an assertion between the description in consideration and the related context through the object property *hasApplicabilityContext*. In the case of dimensions, on the other hand, the context-dependent object is the *relation* between justifications and dimensions. Thus, to represent this, we reify the relation linking a justification to a dimension using an instance of the class *DimensionInContext*, that allows one to specify and the applicability context of such relation. VO formalises this as follows:

```
ObjectProperty: hasApplicabilityContext
   Domain: DescriptionOfNonVagueness or DescriptionOfVagueness or
           DimensionInContext
   Range: ApplicabilityContext
ObjectProperty: hasDimensionInContext
   Domain: Justification that inverse hasJustification only
           (DescriptionOfVagueness that
               (hasVaguenessType value quantitative-vagueness))
   Range: DimensionInContext
Class: DimensionInContext
   SubClassOf: sit:Situation , withDimension exactly 1 ,
               hasApplicabilityContext exactly 1
ObjectProperty: withDimension
   Characteristics: Functional SubPropertyOf: sit:isSettingFor
   Domain: DimensionInContext Range: Dimension
```

According to the above definitions, it is possible to complete the description of the aforementioned example as follows:

```
ex:description-of-vagueness
   :hasApplicabilityContext ex:researcher-hiring-context .
ex:researcher-hiring-context a :ApplicabilityContext .
ex:justification :hasDimensionInContext
   ex:dimension-publications-in-context , ex:dimension-projects-in-context .
ex:dimension-publications-in-context a :DimensionInContext ;
   :withDimension ex:dimension-publications ;
   :hasApplicabilityContext ex:academia-context .
ex:dimension-projects-in-context a :DimensionInContext ;
   :withDimension ex:dimension-projects ;
   :hasApplicabilityContext ex:industry-context .
ex:academia-context a :ApplicabilityContext .
ex:industry-context a :ApplicabilityContext .
```

This approach allows the reuse of the same dimension in different contexts and reasoners to infer automatically all the *hasDimension* assertions starting from the individuals of the class *DimensionInContext* by means of the sub-property chain *hasDimensionInContext o withDimension* defined in the object property *hasDimension*.

4   Vagueness Ontology Evaluation

As an initial assessment of VO's correctness, we used the online tool OOPS![11] [26] to detect potential modelling errors; results indicated no critical errors. Beyond that, we asked from a group of people with a working knowledge of ontologies to use VO to query ontologies that were already annotated with vagueness descriptions. Our goal was to evaluate the comprehensibility and usability of the current version of the ontology and get feedback.

The term "usability" here denotes the easiness by which a user of an ontology that has already been annotated with VO, can access (via SPARQL) and understand this vagueness-related metainformation. To assess this kind of usability, we asked subjects to study VO starting from its sources, documentation and additional material we provided, as well as to use a SPARQL endpoint in order to answer specific competency questions regarding the vagueness of a concrete VO-annotated ontology. The usability of VO in terms of the easiness by which

---

[11] Available at http://oeg-lia3.dia.fi.upm.es/oops/index-content.jsp.

an ontology engineer can annotate vague ontologies is going to be evaluated in future work.

4.1   Experimental Setting

We asked 22 subjects to perform three unsupervised tasks involving querying a SPARQL endpoint containing vagueness information about four entities, three classes and one object property. There were no "administrators" observing the subjects while they were undertaking these tasks, and we made sure that none of the subjects was previously aware of VO. In the end, 10 of these subjects completed the tasks. However, only 6 of which had enough experience in performing proper SPARQL queries, which is a mandatory requirement that subjects had to demonstrate in order to use quantitative data for assessing users' performance in addressing the tasks. Therefore, we used all the 10 subjects' data for analysing the usability of VO as gathered through the questionnaires introduced below, while we considered only the results related to the SPARQL-aware users for evaluating quantitative outcomes.

More specifically, the assessment of the actual subject's experience concerning the ability to provide appropriate SPARQL queries was derived from the answers the subject provided in a preliminary questionnaire, composed by self-assessment questions about subject's preliminary knowledge. In addition, we also analysed the actual SPARQL queries made by the subject during the test, in order to understand if (s)he was able to use basic SPARQL constructs such as *UNION* and *OPTIONAL*, that were necessary for addressing the tasks we proposed properly. In case these requirements were not satisfied, we did not consider the subject's SPARQL queries in the quantitative analysis of such data, in order not to bias the results. Therefore, we used only the queries provided by 6 out of 10 subjects for our quantitative analysis.

On the other hand, we thought that the understandability/learnability of the ontology could be assessed considering all the 10 subjects, since these aspects refer to the subjective perception of people when understanding the ontology and querying ontological data. During the test, we did not tell subjects whether their SPARQL queries were right or not and, thus, the actual correctness of such queries did not bias the subjects' personal perception of the ontology.

In all cases, the ontology we used contained nine annotations, seven of which pointed to descriptions of vagueness (one of those had an applicability context specified), while the remaining two referred to descriptions of non-vagueness (one of those had an applicability context specified). Some of these descriptions referred to seven justifications, while two of these justifications were linked to two dimensions each (in two cases, the justification-dimension relation presented a particular applicability context). The tasks given to the subjects involved the latter translating a natural language query into SPARQL. These queries were designed to ensure that subjects had to use all the entities of VO so as to reach a solution. Both the dataset and the tasks were based on the examples and the informal competency questions we had produced during the development of VO.

Table 1. The three natural language questions of each task (T1, T2 and T3) to translate in SPARQL.

T1	What are all the entities that have been defined either as vague or non-vague and why?
T2	What are all the entities, their related vagueness type, and their related OWL type that have been defined vague by someone?
T3	What entities have quantitative vagueness, in what dimensions and what is the context of their dimensions?

The evaluation session was structured as follows. We first asked subjects to complete a short multiple-choice questionnaire about their background knowledge and skills in OWL, ontology engineering, SPARQL, PROV-O and OADM (max. 2 min). Then, we asked subjects to study VO (max. 25 min), providing them the ontology source in RDF/XML, the complete online documentation with the diagram of Fig. 1, and usage examples. Then, we asked them to complete the three tasks listed in Table 1 (max. 15 min), allowing them to test the SPARQL translations on the dataset, available as SPARQL endpoint. During that, no access to the any exemplar SPARQL queries was given. Finally, we asked subjects to fill in two short questionnaires, one multiple choice and the other textual, to report their experience of using VO (and its related material) to complete these tasks (max. 5 min). All the questionnaires[12] and all the outcomes of the experiments[13] are available online.

4.2 Evaluation

Out of the 30 tasks in total (3 tasks given to each of 10 subjects), 9 were completed successfully (i.e., the right SPARQL queries were given), while 9 had incorrect answers or were not completed at all, giving an overall success rate of 50 %. The remaining 12 ones were not considered in this quantitative analysis since the related 4 users had proved to have not enough experience in performing SPARQL queries. The 9 successes were distributed as follows: 2 (out of 6) in Task 1, 6 in Task 2, and 1 in Task 3. A similar analysis can be done for the actual rows of the 6 users' outcomes matching with the expected results. In this case, we compared the each row returned by executing each user's SPARQL query with the expected rows, listing all the true positives (tp), false positives (fp), and false negatives (fn). We calculated the overall average precision (P) (i.e., $tp/(tp+fp)$) and average recall (R) (i.e., $tp/(tp+fn)$), calculated by considering those obtained by each subject, and we obtained P = 0.61 and R = 0.75. The average precision and recall for each task were P = 0.49 and R = 0.44 in Task 1, P = 1 and R = 1 in Task 2, and P = 0.66 and R = 0.83 in Task 3.

As shown by these quantitative results, the second task was always answered correctly, while issues arose when trying to answer to tasks 1 and 3. On the one

---

[12] Available at http://esurv.org?u=vagueness-ontology.

[13] Available at http://www.essepuntato.it/2013/10/vagueness/evaluation.

hand, in Task 1 we think two users (out of three who provided wrong answers) simply made syntactic mistakes (i.e., one returns the annotation individuals instead of the kinds of descriptions linked by such annotations, while the other named two SPARQL variables in the same way), which could be due to a rushed reading of the task or a distraction when writing the SPARQL query. On the other hand, in Task 3 it seems that subjects' mistakes related to a partial understanding of the ontology, since five of them provided imprecise solutions to the task. This seemed to depend on the possibility of describing dimensions involved in descriptions of quantitative vagueness as contextual objects or not, as introduced in Sect. 3.2. Although we were aware of possible misinterpretation of such part of the ontology, we decided to define dimensions by using the same pattern proposed in PROV-O, where certain relations, for instance between an entity and an agent (e.g., *prov:wasAttributedTo*), can be qualified, if needed, by reifying them as proper classes (e.g., *prov:Attribution*) linking to the entity and the agent in consideration. Of course, in all the above, one needs to consider the constrained time that participants had to study the ontology and perform the tasks.

The usability score for VO (considered together with its documentation and examples) was computed using the *System Usability Scale (SUS)* [7], by using the answers provided by all the 10 users. SUS is a well-known questionnaire used for the perception of the usability of a system, and it has been already used in the past for assessing the usability of ontologies (cf. [9]). SUS has the advantage of being technology independent (it has been tested on hardware, software, Web sites, etc.) and it is reliable even with a very small sample size [29][14]. In addition to the main SUS scale, we also were interested in examining the sub-scales of pure *Usability* and pure *Learnability* of VO, as proposed recently by Lewis and Sauro [20]. The mean SUS score for VO was 67.3 (in a 0–100 range), approaching the target score of 68 to demonstrate a good level of usability [29]. The mean values for the SUS sub-scales Usability and Learnability were 68.8 and 73.4 respectively.

In addition, two sub-scores were calculated for each subject by considering the values of the answers given in the background questionnaire (according to a 0–4 value range for each question). The first sub-score – composed of five questions and, thus, ranging from 0 to 20 – concerned the subject's *experience* with (the development of) ontologies. The other sub-score – composed of three questions and, thus, ranging from 0 to 12 – concerned the subject's personal

---

[14] Even if confidence intervals of the SUS scores will be rather wide (e.g., in our experiment we obtained [56.06, 78.45]), the average SUS score will be surprisingly stable even with a small sample. As stated in [29] and summarised in his blog (see http://www.measuringusability.com/blog/10-things-SUS.php for more details), Sauro "did several computer simulations and showed that [...] the mean from a sample size of just 5 repeated 1000 times [...] was within 6 points of the true SUS score" in the 50 % of the 1000 samples used – note that the true SUS score was calculated using the original big sample Sauro had available. This means that "you get within the ballpark of the actual SUS score in more than half of the cases with very small sample sizes" – e.g., "if the actual SUS score was a 74, average SUS scores from five users will fall between 68 and 80 half of the time".

knowledge about SPARQL, PROV-O and OADM. As shown in Fig. 2, we have plotted these subject's sub-scores (x-axis) with the subject's SUS value and the other sub-scales (y-axis) – and we have also included red dashed lines referring to the related *Least Squares Regression Lines*. Even if we cannot have any statistical significance of such comparisons because of the small size of our sample, it seems that the plots suggest some sort of positive correlation between the experience sub-scores and the SUS values – i.e., the more a subject knew about ontologies in general, the more VO is perceived as usable. The plots referring to the other aspect, namely the relation between the knowledge sub-scores and the SUS values, does not seem to provide enough evidence to speculate on any sort of correlation.

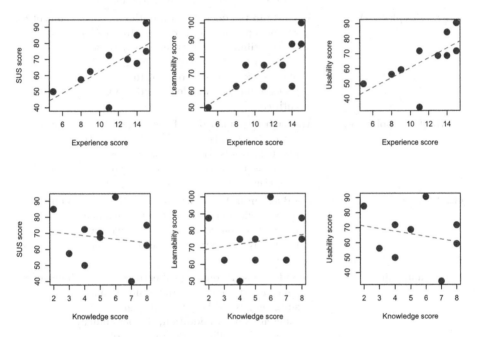

Fig. 2. Six plots showing the relation between subjects' *experience* and *knowledge* scores and the related subjects' SUS values and the other sub-scales. The red dashed lines were calculated by using the Least Squares Regression method.

Axial coding of the personal comments expressed in the final questionnaires [32] revealed a small number of perceived issues. Only 8 of all the subjects tested provided meaningful comments that were used for the study, and, of the 7 terms that were identified as significant in the comments, only 5 (4 positive and 1 negative) were mentioned by more than two individuals (albeit sometimes with different words), as shown in Table 2. The only negative issue mentioned by more than two subjects, i.e., the ambiguities in some ontological terms, was mainly highlighted by the subjects whom answers to tasks were not considered in the quantitative evaluation due to their inexperience in SPARQL. This seems to suggest some sort of (cor)relation between the understandability of VO and the experience users had in using SPARQL.

Table 2. Terms – four positive (+) and one negative (−) – mentioned by more than two individuals in the final questionnaire responses.

Term	Description	Frequency
HTML documentation (+)	The documentation of VO was suggested as one of the success features to help users understanding the ontology and, thus, to consult for writing the SPARQL queries	4 out of 8
ABox examples (+)	Similarly to the documentation, the snippets provided to show how to use VO for describing vagueness entities were a crucial aspect for writing SPARQL queries, since they were used to understand the intended usage of VO entities	3 out of 8
SPARQL endpoint (+)	Although it was not one of the targets of the evaluation, the SPARQL endpoint we provided to test the SPARQL queries was considered useful to assess their correctness and to browse the dataset	3 out of 8
Graffoo diagram (+)	The diagram in Fig. 1, which is part of the ontology documentation, received explicit mention in comments as an effective way to quickly understand the relation among classes	3 out of 8
Entity ambiguities (−)	Users found some ontological entities, e.g., *context* and *dimension*, quite ambiguous since they could be interpreted in different ways. In addition, the possibility of providing different descriptions of vagueness for a certain concept had been perceived as a drawback, as well as the fact that one needs to deal with several levels of indirection in order to express precisely the vagueness of concepts	3 out of 8

5   Discussion Points

5.1   Benefits of Consuming Vagueness-Aware Ontologies

The Vagueness Ontology is to be used by producers and consumers of ontologies and semantic datasets, so as to create and consume vagueness ontology descriptions respectively. Regarding the consumption of a vagueness-aware ontology, a first benefit of a vagueness-aware ontology for potential users is that it makes them aware of the existence of vagueness by explicitly stating the vague elements. This is important as vagueness is not always obvious to people (and certainly never to systems), meaning that it can be easily overlooked and lead to the negative effects described in previous sections. A second benefit is that it enables users to query each of the vague element's metainformation (vagueness dimensions, applicability context etc.) and use it in order to reduce these effects.

To show how this may be done let's revisit the four scenarios of Sect. 1. In the first scenario, involving the structuring of data with an existing vague ontology, the problem is that disagreements may occur on whether particular objects are actually instances or not of vague concepts or relations. If, however, information like the dimensions and applicability conditions and contexts of these elements are made known to the people who perform this task, then the possible interpretation space of them will be reduced. For example, if it is known that in order to classify a given company as a competitor, one needs to consider only the number of common business areas target markets, then other possible dimensions (e.g. the geographical proximity) will be excluded. This exclusion should reduce the number of potential disagreements.

In the second scenario, where vague ontological elements are utilised within some end-user application, the availability of vagueness metainformation can help the system's developers in two ways. First, it will make them aware of the fact that the ontology contains vague information and thus some of the system's output might not be considered accurate by the end-users. Second, they may use the vagueness metainformation to try to deal with that fact. For example, in a recommendation scenario, the applicability context of a vague axiom can be used as part of an explanation to the user of why a particular item was recommended. That might not change the user's opinion on whether this recommendation is accurate, but the potential user's feedback could help pin down the particular element's vagueness as the cause of this inaccuracy and take appropriate action.

In the third scenario, when two or more ontologies need to be integrated, the vagueness metamodel can be used to compare the "compatibility" of these ontologies in terms of vagueness. For example, if the same two vague classes have different vagueness dimensions (e.g. the vague class "Strategic Client"), then the one class's set of instance membership axioms might not be appropriate for the second one's as it might have been defined under a different interpretation of the class's vagueness. A simple query to the two ontologies' vagueness metamodel could reveal this issue. Similarly, in the case of evaluating given ontologies and semantic datasets for reuse purposes, the metamodel can be used to compare the vagueness compatibility of the dataset with the intended domain and application scenario. Table 3 summarises the above use case scenarios and the way the metamodel may be used and benefit each of them.

From the above, it is evident that if the vagueness characteristics that VO specifies (dimensions, context, etc.) were merely part of its documentation and not explicitly represented as metadata, this kind of querying would not be

Table 3. Vagueness metamodel usage and benefits in different scenarios.

Use case scenario	Metamodel usage	Expected benefits
Structuring data with a vague ontology	Communicate the meaning of the vague elements to the domain experts	Make the job of the experts easier and faster and reduce disagreements among them
	Use the metamodel to characterise the created data's vagueness	Enhance the future usability and shareability of the data.
Utilising vague semantic data in an ontology-based system	Check which data is vague	Know a priori which data may affect the system's effectiveness
	Use the properties of the vague elements to provide vagueness-related explanations to the users	
Integrating vague semantic datasets	Compare same-name vague elements across datasets according to their vagueness type and dimensions	Avoid integrating incompatible elements
Evaluating a vague semantic dataset for reuse	Query the metamodel to check the vagueness compatibility of the dataset with the intended domain and application scenario	Avoid re-using (parts of) datasets that are not compatible to own interpretation of vagueness

possible. Moreover, as the VO captures formally the relations that exist between these characteristics (e.g., the relation between a dimension and a context), the same kind of querying would not be possible if these relations were defined using merely the `rdfs:comment` annotation property of OWL. In such a case, if the ontology user would like, for example, to measure the number of different dimensions and contexts a particular entity is vague in, he/she would have to parse, via some NLP method, the entity's `rdfs:comment` value; a process obviously not very effective or easy to perform. On the other hand, with VO as a basis one can access an ontology's vagueness-related metainformation via SPARQL and, potentially, via more high-level services that are suitable not only to ontology engineers but also to domain experts, application developers and data analysts.

5.2 Creating Vagueness-Aware Ontologies

Annotating ontologies with VO is currently a manual task, with knowledge engineers and domain experts having to detect the vague elements, determine the relevant characteristics (type, dimensions, etc.) and instantiate VO. How this may be best facilitated is out of this paper's scope but it is an important aspect of our ongoing and future work. An example of this work is a system we've developed that is able to automatically detect ontology elements that are potentially vague [1]. The system uses a binary classifier that may distinguish between vague and non-vague term word senses and, consequently, between vague and non-vague linguistic definitions of ontology entities. Thus, for example, the definition of the ontology class *"StrategicClient"* as *"A client that has a high value for the company"* would be classified as vague while the definition of *"AmeriCanCompany"* as *"A company that has legal status in the Unites States"* would not. Our goal is to incorporate this classification functionality into an ontology authoring tool that will take as input an ontology, detect automatically its vague entities, guide the user into annotating them with the Vagueness Ontology in a Q&A manner and give as output a vagueness annotation for the given ontology.

5.3 Reasoning with Vagueness-Aware Ontologies

The current version of VO has not been developed with automated reasoning in mind, primarily because we have not yet analysed vagueness in adequate depth so as to define more complex axioms that may facilitate some kind of reasoning. Moreover, some of VO's information, such as dimensions or contexts, is currently described in a textual (and thus imprecise) way, thus making it harder to perform very detailed reasoning. Both these limitations have been purposefully not tackled in this first version of VO, in order to avoid an increased complexity that could discourage people from adopting it and start using it to annotate their ontologies.

In principle, reasoning with VO can be made possible by defining constraints and inference rules that determine how vagueness and its characteristics proliferate when defining more complex OWL axioms, such as complex classes or subsumption relations. A simple example is to say that *"The conjunction of a set of classes is quantitatively vague if all (vague) classes are quantitatively vague whereas it is qualitatively vague if at least one (vague) class is qualitatively vague"*. Then, a vagueness (meta-)reasoner could infer a conjunctive class's vagueness type by considering the types of its constituent classes. Similarly, one could say that *"The inverse of a vague property has the same vagueness characteristics (type, dimensions, contexts, etc.) as the original property"*. On the other hand, it is a matter of further analysis whether and in what way a class's vagueness's characteristics are "transferred" to its subclasses. Such an analysis, that will try to identify and implement a comprehensive set of valid reasoning rules for VO, is left as future work.

As far as the imprecise nature of VO's textual content is concerned, its potentially inhibitive role in reasoning depends on the particular reasoning rule at hand.

For example, the rule in the above paragraph regarding the vagueness type of a conjunctive class is not really affected by imprecision. On the other hand, the implementation of a rule such as *"The conjunction of a set of quantitatively vague classes is quantitatively vague in the superset of all these classes' dimensions"* would require the comparison of vagueness dimensions (and probably contexts) which, when represented as simple strings, can be imprecise. For such cases, a more formal representation of dimensions and contexts (with, e.g., taxonomical relations between contexts) would be probably necessary; nevertheless, such a representation needs to be contemplated along with the specification of VO's reasoning behaviour and, for that, is left as future work.

6   Conclusions and Future Work

In this paper we presented and evaluated the Vagueness Ontology (VO), a metaontology for annotating vague ontological entities with descriptions that describe the nature and characteristics of their vagueness in an explicit way. The metaontology is meant to be used by both producers and consumers of ontologies and semantic datasets, with the former utilising it to annotate the vague part of their produced ontologies and the latter querying this metainformation in order to make a better use of them.

VO's high-level goal is to raise the awareness of human producers and consumers of ontologies and semantic data about vagueness and the potential problems it may cause, and provide them with the means to produce/consume ontologies with a clearer meaning. At the moment, there are neither established practices nor tools in the Semantic Web community for working with vagueness, the result being that vague ontologies and semantic data are created and used without realising the meaning explicitness issues that may arise. Moreover, it should be made clear that our work does not aim to "get rid" of vagueness; on the contrary, we want to highlight it as a central issue in the development of the Semantic Web and, for the scenarios we have identified in this paper, make it more manageable and less problematic by making it explicit, not eliminating it.

Regarding VO's evaluation, our goal was to evaluate how (ontology-savvy) users understood VO. For that, we performed a time-constrained, user-based evaluation of VO showed a satisfying level of clarity and usability. Future experiments will involve domain experts and ontology engineers using VO to annotate ontologies; these experiments, however, will be performed when we have developed appropriate tooling for using VO.

This development will form part of our future work, aiming towards facilitating the easier and seamless usage of VO for the production of vagueness-aware ontologies, not only by ontology engineers but also by domain experts, application developers and data analysts. For that, we are currently developing a semi-automatic framework for generating vagueness descriptions with VO without having to know its implementation details. In another direction, we plan to evolve VO by looking at its potential links with fuzzy ontologies, identifying more sophisticated vagueness distinctions and phenomena and enabling a higher level of automated reasoning.

Acknowledgments. The research has been funded from the People Programme (Marie Curie Actions) of the European Union's 7th Framework Programme P7/2007-2013 under REA grant agreement n° 286348. We also want to thank all the people who helped us with the evaluation of Vagueness Ontology.

References

1. Alexopoulos, P., Pavlopoulos, J.: A vague sense classifier for detecting vague definitions in ontologies. In: Proceedings of the 14th Conference of the European Chapter of the Association for Computational Linguistics: Short Papers, Gothenburg, Sweden, April 2014, vol. 2, pp. 33–37. Association for Computational Linguistics (2014)
2. Alexopoulos, P., Villazon-Terrazas, B., Pan, J.: Towards vagueness-aware semantic data. In: Bobillo, F., Carvalho, R.N., da Costa, P.C.G., d'Amato, C., Fanizzi, N., Laskey, K.B., Laskey, K.J., Lukasiewicz, T., Martin, T., Nickles, M., Pool, M. (eds.) URSW, CEUR Workshop Proceedings, vol. 1073, pp. 40–45. CEUR-WS.org (2013)
3. Bao, J., Tao, J., McGuinness, D.L., Smart, P.: Context representation for the semantic web. In: Web Science Conference, 26–27 April 2010 (2010)
4. Barabucci, G., Di Iorio, A., Peroni, S., Poggi, F., Vitali, F.: Annotations with EAR-MARK in practice: a fairy tale. In: Tomasi, F., Vitali, F., (eds.) Proceedings of the 2013 Workshop on Collaborative Annotations in Shared Environments: Metadata, Vocabularies and Techniques in the Digital Humanities (DH-CASE 2013). ACM Press (2013)
5. Benerecetti, M., Bouquet, P., Ghidini, C.: Contextual reasoning distilled. J. Exp. Theoret. Artif. Intell. **12**(3), 279–305 (2000)
6. Bobillo, F., Straccia, U.: Fuzzy ontology representation using OWL 2. Int. J. Approximate Reasoning **52**(7), 1073–1094 (2011)
7. Brooke, J.: SUS: a quick and dirty usability scale. In: Jordan, P.W., Thomas, B., Weerdmeester, B.A., McClelland, A.L. (eds.) Usability Evaluation in Industry, pp. 189–194. Taylor and Francis, London (1996)
8. Buitelaar, P., Sintek, M., Kiesel, M.: A multilingual/multimedia lexicon model for ontologies. In: Sure, Y., Domingue, J. (eds.) ESWC 2006. LNCS, vol. 4011, pp. 502–513. Springer, Heidelberg (2006)
9. Casellas, N.: Ontology evaluation through usability measures. In: Meersman, R., Herrero, P., Dillon, T. (eds.) OTM 2009 Workshops. LNCS, vol. 5872, pp. 594–603. Springer, Heidelberg (2009)
10. Chandrasekaran, B., Josephson, J., Benjamins, R.: What are ontologies and why do we need them? IEEE Intell. Syst. **14**(1), 20–26 (1999)
11. Falco, R., Gangemi, A., Peroni, S., Vitali, F.: Modeling OWL ontologies with Graffoo. In: Presutti, V., Blomqvist, E., Troncy, R., Sack, H., Papadakis, I., Tordai, A. (eds.) ESWC 2014 Satellite Events. LNCS, vol. 8798, pp. 320–325. Springer, Berlin (2014)
12. Gangemi, A., Lehmann, J., Presutti, V., Nissim, M., Catenacci, C.: C-ODO: an OWL meta-model for collaborative ontology design. In: Alani, H., Noy, N., Stumme, G., Mika, P., Sure, Y., Vrandecic, D. (eds.) Workshop on Social and Collaborative Construction of Structured Knowledge (CKC 2007) at WWW 2007, Banff, Canada (2007)
13. Hartmann, J., Sure, Y., Haase, P., Palma, R., del Carmen Suárez-Figueroa, M.: OMV - Ontology Metadata Vocabulary. In: Welty, C. (ed.) Ontology Patterns for the Semantic Web Workshop, Galway, Ireland (2005)

14. Hellmann, S., Lehmann, J., Auer, S., Brümmer, M.: Integrating NLP using linked data. In: Alani, H., Kagal, L., Fokoue, A., Groth, P., Biemann, C., Parreira, J.X., Aroyo, L., Noy, N., Welty, C., Janowicz, K. (eds.) ISWC 2013, Part II. LNCS, vol. 8219, pp. 98–113. Springer, Heidelberg (2013)

15. Horridge, M., Patel-Schneider, P.F.: OWL 2 Web Ontology Language: Manchester Syntax. W3C working group note, World Wide Web Consortium, 2nd edn., December 2012

16. Hyde, D.: Vagueness, Logic and Ontology. Ashgate New Critical Thinking in Philosophy. Ashgate, Aldershot (2008)

17. Kahan, J., Koivunen, M.-R.: Annotea: an open RDF infrastructure for shared web annotations. In: Proceedings of the 10th International Conference on World Wide Web (WWW 2001), pp. 623–632. ACM Press, New York (2001)

18. Kotis, K., Vouros, A.: Human-centered ontology engineering: the HCOME methodology. Knowl. Inf. Syst. **10**(1), 109–131 (2006)

19. Lebo, T., Sahoo, S., McGuinness, D.: PROV-O: The PROV Ontology. W3C recommendation, World Wide Web Consortium, April 2013

20. Lewis, J.R., Sauro, J.: The factor structure of the system usability scale. In: Kurosu, M. (ed.) HCD 2009. LNCS, vol. 5619, pp. 94–103. Springer, Heidelberg (2009)

21. Lukasiewicz, T., Straccia, U.: Managing uncertainty and vagueness in description logics for the semantic web. J. Web Semant. **6**(4), 291–308 (2008)

22. Montiel-Ponsoda, E., de Cea, G.A., Suarez-Figueroa, M., Palma, R., Peters, W., Gomez-Perez, A.: LexOMV: an OMV extension to capture multilinguality. In: Proceedings of the OntoLex07, pp. 118–127 (2007–06)

23. Noy, N.F., Chugh, A., Liu, W., Musen, M.A.: A framework for ontology evolution in collaborative environments. In: Cruz, I., Decker, S., Allemang, D., Preist, C., Schwabe, D., Mika, P., Uschold, M., Aroyo, L.M. (eds.) ISWC 2006. LNCS, vol. 4273, pp. 544–558. Springer, Heidelberg (2006)

24. Palma, R., Haase, P., Corcho, O., Gómez-Pérez, A.: Change representation for OWL 2 ontologies. In: Hoekstra, R., Patel-Schneider, P.F. (eds.) OWLED, CEUR Workshop Proceedings, vol. 529. CEUR-WS.org (2008)

25. Pan, J.Z., Stamou, G., Stoilos, G., Taylor, S., Thomas, E.: Scalable querying services over Fuzzy ontologies. In: The Proceedings of the 17th International World Wide Web Conference (WWW2008) (2008)

26. Poveda-Villalón, M., Suárez-Figueroa, M.C., Gómez-Pérez, A.: Validating ontologies with OOPS!. In: ten Teije, A., Völker, J., Handschuh, S., Stuckenschmidt, H., d'Acquin, M., Nikolov, A., Aussenac-Gilles, N., Hernandez, N. (eds.) EKAW 2012. LNCS, vol. 7603, pp. 267–281. Springer, Heidelberg (2012)

27. Prud'hommeaux, E., Carothers, G.: Turtle - Terse RDF Triple Language. W3C candidate recommendation, World Wide Web Consortium, February 2013

28. Sanderson, R., Ciccarese, P., Van de Sompel, H.: Designing the W3C open annotation data model. In: Proceedings of the 5th Annual ACM Web Science Conference (WebSci13), pp. 366–375. ACM Press (2013)

29. Sauro, J.: A Practical Guide to the System Usability Scale: Background, Benchmarks & Best Practices. CreateSpace, Denver (2011)

30. Shapiro, S.: Vagueness in Context. Oxford University Press, Oxford (2006)

31. Stoilos, G., Stamou, G., Pan, J.Z., Tzouvaras, V., Horrocks, I.: Reasoning with very expressive Fuzzy description logics. J. Artif. Intell. Res. **30**, 273–320 (2007)

32. Strauss, A., Corbin, J.: Basics of Qualitative Research: Techniques and Procedures for Developing Grounded Theory, 2nd edn. Sage Publications, London (1998)

33. Thomas, C., Sheth, A.: On the expressiveness of the languages for the semantic web - making a case for a 'little more'. In: Sanchez, E. (ed.) Fuzzy Logic and the Semantic Web. Capturing Intelligence, vol. 1, pp. 3–20. Elsevier, Amsterdam (2006)
34. Vrandecic, D., Pinto, H.S., Sure, Y., Tempich, C.: The DILIGENT knowledge processes. J. Knowl. Manag. **9**(5), 85–96 (2005)
35. Zadeh, L.A.: From search engines to question-answering systems - the need for new tools. In: Ruiz, E.M., Segovia, J., Szczepaniak, P.S. (eds.) AWIC 2003. LNAI, vol. 2663, pp. 15–17. Springer, Heidelberg (2003)

Finite Lattices Do Not Make Reasoning in $\mathcal{ALCOI}$ Harder

Stefan Borgwardt[1][(✉)] and Rafael Peñaloza[1,2]

[1] Theoretical Computer Science, TU Dresden, Dresden, Germany
{stefborg,penaloza}@tcs.inf.tu-dresden.de
[2] Center for Advancing Electronics Dresden, Dresden, Germany

Abstract. We consider the fuzzy description logic $\mathcal{ALCOI}$ with semantics based on a finite residuated De Morgan lattice. We show that reasoning in this logic is EXPTIME-complete w.r.t. general TBoxes. In the sublogics $\mathcal{ALCI}$ and $\mathcal{ALCO}$, it is PSPACE-complete w.r.t. acyclic TBoxes. This matches the known complexity bounds for reasoning in classical description logics between $\mathcal{ALC}$ and $\mathcal{ALCOI}$.

1 Introduction

OWL 2, the current standard ontology language for the semantic web, is a syntactic variant of the crisp description logic (DL) $\mathcal{SROIQ(D)}$. The knowledge of an application domain can be formalized in such an ontology, and then reasoning problems such as ontology consistency, concept satisfiability, and concept subsumption can be used to infer new knowledge. As all crisp logics, this language is not well suited for expressing vague or imprecise notions that can be found in numerous domains. For instance, in the biomedical areas it is common to encounter concepts, such as HighTemperature or Large, that cannot be precisely represented using a classical logic.

Fuzzy extensions of DLs have been studied for the last two decades, and the literature on the topic is very extensive (see the surveys [27,34]). Most of those approaches are based on the very simple *Zadeh semantics*, where the conjunction of two statements is interpreted as the minimum of their truth values; these values range over the interval $[0,1]$ of rational numbers. The last decade has seen a shift towards more general semantics for treating vagueness, motivated by the development of mathematical fuzzy logic [23]. On the one hand, the use of continuous t-norms as the underlying interpretation function for conjunction was proposed in [24]. On the other hand, [33] considers incomparable truth degrees, which are structured into a lattice. However, this latter work still restricts to Zadeh-like semantics.

Most of the work on fuzzy DLs since then has focused on t-norm-based semantics over the unit interval; yet, even in those cases, ontologies are usually restricted to be unfoldable or acyclic [7–9]. Indeed, it has been shown that

---

Partially supported by the DFG under grant BA 1122/17-1, in the research training group 1763 (QuantLA), and in the Cluster of Excellence 'cfAED'.

F. Bobillo et al. (Eds.): URSW 2011-2013, LNAI 8816, pp. 122–141, 2014.
DOI: 10.1007/978-3-319-13413-0_7

general concept inclusion axioms (GCIs) can cause undecidability even in fuzzy extensions of the basic DL $\mathcal{ALC}$ [4,5,15,16,20]. In order to allow the expressivity of GCIs within the knowledge base, while retaining decidability, it is necessary to restrict the expressivity of the logic in other ways. This has motivated the study of fuzzy DLs over finitely-valued semantics [19]. The notion of t-norm can be rephrased in terms of finite lattices to preserve the relationship with mathematical fuzzy logic.

If one considers the Łukasiewicz t-norm over finitely many values, then reasoning is decidable even for very expressive DLs, as shown in [10] through a reduction to crisp reasoning. When restricted to $\mathcal{ALC}$ without terminological axioms, concept satisfiability is PSPACE-complete as in the crisp case [18].[1] In the presence of general TBoxes, this problem becomes EXPTIME-complete [13,14], again matching the complexity of the crisp case, even if arbitrary (finite) lattices and t-norms are allowed. However, the complexity of subsumption of concepts was left as an open problem, as the standard reduction used in crisp DLs does not work with general t-norm semantics.

In [11,12,17], matching complexity bounds were shown for other reasoning tasks and for logics up to $\mathcal{SHI}$, which extends $\mathcal{ALC}$ with transitive and inverse roles, and allows for role inclusion axioms. More precisely, it was shown that all standard reasoning tasks are EXPTIME-complete in lattice-valued $\mathcal{SHI}$ w.r.t. general TBoxes. If restricted to acyclic TBoxes, then the complexity reduces to PSPACE in $\mathcal{ALCHI}$; the same holds for $\mathcal{SI}$ under a restriction on the interpretation of the roles. For $\mathcal{SH}$, reasoning is EXPTIME-complete, even if the TBox is empty.

In this paper we complement those complexity results by showing that finite lattices do not affect the complexity of reasoning even if nominals are allowed, and provide tight complexity bounds for the fuzzy logic L-$\mathcal{ALCOI}$ over a finite lattice L. More precisely, we show that in this logic concept satisfiability is EXPTIME-complete w.r.t. general TBoxes and acyclic TBoxes. It was shown in [11,17] that the restriction to acyclic TBoxes and the sublogic L-$\mathcal{ALCI}$ leads to a PSPACE-complete satisfiability problem. We show here that this is also the case in L-$\mathcal{ALCO}$. Moreover, the same complexity bounds hold for deciding subsumption between concepts. These results are in accordance with the complexity of reasoning in the classical DLs underlying these logics.

2 Preliminaries

We first recall some results about automata on infinite trees from [2] that will allow us to obtain tight upper bounds for our reasoning problems. Afterwards, we briefly introduce residuated lattices, which will be used for the semantics of our logic. For a more comprehensive view on residuated lattices, in particular in connection with mathematical fuzzy logic, we refer the reader to [21–23].

---

[1] The paper [18] considers the fuzzy modal logic K, which can be seen as a syntactic variant of fuzzy $\mathcal{ALC}$ with only one role.

2.1   Looping Automata on Infinite Trees

To obtain upper bounds for the complexity of reasoning in $L\text{-}\mathcal{ALCOI}$, we describe in Sect. 4 a reduction to the emptiness problem of looping automata on infinite trees. Such automata receive as input the unlabeled infinite k-ary tree for a fixed $k \in \mathbb{N}$. The *nodes* of this tree are represented by words in K^*, where $K := \{1, \ldots, k\}$: the empty word ε represents the root node, and ui represents the i-th successor of the node u. An *ancestor* of a node $u \in K^*$ is a node $u' \in K^*$ for which there exists a $u'' \in K^*$ such that $u = u'u''$. A *path* is a sequence $v_1, \ldots, v_m$ of nodes such that $v_1 = \varepsilon$ and each v_{i+1} is a successor of v_i for every $i, 1 \le i < m$.

Definition 1 (looping automaton). *A* looping (tree) automaton *is a tuple* $A = (Q, I, \Delta)$ *where Q is a finite set of* states, $I \subseteq Q$ *is a set of* initial states, *and $\Delta \subseteq Q^{k+1}$ is the* transition relation. *A* run *of* A *is a mapping* $r\colon K^* \to Q$ *assigning states to each node of K^* such that $r(\varepsilon) \in I$ and for every $u \in K^*$,* $(r(u), r(u1), \ldots, r(uk)) \in \Delta$. *The* emptiness problem *for looping automata is to decide whether a given looping automaton has a run.*

The emptiness of looping automata can be decided in polynomial time using a bottom-up approach that finds all states that can appear in a run [37]. Alternatively, one can use a top-down approach, which relies on the fact that if there is a run, then there is also a periodic run. To speed up this search, the period should be as short as possible. This motivates the notion of *blocking automata*.

Definition 2 (m-blocking). *Let* $A = (Q, I, \Delta)$ *be a looping automaton. We say that* A *is m-blocking for $m \in \mathbb{N}$ if every path $v_1, \ldots, v_m$ of length m in a run r of* A *contains two nodes v_i and v_j ($i < j$) such that $r(v_i) = r(v_j)$.*

Clearly, every looping automaton is m-blocking for all $m > |Q|$. However, the main interest in blocking automata arises when one can find a smaller bound on m. One way to reduce this is through a so-called *faithful* family of functions.

Definition 3 (faithful). *Let* $A = (Q, I, \Delta)$ *be a looping automaton. The family of functions $f_q\colon Q \to Q$, $q \in Q$, is* faithful *w.r.t.* A *if for all $q, q_0, q_1, \ldots, q_k \in Q$,*

- *if $(q, q_1, \ldots, q_k) \in \Delta$, then $(q, f_q(q_1), \ldots, f_q(q_k)) \in \Delta$; and*
- *if $(q_0, q_1, \ldots, q_k) \in \Delta$, then $(f_q(q_0), f_q(q_1), \ldots, f_q(q_k)) \in \Delta$.*

The subautomaton $A^S = (Q, I, \Delta^S)$ *of* A *induced by this family has the transition relation $\Delta^S = \{(q, f_q(q_1), \ldots, f_q(q_k)) \mid (q, q_1, \ldots, q_k) \in \Delta\}$.*

Lemma 4 [2]. *Let* A *be a looping automaton and A^S its subautomaton induced by a faithful family of functions. Then* A *has a run iff A^S has a run.*

The construction in Sect. 4 produces automata that are exponential in the size of the input. For such cases, it has been shown that if the automata are m-blocking for some m bounded polynomially in the size of the input (that is, logarithmically in the size of the automaton), then the emptiness test requires only polynomial space.

Definition 5 (PSPACE on-the-fly construction). *Consider a set $\mathfrak{I}$ of inputs and a construction that yields, for every $i \in \mathfrak{I}$, an m_i-blocking looping automaton $A_i = (Q_i, I_i, \Delta_i)$ on k_i-ary trees. This construction is a* PSPACE *on-the-fly construction if there is a polynomial P such that, for every input i of size n,*

(i) $m_i \leq P(n)$ and $k_i \leq P(n)$,
(ii) every element of Q_i has size bounded by $P(n)$, and
(iii) one can nondeterministically guess in time bounded by $P(n)$ an element of I_i, and, for a state $q \in Q_i$, a transition from Δ_i with first component q.

As hinted at by the name, these conditions guarantee the following complexity result for checking emptiness of the constructed automata.

Theorem 6 [2]. *If the looping automata A_i are obtained from the inputs $i \in \mathfrak{I}$ by a* PSPACE *on-the-fly construction, then the emptiness of A_i can be checked in space polynomial in the size of i.*

In Sect. 5, we will use this theorem to give PSPACE upper bounds on the complexity of reasoning in sublogics of $L\text{-}\mathcal{ALCOI}$.

2.2 Residuated Lattices

A *lattice* is an algebraic structure $(L, \vee, \wedge)$ over a *carrier set* L with two idempotent, associative, and commutative binary operations *join* $\vee$ and *meet* $\wedge$ that satisfy the absorption laws $\ell_1 \vee (\ell_1 \wedge \ell_2) = \ell_1 = \ell_1 \wedge (\ell_1 \vee \ell_2)$ for all $\ell_1, \ell_2 \in L$. The order $\leq$ on L is uniquely defined by $\ell_1 \leq \ell_2$ iff $\ell_1 \wedge \ell_2 = \ell_1$ for all $\ell_1, \ell_2 \in L$. A lattice L is *distributive* if $\vee$ and $\wedge$ distribute over each other, *finite* if L is finite, and *bounded* if it has a *minimum* and a *maximum* element, denoted as $\mathbf{0}$ and $\mathbf{1}$, respectively. It is *complete* if joins and meets of arbitrary subsets $T \subseteq L$, denoted by $\bigvee_{t \in T} t$ and $\bigwedge_{t \in T} t$ respectively, exist. Every finite lattice is also bounded and complete. Whenever it is clear from the context, we simply use the carrier set L to represent the lattice $(L, \vee, \wedge)$.

A *De Morgan lattice* is a distributive lattice with an involutive and antimonotonic unary operator $\sim$, called *(De Morgan) negation*, satisfying the De Morgan laws $\sim(\ell_1 \vee \ell_2) = \sim\ell_1 \wedge \sim\ell_2$ and $\sim(\ell_1 \wedge \ell_2) = \sim\ell_1 \vee \sim\ell_2$ for all $\ell_1, \ell_2 \in L$. For example, for every $n \in \mathbb{N}$, let $L_n = \{k/n \mid 0 \leq k \leq n\}$. Then $(L_n, \max, \min)$ is a distributive lattice which, together with the negation $\sim\ell = 1 - \ell$, forms a De Morgan lattice.

An important notion in mathematical fuzzy logic is that of a *triangular norm*, or *t-norm* for short. We define this for arbitrary (bounded) lattices, although in the literature the term is usually only used in the context of the real interval $[0, 1]$ or finite chains [22, 23, 26].

Definition 7 (t-norm, residuum). *Given a bounded lattice L, a (generalized) t-norm is an associative and commutative binary operator on L that is monotone w.r.t. the lattice order and has unit $\mathbf{1}$. A residuated lattice is a bounded lattice L extended with a t-norm $\otimes$ and a binary operator $\Rightarrow$ (called (generalized) residuum) such that, for all $\ell_1, \ell_2, \ell_3 \in L$, we have $\ell_1 \otimes \ell_2 \leq \ell_3$ iff $\ell_2 \leq \ell_1 \Rightarrow \ell_3$.*

Notice that what we call a residuated lattice corresponds to a commutative, distributive, integral, zero-bounded FL-algebra from [22]. We chose to call it a residuated lattice to keep the relation with mathematical fuzzy logic explicit. A simple consequence of Definition 7 is that, for every $\ell_1, \ell_2 \in L$,

- $1 \Rightarrow \ell_1 = \ell_1$, and
- $\ell_1 \leq \ell_2$ iff $\ell_1 \Rightarrow \ell_2 = 1$.

For a t-norm $\otimes$ over a complete lattice L, there is a binary operator $\Rightarrow$ that satisfies the residuation property w.r.t. $\otimes$ iff the t-norm is *join-preserving* [22], i.e. for all $\ell \in L$ and $T \subseteq L$ we have

$$\ell \otimes \left(\bigvee_{\ell' \in T} \ell' \right) = \bigvee_{\ell' \in T} (\ell \otimes \ell').$$

If $\otimes$ also preserves arbitrary meets in the dual way, then $\Rightarrow$ is uniquely determined as the function that satisfies, for all $\ell_1, \ell_2 \in L$,

$$\ell_1 \Rightarrow \ell_2 = \bigvee \{x \mid \ell_1 \otimes x \leq \ell_2\}.$$

Using this result, we often characterize a complete residuated lattice through its (join- and meet-preserving) t-norm, without explicitly mentioning its residuum. The t-norm and the De Morgan negation also uniquely determine the *t-conorm* $\ell_1 \oplus \ell_2 := \sim(\sim\ell_1 \otimes \sim\ell_2)$ and the *residual negation* $\ominus\ell := \ell \Rightarrow \mathbf{0}$.

For the rest of this paper, we fix a complete residuated De Morgan lattice L with De Morgan negation $\sim$ and a join- and meet-preserving t-norm $\otimes$. The operators $\Rightarrow, \oplus$, and $\ominus$ are then given by the above equations.

3   L-$\mathcal{ALCOI}$

We now describe the fuzzy description logic L-$\mathcal{ALCOI}$, whose semantics is based on the operators of L. It generalizes the classical DL $\mathcal{ALCOI}$ by using the elements of L as truth values, instead of just the Boolean *true* and *false*. The syntax of L-$\mathcal{ALCOI}$ is very similar to that of $\mathcal{ALCOI}$. It is based on non-empty and pairwise disjoint sets N_C, N_R, and N_I of *concept names*, *role names*, and *individual names*, respectively.

Definition 8 (syntax). *A (complex) role is of the form r or r^- for $r \in N_R$. (Complex) concepts are constructed from concept names using the constructors $\top$ (top), $\{a\}$ (nominal for $a \in N_I$), $\neg C$ (negation), $C \sqcap D$ (conjunction), $C \to D$ (implication), $\exists s.C$ (existential restriction for a complex role s), and $\forall s.C$ (value restriction).*

For a complex role s, the *inverse of s* (denoted by $\bar{s}$) is s^- if $s \in N_R$ and r if $s = r^-$. The main difference to the syntax of classical $\mathcal{ALCOI}$ is the explicit presence of the implication constructor.

The semantics of this logic is based on interpretation functions that map every concept C to a *fuzzy set* over the truth degrees from L, i.e. a function specifying the membership degree of every domain element to C.

Definition 9 (semantics). *An* interpretation *is a pair* $\mathcal{I} = (\Delta^{\mathcal{I}}, \cdot^{\mathcal{I}})$, *where* $\Delta^{\mathcal{I}}$ *is a non-empty set, called the* domain *of* $\mathcal{I}$, *and* $\cdot^{\mathcal{I}}$ *is an* interpretation function *that maps every concept name A to a function $A^{\mathcal{I}} : \Delta^{\mathcal{I}} \to L$, every role name r to a function $r^{\mathcal{I}} : \Delta^{\mathcal{I}} \times \Delta^{\mathcal{I}} \to L$, and every individual name a to an element $a^{\mathcal{I}} \in \Delta^{\mathcal{I}}$. This function is extended to complex roles and concepts for all $x, y \in \Delta^{\mathcal{I}}$ as follows:*

- $(r^-)^{\mathcal{I}}(x, y) := r^{\mathcal{I}}(y, x)$,
- $\top^{\mathcal{I}}(x) := \mathbf{1}$,
- $\{a\}^{\mathcal{I}}(x) := \mathbf{1}$ *if* $x = a^{\mathcal{I}}$, *and* $\{a\}^{\mathcal{I}}(x) := \mathbf{0}$ *otherwise*,
- $(\neg C)^{\mathcal{I}}(x) := {\sim} C^{\mathcal{I}}(x)$,
- $(C \sqcap D)^{\mathcal{I}}(x) := C^{\mathcal{I}}(x) \otimes D^{\mathcal{I}}(x)$,
- $(C \to D)^{\mathcal{I}}(x) := C^{\mathcal{I}}(x) \Rightarrow D^{\mathcal{I}}(x)$,
- $(\exists s.C)^{\mathcal{I}}(x) := \bigvee_{z \in \Delta^{\mathcal{I}}} s^{\mathcal{I}}(x, z) \otimes C^{\mathcal{I}}(z)$,
- $(\forall s.C)^{\mathcal{I}}(x) := \bigwedge_{z \in \Delta^{\mathcal{I}}} s^{\mathcal{I}}(x, z) \Rightarrow C^{\mathcal{I}}(z)$.

Note that we did not include the disjunction constructor, usually interpreted by the t-conorm, as it can be expressed using conjunction and negation. Likewise, the residual negation can be simulated by the implication, negation, and top. However, unlike in classical DLs, existential and universal quantifiers are not dual to each other, i.e. in general it does not hold that $(\neg \exists s.C)^{\mathcal{I}}(x) = (\forall s.\neg C)^{\mathcal{I}}(x)$.

The axioms of this logic also have an associated lattice value, which expresses the degree to which the restriction must be satisfied.

Definition 10 (axioms). *An* axiom *is an* assertion $\langle a{:}C \bowtie \ell \rangle$, *a* concept definition $\langle A \doteq C \geq \ell \rangle$, *or a* general concept inclusion *(GCI)* $\langle C \sqsubseteq D \geq \ell \rangle$, *where* $A \in \mathsf{N_C}$, $a \in \mathsf{N_I}$, $\ell \in L$, C, D *are concepts, and* $\bowtie \in \{<, \leq, =, \geq, >\}$. *An* ABox *is a finite set of assertions. A* general TBox *is a finite set of GCIs. An* acyclic TBox *is a finite set of concept definitions such that every concept name occurs at most once on the left-hand side of an axiom, and there is no cyclic dependency between definitions. A* TBox *is either a general TBox or an acyclic TBox.*[2] *An* ontology *is a pair* $\mathcal{O} = (\mathcal{A}, \mathcal{T})$ *where $\mathcal{A}$ is an ABox and $\mathcal{T}$ is a TBox.*

The interpretation $\mathcal{I}$ satisfies *(or is a* model of*)*

- *an assertion $\langle a{:}C \bowtie \ell \rangle$ if $C^{\mathcal{I}}(a^{\mathcal{I}}) \bowtie \ell$;*
- *a concept definition $\langle A \doteq C \geq \ell \rangle$ if for every element $x \in \Delta^{\mathcal{I}}$ it holds that $\left(A^{\mathcal{I}}(x) \Rightarrow C^{\mathcal{I}}(x) \right) \otimes \left(C^{\mathcal{I}}(x) \Rightarrow A^{\mathcal{I}}(x) \right) \geq \ell$;*
- *a GCI $\langle C \sqsubseteq D \geq \ell \rangle$ if for every $x \in \Delta^{\mathcal{I}}$ we have $C^{\mathcal{I}}(x) \Rightarrow D^{\mathcal{I}}(x) \geq \ell$;*
- *an ABox, TBox, or ontology if it satisfies all axioms in it.*

If $\mathcal{T}$ is an acyclic TBox, then all concept names occurring on the left-hand side of some definition in $\mathcal{T}$ are called *defined*, all others are called *primitive*. If $\mathcal{T}$ is a general TBox, then all concept names appearing in it are *primitive*.

---

[2] We do not consider mixed TBoxes. We could allow axioms of the form $\langle A \sqsubseteq C \geq \ell \rangle$ in acyclic TBoxes, as long as they do not introduce cyclic dependencies. To avoid overloading the notation, we exclude this case.

Usually, ABoxes also contain *role assertions* of the form $\langle (a,b):r \bowtie \ell \rangle$, expressing that $r^{\mathcal{I}}(a^{\mathcal{I}}, b^{\mathcal{I}}) \bowtie \ell$ should hold. We do not consider such axioms here, since they can be simulated by the concept assertion $\langle a{:}\exists r.\{b\} \bowtie \ell \rangle$ (or $\langle b{:}\exists r^-.\{a\} \bowtie \ell \rangle$).

We emphasize that $\mathcal{ALCOI}$ is a special case of $L\text{-}\mathcal{ALCOI}$, where the underlying lattice contains only the elements $\mathbf{0}$ and $\mathbf{1}$, which may be interpreted as *false* and *true*, respectively, and the t-norm is simply the classical conjunction. Accordingly, one can generalize the reasoning problems for $\mathcal{ALCOI}$ to lattices.

Definition 11 (reasoning). *Let C, D be concepts, $\mathcal{O}$ an ontology, and $\ell \in L$.*

- *$\mathcal{O}$ is consistent if it has a model.*
- *C is ℓ-satisfiable w.r.t. $\mathcal{O}$ if there is a model $\mathcal{I}$ of $\mathcal{O}$ and an element $x \in \Delta^{\mathcal{I}}$ such that $C^{\mathcal{I}}(x) \geq \ell$.*
- *C is ℓ-subsumed by D w.r.t. $\mathcal{O}$ if every model of $\mathcal{O}$ is also a model of $\langle C \sqsubseteq D \geq \ell \rangle$.*
- *The best satisfiability degree for C w.r.t. $\mathcal{O}$ is the supremum of all $\ell' \in L$ such that C is ℓ'-satisfiable w.r.t. $\mathcal{O}$.*
- *The best subsumption degree of C and D w.r.t. $\mathcal{O}$ is the supremum of all $\ell' \in L$ such that C is ℓ'-subsumed by D w.r.t. $\mathcal{O}$.*

Unfortunately, consistency and satisfiability even in the smaller logic $L\text{-}\mathcal{ALC}$ are undecidable in general [1,4,5,15,16,20]. Here we analyze the complexity of these problems under the assumption that L is *finite* and given as a list of its elements, and all lattice operations are computable in polynomial time in the size of their operands.

Observe that C is ℓ-satisfiable w.r.t. $(\mathcal{A}, \mathcal{T})$ iff $(\mathcal{A} \cup \{\langle a{:}C \geq \ell \rangle\}, \mathcal{T})$ is consistent, where a is a fresh individual name. Likewise, C is *not* ℓ-subsumed by D w.r.t. $(\mathcal{A}, \mathcal{T})$ iff $(\mathcal{A} \cup \{\langle a{:}C \to D < \ell \rangle\}, \mathcal{T})$ is consistent. To obtain the *best* degrees to which these inferences hold, one has to solve at most polynomially many consistency problems. For more details on these reductions, see e.g. [12]. We can thus focus on deciding consistency of ontologies to solve all other reasoning problems.

We show in Sect. 4 that the complexity of this problem is the same as for classical $\mathcal{ALCOI}$: it is ExpTime-complete w.r.t. both general and acyclic TBoxes [29,30,35,36]. However, in the sublogics $\mathcal{ALCO}$ (without inverse roles) and $\mathcal{ALCI}$ (without nominals), consistency w.r.t. *acyclic TBoxes* is decidable in PSpace [2,3], and PSpace-hard already in $\mathcal{ALC}$ [31]. We show in Sect. 5 that these bounds also apply to $L\text{-}\mathcal{ALCO}$ for arbitrary finite lattices L. The same holds for $L\text{-}\mathcal{ALCI}$ [11,17].

4   Consistency

We now show that consistency of $L\text{-}\mathcal{ALCOI}$ ontologies is in ExpTime. To achieve this, we adapt the approach from [2], where reasoning in classical DLs is reduced to the emptiness of an exponentially large looping automaton. To handle nominals and inverse roles correctly, we adapt ideas from [3,25].

Recall that the semantics of the quantifiers requires the computation of a supremum or infimum of the membership degrees of a possibly infinite set of elements of the domain. To obtain an effective decision procedure, reasoning is usually restricted to witnessed models [24].

Definition 12 (n-witnessed). *An interpretation $\mathcal{I}$ is n-witnessed, $n \in \mathbb{N}$, if for every $x \in \Delta^{\mathcal{I}}$ and every concept of the form $\exists s.C$ there exist n elements $x_1, \ldots, x_n \in \Delta^{\mathcal{I}}$ such that*

$$(\exists s.C)^{\mathcal{I}}(x) = \bigvee_{i=1}^{n} s^{\mathcal{I}}(x, x_i) \otimes C^{\mathcal{I}}(x_i),$$

and analogously for all value restrictions $\forall s.C$.

In particular, if $n = 1$, then the suprema and infima from the semantics of $\exists r.C$ and $\forall r.C$ become maxima and minima, respectively. In this case, we simply say that $\mathcal{I}$ is *witnessed*.

It was shown in [17] that every interpretation over the finite lattice L is n-witnessed for some n bounded by the cardinality of L. To simplify the description of the algorithm, in the following we consider only the case of $n = 1$. All constructions can easily be adapted for any other $n \in \mathbb{N}$.

Our algorithm for deciding consistency exploits the fact that an ontology $\mathcal{O}$ has a model iff it has a well-structured *forest model*, consisting of interconnected tree-like structures rooted in the named individuals. We model them using so-called *Hintikka trees* that abstract from the complexity of a full model by only expressing the membership degrees for all relevant concepts. We construct automata that have exactly these Hintikka trees as their runs, and use the initial states to verify the restrictions imposed by the ABox. Reasoning is hence reduced to a polynomial guessing step and polynomially many emptiness tests for these automata.

In the following, we consider an ontology $\mathcal{O} = (\mathcal{A}, \mathcal{T})$ for which we want to decide consistency. We denote by $\mathsf{sub}(\mathcal{O})$ the set of all subconcepts occurring in $\mathcal{O}$, and by $\mathsf{Ind}(\mathcal{O})$ the set of all individual names occurring in $\mathcal{O}$. Similarly, the set $\mathsf{Rol}(\mathcal{O})$ contains all role names used in $\mathcal{O}$, and $\mathsf{Rol}^{-}(\mathcal{O})$ contains all complex roles occurring in $\mathcal{O}$. The nodes of the Hintikka trees are labeled with so-called Hintikka functions over the domain $\mathsf{sub}(\mathcal{O}) \cup \{\rho\}$, where ρ is an arbitrary new element that will be used to express the degree with which the role relation to the parent node holds.

Definition 13 (Hintikka function). *A Hintikka function for $\mathcal{O}$ is a partial function $H: \mathsf{sub}(\mathcal{O}) \cup \{\rho\} \to L$ satisfying the following conditions:*

(i) $H(\rho)$ *is defined;*
(ii) *if $H(\top)$ is defined, then $H(\top) = \mathbf{1}$;*
(iii) *if $H(\{a\})$ is defined, then $H(\{a\}) \in \{\mathbf{0}, \mathbf{1}\}$;*
(iv) *if $H(\neg D)$ is defined, then $H(D)$ is defined and $H(\neg D) = {\sim}H(D)$;*
(v) *if $H(C \sqcap D)$ is defined, then $H(C)$ and $H(D)$ are also defined and it holds that $H(C \sqcap D) = H(C) \otimes H(D)$; and*
(vi) *if $H(C \to D)$ is defined, then $H(C)$ and $H(D)$ are also defined and it holds that $H(C \to D) = H(C) \Rightarrow H(D)$.*

This function is compatible *with*

- *an assertion* $\langle a{:}C \bowtie \ell \rangle$ *if* $H(C)$ *is defined and* $H(C) \bowtie \ell$;
- *a concept definition* $\langle A \doteq C \geq \ell \rangle$ *if, whenever* $H(A)$ *is defined, then* $H(C)$ *is defined and* $(H(A) \Rightarrow H(C)) \otimes (H(C) \Rightarrow H(A)) \geq \ell$;[3]
- *a GCI* $\langle C \sqsubseteq D \geq \ell \rangle$ *if* $H(C)$ *and* $H(D)$ *are defined and* $H(C) \Rightarrow H(D) \geq \ell$;
- *an ABox/TBox if it is compatible with all axioms in it.*

The support *of* H *is the set* $\mathsf{supp}(H)$ *of all* $C \in \mathsf{sub}(\mathcal{O})$ *for which* H *is defined, and* $\mathsf{Ind}(H)$ *is the set of all* $a \in \mathsf{Ind}(\mathcal{O})$ *for which* $\{a\} \in \mathsf{supp}(H)$ *and* $H(\{a\}) = 1$.

We denote by $H|_{\mathsf{sub}(\mathcal{O})}$ the restriction of a Hintikka function H to $\mathsf{sub}(\mathcal{O})$.

The first step of our decision procedure is to guess Hintikka functions for all named individuals. Since a domain element can have several names, we first need a partition $\mathcal{P}$ of $\mathsf{Ind}(\mathcal{O})$ that groups together all names referring to the same element. Given $\mathcal{P}$, we denote by $[a]_\mathcal{P}$ the class of $\mathcal{P}$ that contains $a \in \mathsf{Ind}(\mathcal{O})$. Then, we guess, for each $X \in \mathcal{P}$, one Hintikka function describing the behavior of the individual designated by the names in X. This is similar to the approach used in [3] to decide concept satisfiability in classical $\mathcal{ALCOQ}$ with acyclic TBoxes. We additionally guess the values of the role connections between all named elements using fuzzy binary relations $r_\mathcal{P}$ on $\mathcal{P}$ for every role name $r \in \mathsf{Rol}(\mathcal{O})$.

Definition 14 (pre-completion). *A* pre-completion *for the ontology* $\mathcal{O}$ *is a triple* $(\mathcal{P}, \mathcal{H}_\mathcal{P}, \mathcal{R}_\mathcal{P})$, *where* $\mathcal{P}$ *is a partition of* $\mathsf{Ind}(\mathcal{O})$, $\mathcal{H}_\mathcal{P} = (H_X)_{X \in \mathcal{P}}$ *is a family of Hintikka functions for* $\mathcal{O}$, *and* $\mathcal{R}_\mathcal{P} = (r_\mathcal{P})_{r \in \mathsf{Rol}(\mathcal{O})}$ *is a family of fuzzy binary relations* $r_\mathcal{P} : \mathcal{P} \times \mathcal{P} \to L$ *such that, for all* $X \in \mathcal{P}$,

- $\mathsf{Ind}(H_X) = X$;
- H_X *is compatible with* $\mathcal{T}$; *and*
- H_X *is compatible with* $\mathcal{A}_X := \{\langle a{:}C \bowtie \ell \rangle \in \mathcal{A} \mid a \in X\}$.

A Hintikka function H *for* $\mathcal{O}$ *is* compatible with this pre-completion *if for all* $a \in \mathsf{Ind}(H)$, *we have* $H|_{\mathsf{sub}(\mathcal{O})} = H_{[a]_\mathcal{P}}|_{\mathsf{sub}(\mathcal{O})}$.

Each H_X is compatible with the pre-completion $(\mathcal{P}, \mathcal{H}_\mathcal{P}, \mathcal{R}_\mathcal{P})$ since, for every $a \in \mathsf{Ind}(H_X) = X$ we have $[a]_\mathcal{P} = X$. We extend the family $\mathcal{R}_\mathcal{P}$ to complex roles by setting $r_\mathcal{P}^-(X, Y) := r_\mathcal{P}(Y, X)$ for all $X, Y \in \mathcal{P}$.

Hintikka trees are k-ary trees labelled with Hintikka functions, where k is the number of existential and value restrictions in $\mathsf{sub}(\mathcal{O})$. Intuitively, each successor acts as the witness for one of these restrictions. As in Sect. 2.1, we define $K := \{1, \ldots, k\}$. Since we need to know which successor in the tree corresponds to which restriction, we fix an arbitrary bijection

$$\varphi \colon \{C \mid C \in \mathsf{sub}(\mathcal{O}) \text{ is of the form } \exists s.D \text{ or } \forall s.D\} \to K,$$

and denote by $\varphi_s(\mathcal{O})$ for a role s the set of all indices $i \in K$ such that $i = \varphi(C)$ for a $C \in \mathsf{sub}(\mathcal{O})$ of the form $\exists s.D$ or $\forall s.D$.

---

[3] This method, called *lazy unfolding*, is only correct for acyclic TBoxes.

Definition 15 (Hintikka condition). *The tuple* $(H_0, H_1, \ldots, H_k)$ *of Hintikka functions for* $\mathcal{O}$ *satisfies the* Hintikka condition *if the following hold:*

(i) *For every existential restriction* $\exists s.C \in \mathsf{sub}(\mathcal{O})$:

- *If* $\exists s.C \in \mathsf{supp}(H_0)$ *and* $i = \varphi(\exists s.C)$, *then we have* $C \in \mathsf{supp}(H_i)$ *and* $H_0(\exists s.C) = H_i(\rho) \otimes H_i(C)$.
- *If* $\exists s.C \in \mathsf{supp}(H_0)$, *then for all* $i \in \varphi_s(\mathcal{O})$ *we have* $C \in \mathsf{supp}(H_i)$ *and* $H_0(\exists s.C) \geq H_i(\rho) \otimes H_i(C)$.
- *For all* $i \in \varphi_{\bar{s}}(\mathcal{O})$ *with* $\exists s.C \in \mathsf{supp}(H_i)$, *we have* $C \in \mathsf{supp}(H_0)$ *and* $H_i(\exists s.C) \geq H_i(\rho) \otimes H_0(C)$.

(ii) *For every universal restriction* $\forall s.C \in \mathsf{sub}(\mathcal{O})$:

- *If* $\forall s.C \in \mathsf{supp}(H_0)$ *and* $i = \varphi(\forall s.C)$, *then we have* $C \in \mathsf{supp}(H_i)$ *and* $H_0(\forall s.C) = H_i(\rho) \Rightarrow H_i(C)$.
- *If* $\forall s.C \in \mathsf{supp}(H_0)$, *then for all* $i \in \varphi_s(\mathcal{O})$ *we have* $C \in \mathsf{supp}(H_i)$ *and* $H_0(\forall s.C) \leq H_i(\rho) \Rightarrow H_i(C)$.
- *For all* $i \in \varphi_{\bar{s}}(\mathcal{O})$ *with* $\forall s.C \in \mathsf{supp}(H_i)$, *we have* $C \in \mathsf{supp}(H_0)$ *and* $H_i(\forall s.C) \geq H_i(\rho) \Rightarrow H_0(C)$.

(iii) *For all* $s \in \mathsf{Rol}^-(\mathcal{O})$, $i, j \in \varphi_s(\mathcal{O})$, $a \in \mathsf{Ind}(H_i)$, *and* $b \in \mathsf{Ind}(H_j)$ *with* $[a]_{\mathcal{P}} = [b]_{\mathcal{P}}$, *we have* $H_i(\rho) = H_j(\rho)$.

(iv) *For all* $a \in \mathsf{Ind}(H_0)$, $s \in \mathsf{Rol}^-(\mathcal{O})$, $i \in \varphi_s(\mathcal{O})$, *and* $b \in \mathsf{Ind}(H_i)$, *we have* $H_i(\rho) = s_{\mathcal{P}}([a]_{\mathcal{P}}, [b]_{\mathcal{P}})$.

Condition (i) makes sure that an existential restriction $\exists s.C$ is witnessed by its designated successor $\varphi(\exists s.C)$ and all other s-successors do not contradict the witness; this in particular includes possible $\bar{s}$-predecessors.[4] Condition (ii) treats the universal restrictions analogously. Condition (iii) ensures that the value of a role connection to a named individual remains the same regardless of which index i is used to express it, and by Condition (iv) we enforce the previously guessed role connections between named elements.

Definition 16 (Hintikka tree). *Let* $(\mathcal{P}, \mathcal{H}_{\mathcal{P}}, \mathcal{R}_{\mathcal{P}})$ *be a pre-completion for* $\mathcal{O}$ *and* $X \in \mathcal{P}$. *A* Hintikka tree *for* $\mathcal{O}$ *starting with* H_X *is a mapping* $\mathbf{T}$ *that assigns to each node* $u \in K^*$ *a Hintikka function for* $\mathcal{O}$ *such that*

(i) $\mathbf{T}(\varepsilon) = H_X$;
(ii) *for every* $u \in K^*$, $\mathbf{T}(u)$ *is compatible with* $\mathcal{T}$ *and the pre-completion; and*
(iii) *for every* $u \in K^*$, $(\mathbf{T}(u), \mathbf{T}(u1), \ldots, \mathbf{T}(uk))$ *satisfies the Hintikka condition.*

The definition of compatibility ensures that all axioms of $\mathcal{T}$ are satisfied at every node of the Hintikka tree, while the Hintikka condition makes sure that the tree satisfies the witnessing conditions for all relevant quantified concepts.

The proof of the following theorem uses arguments similar to those in [2,17]. The main difference to the classical case is the presence of successors witnessing the universal restrictions. We additionally have to deal here with the side condition of compatibility with the pre-completion.

---

[4] There is at most one such predecessor, namely the parent node.

Lemma 17. $\mathcal{O}$ *is consistent iff there are a pre-completion* $(\mathcal{P}, \mathcal{H}_{\mathcal{P}}, \mathcal{R}_{\mathcal{P}})$ *for* $\mathcal{O}$ *and, for each* $X \in \mathcal{P}$, *a Hintikka tree for* $\mathcal{O}$ *starting with* H_X.

Proof. Assume that a pre-completion and the required Hintikka trees $\mathbf{T}_X$ for $\mathcal{O}$ starting with H_X exist. We first remove irrelevant nodes in these Hintikka trees. A node $u \in K^*$ is *relevant* in $\mathbf{T}_X$ if $\mathsf{Ind}(\mathbf{T}_X(u')) = \emptyset$ for all (non-empty) ancestors $u' \in K^+$ of u. The idea is that if $a \in \mathsf{Ind}(\mathbf{T}_X(u'))$, then by the compatibility with the pre-completion $\mathbf{T}_X(u')$ agrees with $H_{[a]_{\mathcal{P}}} = \mathbf{T}_{[a]_{\mathcal{P}}}(\varepsilon)$ on the values of all concepts in $\mathsf{sub}(\mathcal{O})$, and thus $\mathbf{T}_X(u')$ can be replaced with $\mathbf{T}_{[a]_{\mathcal{P}}}(\varepsilon)$. The root nodes are always relevant since they are needed to represent the named individuals. We now define the interpretation $\mathcal{I}$ with domain

$$\Delta^{\mathcal{I}} := \{(X, u) \in \mathcal{P} \times K^* \mid u \text{ is relevant in } \mathbf{T}_X\}.$$

We set $a^{\mathcal{I}} := ([a]_{\mathcal{P}}, \varepsilon)$ for all $a \in \mathsf{Ind}(\mathcal{O})$. For $r \in \mathsf{N_R}$ and $(X, u), (Y, v) \in \Delta^{\mathcal{I}}$,

- $r^{\mathcal{I}}((X, u), (Y, v)) := \mathbf{T}_X(ui)(\rho)$ if there is an index $i \in \varphi_r(\mathcal{O})$ such that either (i) $(Y, v) = (X, ui)$ or (ii) $v = \varepsilon$ and $\mathsf{Ind}(\mathbf{T}_X(ui)) \cap Y \neq \emptyset$;
- $r^{\mathcal{I}}((X, u), (Y, v)) := \mathbf{T}_Y(vi)(\rho)$ if there is an index $i \in \varphi_{r^-}(\mathcal{O})$ such that either (i) $(X, u) = (Y, vi)$ or (ii) $u = \varepsilon$ and $\mathsf{Ind}(\mathbf{T}_Y(vi)) \cap X \neq \emptyset$; and
- $r^{\mathcal{I}}((X, u), (Y, v)) := \mathbf{0}$ otherwise.

To see that this is well-defined, consider the following three cases.

- If there are $i, j \in \varphi_r(\mathcal{O})$ such that $v = \varepsilon$ and both $\mathsf{Ind}(\mathbf{T}_X(ui)) \cap Y$ and $\mathsf{Ind}(\mathbf{T}_X(uj)) \cap Y$ are non-empty, then from Condition (iii) of Definition 15 we obtain $\mathbf{T}_X(ui)(\rho) = \mathbf{T}_X(uj)(\rho)$.
- For the dual case of $i, j \in \varphi_{r^-}(\mathcal{O})$ with $u = \varepsilon$, $\mathsf{Ind}(\mathbf{T}_Y(vi)) \cap X \neq \emptyset$, and $\mathsf{Ind}(\mathbf{T}_Y(vj)) \cap X \neq \emptyset$, we have $\mathbf{T}_Y(vi)(\rho) = \mathbf{T}_Y(vj)(\rho)$ by the same condition.
- If $u = v = \varepsilon$ and there are $i \in \varphi_r(\mathcal{O})$, $j \in \varphi_{r^-}(\mathcal{O})$, $a \in \mathsf{Ind}(\mathbf{T}_X(i)) \cap Y$, and $b \in \mathsf{Ind}(\mathbf{T}_Y(j)) \cap X$, then we have $Y = [a]_{\mathcal{P}}$ and $X = [b]_{\mathcal{P}}$. By Condition (iv) of Definition 15, this implies $\mathbf{T}_X(i)(\rho) = r_{\mathcal{P}}(X, Y) = r_{\mathcal{P}}^-(Y, X) = \mathbf{T}_Y(j)(\rho)$.

We now define the interpretation of concept names. For a primitive concept name A, we simply set $A^{\mathcal{I}}(X, u) := \mathbf{T}_X(u)(A)$ for all $(X, u) \in \Delta^{\mathcal{I}}$. $\mathcal{I}$ is extended to the defined concept names while showing the following claim:

for all $(X, u) \in \Delta^{\mathcal{I}}$ and all $C \in \mathsf{sub}(\mathcal{O})$ for which $\mathbf{T}_X(u)(C)$ is defined, we have $C^{\mathcal{I}}(X, u) = \mathbf{T}_X(u)(C)$. $\qquad(1)$

We prove this by induction on the (non-negative) *weight* $o(C)$, which is defined inductively as follows:

- $o(A) := o(\top) := o(\{a\}) := 0$ for every primitive concept name A and every $a \in \mathsf{N_I}$;
- $o(A) := o(C) + 1$ for every definition $\langle A \doteq C \geq \ell \rangle \in \mathcal{T}$;
- $o(\neg C) := o(C) + 1$;
- $o(C \sqcap D) := o(C \to D) := \max\{o(C), o(D)\} + 1$; and
- $o(\exists s.C) := o(\forall s.C) := o(C) + 1$.

This weight is well-defined for general and acyclic TBoxes.

For $C = \top$, Claim (1) follows immediately from Definition 13. For a primitive concept name A, it holds by the definition of $A^{\mathcal{I}}$ above.

If $\mathbf{T}_X(u)(\{a\})$ is defined for some $a \in \mathsf{Ind}(\mathcal{O})$, then by Definition 13 this value is either $\mathbf{0}$ or $\mathbf{1}$. If it is $\mathbf{0}$, then we cannot have $\mathbf{T}_X(u) = H_{[a]_{\mathcal{P}}}$ by Definition 14. Thus, $a^{\mathcal{I}} = ([a]_{\mathcal{P}}, \varepsilon) \neq (X, u)$, and hence $\{a\}^{\mathcal{I}}(X, u) = \mathbf{0} = \mathbf{T}_X(u)(\{a\})$. Otherwise, we have $\mathbf{T}_X(u)(\{a\}) = \mathbf{1}$, i.e. $a \in \mathsf{Ind}(\mathbf{T}_X(u))$. Since u is relevant in $\mathbf{T}_X$, we infer that $u = \varepsilon$. By Definition 14, we get $a \in \mathsf{Ind}(\mathbf{T}_X(u)) = \mathsf{Ind}(H_X) = X$, and thus $a^{\mathcal{I}} = ([a]_{\mathcal{P}}, \varepsilon) = (X, u)$. We conclude $\{a\}^{\mathcal{I}}(X, u) = \mathbf{1} = \mathbf{T}_X(u)(\{a\})$.

Consider now a defined concept name A with the definition $\langle A \doteq C \geq \ell \rangle \in \mathcal{T}$. If $\mathbf{T}_X(u)(A)$ is defined, then by the compatibility with $\mathcal{T}$ the value $\mathbf{T}_X(u)(C)$ is also defined and $\big(\mathbf{T}_X(u)(A) \Rightarrow \mathbf{T}_X(u)(C)\big) \otimes \big(\mathbf{T}_X(u)(C) \Rightarrow \mathbf{T}_X(u)(A)\big) \geq \ell$. Since $o(C) < o(A)$, we get $C^{\mathcal{I}}(X, u) = \mathbf{T}_X(u)(C)$ by induction. Thus, we can define $A^{\mathcal{I}}(X, u) := \mathbf{T}_X(u)(A)$ to ensure that $\mathcal{I}$ satisfies $\langle A \doteq C \geq \ell \rangle$ at (X, u). Whenever $\mathbf{T}_X(u)(A)$ is undefined, we can set $A^{\mathcal{I}}(X, u) := C^{\mathcal{I}}(X, u)$ to satisfy this concept definition without violating the claim.

If $\mathbf{T}_X(u)(\neg C)$ is defined, then $\mathbf{T}_X(u)(C)$ is also defined. By induction, we obtain $(\neg C)^{\mathcal{I}}(X, u) = {\sim} C^{\mathcal{I}}(X, u) = {\sim} \mathbf{T}_X(u)(C) = \mathbf{T}_X(u)(\neg C)$. Similar arguments show Claim (1) for conjunctions and implications.

Assume now that $\ell := \mathbf{T}_X(u)(\exists s.C)$ is defined for a complex role s and a concept C, and let $i := \varphi(\exists s.C)$. We first prove the existence of a witness $(Y, v) \in \Delta^{\mathcal{I}}$ such that $s^{\mathcal{I}}((X, u), (Y, v)) \otimes C^{\mathcal{I}}(Y, v) = \ell$. By the Hintikka condition, we know that $\mathbf{T}_X(ui)(C)$ is defined and $\ell = \mathbf{T}_X(ui)(\rho) \otimes \mathbf{T}_X(ui)(C)$. Since u is relevant in $\mathbf{T}_X$, ui can only be irrelevant in $\mathbf{T}_X$ if $\mathsf{Ind}(\mathbf{T}_X(ui)) \neq \emptyset$. We make a case distinction on whether ui is relevant or not. (1) If there exists an $a \in \mathsf{Ind}(\mathbf{T}_X(ui))$, then the compatibility of $\mathbf{T}_X(ui)$ with the pre-completion implies that $\mathbf{T}_{[a]_{\mathcal{P}}}(\varepsilon)(C) = H_{[a]_{\mathcal{P}}}(C) = \mathbf{T}_X(ui)(C)$ is defined. Since the root node ε is relevant in $\mathbf{T}_{[a]_{\mathcal{P}}}$, by induction we get $C^{\mathcal{I}}([a]_{\mathcal{P}}, \varepsilon) = \mathbf{T}_{[a]_{\mathcal{P}}}(\varepsilon)(C)$. Furthermore, by the definition of $s^{\mathcal{I}}$ we know that $s^{\mathcal{I}}((X, u), ([a]_{\mathcal{P}}, \varepsilon)) = \mathbf{T}_X(ui)(\rho)$, and thus we can choose the witness $(Y, v) := ([a]_{\mathcal{P}}, \varepsilon)$. (2) Otherwise, $\mathsf{Ind}(\mathbf{T}_X(ui)) = \emptyset$ and $(X, ui) \in \Delta^{\mathcal{I}}$. By induction, we have $C^{\mathcal{I}}(X, ui) = \mathbf{T}_X(ui)(C)$, and from the definition of $s^{\mathcal{I}}$ we obtain $s^{\mathcal{I}}((X, u), (X, ui)) = \mathbf{T}_X(ui)(\rho)$, which allows us to choose $(Y, v) := (X, ui)$. It remains to show that $s^{\mathcal{I}}((X, u), (Z, w)) \otimes C^{\mathcal{I}}(Z, w) \leq \ell$ holds for all other $(Z, w) \in \Delta^{\mathcal{I}}$, which implies that $(\exists s.C)^{\mathcal{I}}(X, u) = \ell$, as desired. In the case that $s^{\mathcal{I}}((X, u), (Z, w)) = \mathbf{0}$, the claim is trivial. Otherwise, one of the following two alternatives must hold:

- There is an index $i \in \varphi_s(\mathcal{O})$ with $s^{\mathcal{I}}((X, u), (Z, w)) = \mathbf{T}_X(ui)(\rho)$ and (i) $(Z, w) = (X, ui)$ or (ii) $w = \varepsilon$ and $\mathsf{Ind}(\mathbf{T}_X(ui)) \cap Z \neq \emptyset$. From the Hintikka condition we know that the value $\mathbf{T}_X(ui)(C)$ is defined and satisfies $\ell = \mathbf{T}_X(u)(\exists s.C) \geq \mathbf{T}_X(ui)(\rho) \otimes \mathbf{T}_X(ui)(C)$. It thus suffices to show that $C^{\mathcal{I}}(Z, w) = \mathbf{T}_X(ui)(C)$. In the first case, $\mathbf{T}_Z(w)(C) = \mathbf{T}_X(ui)(C)$ is defined, and thus induction yields that $C^{\mathcal{I}}(Z, w) = \mathbf{T}_Z(w)(C) = \mathbf{T}_X(ui)(C)$. In Case (ii), we know that $\mathbf{T}_Z(\varepsilon)(C) = H_Z(C) = \mathbf{T}_X(ui)(C)$ by the compatibility of $\mathbf{T}_X(ui)$ with the pre-completion. By induction we thus obtain $C^{\mathcal{I}}(Z, w) = C^{\mathcal{I}}(Z, \varepsilon) = \mathbf{T}_Z(\varepsilon)(C) = \mathbf{T}_X(ui)(C)$.

– There is an index $i \in \varphi_{\overline{s}}(\mathcal{O})$ with $s^{\mathcal{I}}((X,u),(Z,w)) = \mathbf{T}_Z(wi)(\rho)$ and
(i') $(X,u) = (Z,wi)$ or (ii') $u = \varepsilon$ and $\mathsf{Ind}(\mathbf{T}_Z(wi)) \cap X \neq \emptyset$. In Case (i'),
we immediately get $\ell = \mathbf{T}_X(u)(\exists s.C) = \mathbf{T}_Z(wi)(\exists s.C)$. In the latter case,
from the compatibility of $\mathbf{T}_Z(wi)$ with the pre-completion we also get that
$\ell = \mathbf{T}_X(u)(\exists s.C) = \mathbf{T}_X(\varepsilon)(\exists s.C) = H_X(\exists s.C) = \mathbf{T}_Z(wi)(\exists s.C)$. Thus, in
both cases the Hintikka condition yields that $\mathbf{T}_Z(w)(C)$ is defined and we
have $\ell = \mathbf{T}_Z(wi)(\exists s.C) \geq \mathbf{T}_Z(wi)(\rho) \otimes \mathbf{T}_Z(w)(C)$. By induction, we obtain
$\mathbf{T}_Z(w)(C) = C^{\mathcal{I}}(Z,w)$, which proves the claim.

Claim (1) can be shown for value restrictions using similar arguments.

We have thus defined an interpretation $\mathcal{I}$ that satisfies all concept definitions
in $\mathcal{T}$. In the case that $\mathcal{T}$ is a general TBox, consider any GCI $\langle C \sqsubseteq D \geq \ell \rangle \in \mathcal{T}$
and $(X,u) \in \Delta^{\mathcal{I}}$. By the compatibility of $\mathbf{T}_X(u)$ with $\mathcal{T}$, we know that $\mathbf{T}_X(u)(C)$
and $\mathbf{T}_X(u)(D)$ are defined and $\mathbf{T}_X(u)(C) \Rightarrow \mathbf{T}_X(u)(D) \geq \ell$. By Claim (1), we
thus have $C^{\mathcal{I}}(X,u) \Rightarrow D^{\mathcal{I}}(X,u) \geq \ell$, which shows that $\mathcal{I}$ satisfies the GCI.
Finally, consider an assertion $\langle a{:}C \bowtie \ell \rangle \in \mathcal{A}$. By the compatibility of $H_{[a]_\mathcal{P}}$ with
$\mathcal{A}_{[a]_\mathcal{P}}$ (see Definition 14), we know that $H_{[a]_\mathcal{P}}(C)$ is defined and $H_{[a]_\mathcal{P}}(C) \bowtie \ell$.
By Claim (1), we conclude $C^{\mathcal{I}}(a^{\mathcal{I}}) = C^{\mathcal{I}}([a]_\mathcal{P}, \varepsilon) = \mathbf{T}_{[a]_\mathcal{P}}(\varepsilon)(C) = H_{[a]_\mathcal{P}}(C) \bowtie \ell$;
that is, $\mathcal{I}$ satisfies the assertion.

Conversely, assume that there is a model $\mathcal{I}$ of $\mathcal{O}$. We define a pre-completion
$(\mathcal{P}, \mathcal{H}_\mathcal{P}, \mathcal{R}_\mathcal{P})$ for $\mathcal{O}$ based on the partition $\mathcal{P} := \{\{b \in \mathsf{Ind} \mid a^{\mathcal{I}} = b^{\mathcal{I}}\} \mid a \in \mathsf{Ind}\}$. For every $r \in \mathsf{Rol}(\mathcal{O})$ and $X, Y \in \mathcal{P}$, we set $r_\mathcal{P}(X,Y) := r^{\mathcal{I}}(a^{\mathcal{I}}, b^{\mathcal{I}})$,
where (a,b) is an arbitrary element of $X \times Y$. Similarly, we set $H_X(\rho) := \mathbf{0}$ and
$H_X(C) := C^{\mathcal{I}}(a^{\mathcal{I}})$ for every $C \in \mathsf{sub}(\mathcal{O})$ to define the family $\mathcal{H}_\mathcal{P} = (H_X)_{X \in \mathcal{P}}$.
Since $\mathcal{I}$ satisfies $\mathcal{T}$, this obviously defines Hintikka functions that are compatible
with $\mathcal{T}$, and we also have $\mathsf{Ind}(H_X) = X$ for every $X \in \mathcal{P}$. Furthermore, for every
$\langle a{:}C \bowtie \ell \rangle \in \mathcal{A}$, we have $C^{\mathcal{I}}(a^{\mathcal{I}}) \bowtie \ell$, and thus $H_{[a]_\mathcal{P}}(C) \bowtie \ell$, which shows that
what we have defined above is indeed a pre-completion for $\mathcal{O}$.

For a given $X \in \mathcal{P}$, we now define the Hintikka tree $\mathbf{T}_X$ starting with H_X by
inductively constructing a mapping $g_X \colon K^* \to \Delta^{\mathcal{I}}$ that specifies which elements
of $\Delta^{\mathcal{I}}$ represent the nodes of $\mathbf{T}_X$ and satisfies the following property:

$$\text{For all } u \in K^*, C \in \mathsf{sub}(\mathcal{O}), s \in \mathsf{Rol}^-(\mathcal{O}), \text{ and } i \in \varphi_s(\mathcal{O}), \text{ we have} \\ \mathbf{T}_X(u)(C) = C^{\mathcal{I}}(g_X(u)) \text{ and } \mathbf{T}_X(ui)(\rho) = s^{\mathcal{I}}(g_X(u), g_X(ui)). \tag{2}$$

This in particular ensures that all constructed Hintikka functions are compatible
with $\mathcal{T}$ and with the pre-completion.

We start the construction by setting $\mathbf{T}_X(\varepsilon) := H_X$ and $g_X(\varepsilon) := a^{\mathcal{I}}$, where
a is an arbitrary element of X. Thus, $\mathbf{T}_X$ starts with H_X and Claim (2) is satisfied at ε. Let now $u \in K^*$ be any node for which $\mathbf{T}_X$ and g_X have already
been defined while satisfying Claim (2), and consider any existential restric-
tion $\exists s.C \in \mathsf{sub}(\mathcal{O})$ and $i := \varphi(\exists s.C)$. Since $\mathcal{I}$ is witnessed, there must be
a $y \in \Delta^{\mathcal{I}}$ such that $(\exists s.C)^{\mathcal{I}}(g_X(u)) = s^{\mathcal{I}}(g_X(u), y) \otimes C^{\mathcal{I}}(y)$. We now set
$g_X(ui) := y$, $\mathbf{T}_X(ui)(\rho) := s^{\mathcal{I}}(g_X(u), y)$, and $\mathbf{T}_X(ui)(C) := C^{\mathcal{I}}(y)$ for all
$C \in \mathsf{sub}(\mathcal{O})$ to satisfy Claim (2) at ui. Likewise, for any $\forall s.C \in \mathsf{sub}(\mathcal{O})$ there
must be a $y \in \Delta^{\mathcal{I}}$ with $(\forall s.C)^{\mathcal{I}}(g_X(u)) = s^{\mathcal{I}}(g_X(u), y) \Rightarrow C^{\mathcal{I}}(y)$, and we proceed
as above to define $\mathbf{T}_X$ and g_X at ui for $i := \varphi(\forall s.C)$.

We now show that every tuple $(\mathbf{T}_X(u), \mathbf{T}_X(u1), \ldots, \mathbf{T}_X(uk))$ with $u \in K^*$ satisfies the Hintikka condition. The first point of Condition (i) from Definition 15 is obviously satisfied by the above construction. Consider now any $\exists s.C \in \mathsf{sub}(\mathcal{O})$ and $i \in \varphi_s(\mathcal{O})$. By Claim (2) and the semantics of existential restrictions, we obtain

$$\mathbf{T}_X(u)(\exists s.C) = (\exists s.C)^{\mathcal{I}}(g_X(u))$$
$$\geq s^{\mathcal{I}}(g_X(u), g_X(ui)) \otimes C^{\mathcal{I}}(g_X(ui))$$
$$= \mathbf{T}_X(ui)(\rho) \otimes \mathbf{T}_X(ui)(C).$$

Similarly, for all $i \in \varphi_{\bar{s}}(\mathcal{O})$, we have

$$\mathbf{T}_X(ui)(\exists s.C) = (\exists s.C)^{\mathcal{I}}(g_X(ui))$$
$$\geq s^{\mathcal{I}}(g_X(ui), g_X(u)) \otimes C^{\mathcal{I}}(g_X(u))$$
$$= \mathbf{T}_X(ui)(\rho) \otimes \mathbf{T}_X(u)(C).$$

Condition (ii) can be shown by analogous arguments. For Condition (iii), let $u \in K^*$, $s \in \mathsf{Rol}^-(\mathcal{O})$, $i, j \in \varphi_s(\mathcal{O})$, $a \in \mathsf{Ind}(\mathbf{T}_X(ui))$, and $b \in \mathsf{Ind}(\mathbf{T}_X(uj))$ with $[a]_{\mathcal{P}} = [b]_{\mathcal{P}}$. Then Claim (2) yields $g_X(ui) = a^{\mathcal{I}} = b^{\mathcal{I}} = g_X(uj)$, and thus $\mathbf{T}_X(ui)(\rho) = s^{\mathcal{I}}(g_X(u), a^{\mathcal{I}}) = \mathbf{T}_X(uj)(\rho)$.

For Condition (iv) consider $u \in K^*$, $a \in \mathsf{Ind}(\mathbf{T}_X(u))$, $s \in \mathsf{Rol}^-(\mathcal{O})$, $i \in \varphi_s(\mathcal{O})$, and $b \in \mathsf{Ind}(\mathbf{T}_X(ui))$. By Claim (2), we get $g_X(u) = a^{\mathcal{I}}$, $g_X(ui) = b^{\mathcal{I}}$, and $\mathbf{T}_X(ui)(\rho) = s^{\mathcal{I}}(g_X(u), g_X(ui)) = s^{\mathcal{I}}(a^{\mathcal{I}}, b^{\mathcal{I}}) = s_{\mathcal{P}}([a]_{\mathcal{P}}, [b]_{\mathcal{P}})$.   □

From this lemma it follows that consistency in $L\text{-}\mathcal{ALCOI}$ can be reduced to deciding the existence of suitable family of Hintikka trees. By building looping automata whose runs correspond exactly to those Hintikka trees, we further reduce it to the emptiness problem for this class of automata.

Definition 18 (Hintikka automaton). *Let $(\mathcal{P}, \mathcal{H}_{\mathcal{P}}, \mathcal{R}_{\mathcal{P}})$ be a pre-completion for $\mathcal{O}$ and $X \in \mathcal{P}$. The* Hintikka automaton *for $\mathcal{O}$ and H_X is the looping automaton $\mathsf{A}_{\mathcal{O}, H_X} = (Q_{\mathcal{O}}, I_{\mathcal{O}, H_X}, \Delta_{\mathcal{O}})$, where*

- *$Q_{\mathcal{O}}$ is the set of all Hintikka functions for $\mathcal{O}$ that are compatible with $\mathcal{T}$ and the pre-completion,*
- *$I_{\mathcal{O}, H_X} := \{H_X\}$, and*
- *$\Delta_{\mathcal{O}}$ is the set of all elements of $Q_{\mathcal{O}}^{k+1}$ that satisfy the Hintikka condition.*

The runs of $\mathsf{A}_{\mathcal{O}, H_X}$ are exactly the Hintikka trees for $\mathcal{O}$ starting with H_X. Thus, $\mathcal{O}$ is consistent iff there is a pre-completion $(\mathcal{P}, \mathcal{H}_{\mathcal{P}}, \mathcal{R}_{\mathcal{P}})$ for $\mathcal{O}$ such that $\mathsf{A}_{\mathcal{O}, H_X}$ is not empty for all $X \in \mathcal{P}$.

Note that the number of partitions of $\mathsf{Ind}(\mathcal{O})$ is bounded by $2^{|\mathsf{Ind}(\mathcal{O})|^2}$, the number of Hintikka functions for $\mathcal{O}$ is bounded by $(|L|+1)^{|\mathsf{sub}(\mathcal{O})|+1}$, and the number of fuzzy binary relations on a partition of $\mathsf{Ind}(\mathcal{O})$ is bounded by $|L|^{|\mathsf{Ind}(\mathcal{O})|^2}$. Thus, the number of pre-completions for $\mathcal{O}$ is bounded exponentially in the size of $\mathcal{O}$ and polynomially in the size of L. However, each pre-completion is only of size polynomial in the size of the input. We can thus enumerate all pre-completions in exponential time and for each of them check emptiness of polynomially many looping automata. Since the size of these automata is exponential in the size of $\mathcal{O}$ and emptiness of looping automata is decidable in polynomial time in the size of the automaton [37], the overall runtime of this algorithm is bounded exponentially in the size of $\mathcal{O}$.

This gives a tight upper bound for the complexity of consistency in $L\text{-}\mathcal{ALCOI}$ since this problem is already ExpTime-hard for classical $\mathcal{ALCOI}$, even for empty TBoxes [30,35].

Theorem 19. *In all fuzzy DLs between $L\text{-}\mathcal{ALC}$ and $L\text{-}\mathcal{ALCOI}$, deciding consistency w.r.t. general TBoxes is ExpTime-complete.*

When restricting to acyclic TBoxes, reasoning in classical $\mathcal{ALCO}$ is PSpace-complete [3,31]. We show in the following section that this remains true under finite lattice semantics. A similar approach was used in [17] to prove the same for $L\text{-}\mathcal{ALCI}$, even in the presence of a role hierarchy or (crisp) transitive roles.

5   Consistency w.r.t. Acyclic TBoxes

Consider now an ontology $\mathcal{O} = (\mathcal{A}, \mathcal{T})$ from $L\text{-}\mathcal{ALCO}$, i.e. that does not use inverse roles, where $\mathcal{T}$ is an acyclic TBox. Notice that we can guess a partition $\mathcal{P}$ of $\mathsf{Ind}(\mathcal{O})$ and families $(H_X)_{X \in \mathcal{P}}$ and $(r_\mathcal{P})_{r \in \mathsf{Rol}(\mathcal{O})}$ and verify the conditions of Definition 14 in (non-deterministic) polynomial space. Thus, if we can show that emptiness of the polynomially many Hintikka automata $\mathsf{A}_{\mathcal{O},H_X}$ can be decided in polynomial space, then we would obtain a PSpace upper bound for deciding consistency in this logic. The idea is to modify the construction of these automata into a PSpace on-the-fly construction. It is easy to see that the automata $\mathsf{A}_{\mathcal{O},H_X}$ already satisfy all but one of the conditions from Definition 5:

(i) the arity k of the automata is given by the number of existential and value restrictions in $\mathsf{sub}(\mathcal{O})$;
(ii) every Hintikka function (i.e. every state of the automaton) has size bounded by $|L|(|\mathsf{sub}(\mathcal{O})| + 1)$ since it consists of $|\mathsf{sub}(\mathcal{O})| + 1$ lattice values;
(iii) building a state or a transition of the automaton requires only to guess $|\mathsf{sub}(\mathcal{O})|+1$ or $k(|\mathsf{sub}(\mathcal{O})|+1)$ lattice values, respectively, and then verifying that this is indeed a valid state or transition of the automaton, which can be done in time polynomial in the size of L and $\mathcal{O}$.

However, it is possible to build runs of $\mathsf{A}_{\mathcal{O},H_X}$ where blocking occurs only after exponentially many transitions, violating the first condition of PSpace on-the-fly constructions. We will use a faithful family of functions to obtain reduced automata that guarantee blocking after at most polynomially many transitions, thus obtaining the PSpace upper bound.

For our construction to work, we need to make a small change to the definition of compatibility of a Hintikka function H with the pre-completion $(\mathcal{P}, \mathcal{H}_\mathcal{P}, \mathcal{R}_\mathcal{P})$

that we have guessed: for every $a \in \mathsf{Ind}(H)$, we only require the following weaker condition: For all $C \in \mathsf{sub}(\mathcal{O})$ for which $H(C)$ is defined, $H_{[a]_\mathcal{P}}(C)$ is also defined and we have $H(C) = H_{[a]_\mathcal{P}}(C)$.

It can be seen that this condition would yield an incorrect construction in the presence of inverse roles. However, if inverse roles are disallowed, then all results obtained so far remain true under this modification. More precisely, the only changes in the proof of Lemma 17 are in two places that refer to the compatibility with the pre-completion (without using inverse roles). These are part of the proof of Claim (1) for existential and value restrictions. In both places, it is enough to be able to infer from the fact that a Hintikka function H is compatible with the pre-completion, $a \in \mathsf{Ind}(H)$, and $C \in \mathsf{supp}(H)$ that we have $C \in \mathsf{supp}(H_{[a]_\mathcal{P}})$ and $H_{[a]_\mathcal{P}}(C) = H(C)$. This is precisely the new condition that we are now considering.

We now describe a faithful family of functions for $\mathsf{A}_{\mathcal{O},H_X}$ that allows us to obtain a PSPACE-on-the-fly construction. The idea is that it suffices to consider transitions that reduce the maximal role depth (w.r.t. $\mathcal{T}$) in the support of the states. The *role depth w.r.t.* $\mathcal{T}$ ($\mathsf{rd}_\mathcal{T}$) of concepts is recursively defined as follows:

- $\mathsf{rd}_\mathcal{T}(A) := \mathsf{rd}_\mathcal{T}(\top) := \mathsf{rd}_\mathcal{T}(\{a\}) := 0$ for every primitive concept name A and every $a \in \mathsf{N_I}$;
- $\mathsf{rd}_\mathcal{T}(A) := \mathsf{rd}_\mathcal{T}(C)$ for every definition $\langle A \doteq C \geq \ell \rangle \in \mathcal{T}$;
- $\mathsf{rd}_\mathcal{T}(\neg C) := \mathsf{rd}_\mathcal{T}(C)$;
- $\mathsf{rd}_\mathcal{T}(C \sqcap D) := \mathsf{rd}_\mathcal{T}(C \to D) := \max\{\mathsf{rd}_\mathcal{T}(C), \mathsf{rd}_\mathcal{T}(D)\}$; and
- $\mathsf{rd}_\mathcal{T}(\exists r.C) := \mathsf{rd}_\mathcal{T}(\forall r.C) := \mathsf{rd}_\mathcal{T}(C) + 1$.

This is well-defined since $\mathcal{T}$ is an acyclic TBox.

Given a Hintikka function H for $\mathcal{O}$, we define $\mathsf{rd}_\mathcal{T}(H)$ as the maximal role depth $\mathsf{rd}_\mathcal{T}(C)$ of a concept $C \in \mathsf{supp}(H)$. For $n \geq 0$, we further denote by $\mathsf{sub}^{\leq n}(\mathcal{O})$ the set of all concepts $C \in \mathsf{sub}(\mathcal{O})$ with $\mathsf{rd}_\mathcal{T}(C) \leq n$.

Definition 20 (functions f_H). *Let H and H' be two states of $\mathsf{A}_{\mathcal{O},H_X}$ and $n := \mathsf{rd}_\mathcal{T}(H)$. We define the function $f_H(H')$ as follows for every $C \in \mathsf{sub}(\mathcal{O})$:*

$$f_H(H')(C) := \begin{cases} H'(C) & \text{if } C \in \mathsf{sub}^{\leq n-1}(\mathcal{O}), \\ \text{undefined} & \text{otherwise,} \end{cases}$$

$$f_H(H')(\rho) := \begin{cases} 0 & \text{if } \mathsf{supp}(H) = \emptyset, \\ H'(\rho) & \text{otherwise.} \end{cases}$$

Since $\mathcal{T}$ is acyclic, the function $f_H(H')$ defined above is still a Hintikka function for $\mathcal{O}$ compatible with all axioms of $\mathcal{T}$. It also remains compatible with the pre-completion (according to the modified definition of this notion) since it only discards some values from H'. Thus, it is again an element of $Q_\mathcal{O}$.

Lemma 21. *In L-$\mathcal{ALCO}$, the family of functions f_H is faithful w.r.t. $\mathsf{A}_{\mathcal{O},H_X}$.*

Proof. Consider $H, H_0, H_1, \ldots, H_k \in Q_\mathcal{O}$ and let $n := \mathsf{rd}_\mathcal{T}(H)$ and $H_i' := f_H(H_i)$ for all i, $0 \leq i \leq k$. We first verify that if $(H, H_1, \ldots, H_k)$ satisfies the Hintikka condition, then $(H, H_1', \ldots, H_k')$ also satisfies it.

If $\exists s.C \in \mathsf{supp}(H)$, then $C \in \mathsf{supp}(H_i)$ and $H(\exists s.C) = H_i(\rho) \otimes H_i(C)$, where $i := \varphi(\exists s.C)$. Since we have $\mathsf{rd}_{\mathcal{T}}(C) < \mathsf{rd}_{\mathcal{T}}(\exists s.G) \leq \mathsf{rd}_{\mathcal{T}}(H) = n$, this implies that $C \in \mathsf{supp}(H_i')$ and $H_i'(C) = H_i(C)$. Moreover, we know that $\mathsf{supp}(H) \neq \emptyset$, and thus $H_i'(\rho) = H_i(\rho)$. This shows that the required equality $H(\exists s.C) = H_i(\rho) \otimes H_i(C) = H_i'(\rho) \otimes H_i'(C)$ remains satisfied.

Let now $\exists s.C \in \mathsf{supp}(H)$ and $i \in \varphi_s(\mathcal{O})$. By the same arguments as above, $H_i'(C)$ and $H_i'(\rho)$ are defined and equal to $H_i(C)$ and $H_i(\rho)$, respectively. Thus, the required inequality is still satisfied after applying f_H. Since in $L\text{-}\mathcal{ALCO}$ there are no inverse roles, the rest of Condition (i) of Definition 15 is trivially satisfied. Condition (ii) follows by similar arguments.

For Condition (iii), consider any role name $r \in \mathsf{Rol}(\mathcal{O})$ and $i, j \in \varphi_s(\mathcal{O})$. If there are $a \in \mathsf{Ind}(H_i')$ and $b \in \mathsf{Ind}(H_j')$ with $[a]_{\mathcal{P}} = [b]_{\mathcal{P}}$, then this must already have been the case for H_i and H_j. Since then $\mathsf{supp}(H)$ cannot be empty, we still have $H_i'(\rho) = H_i(\rho) = H_j(\rho) = H_j'(\rho)$. A similar argument shows that Condition (iv) remains satisfied.

For the second condition of Definition 3, we assume that $(H_0, H_1, \ldots, H_k)$ satisfies the Hintikka condition and verify it for $(H_0', H_1', \ldots, H_k')$.

Consider any $\exists s.C \in \mathsf{supp}(H_0')$ and $i := \varphi(\exists s.C)$. By the definition of f_H, we get $H_0(\exists s.C) = H_0'(\exists s.C)$ and $\mathsf{rd}_{\mathcal{T}}(C) < \mathsf{rd}_{\mathcal{T}}(\exists s.C) < \mathsf{rd}_{\mathcal{T}}(H)$. Thus, $H_i(C)$ is defined and equal to $H_i'(C)$. Moreover, $\mathsf{supp}(H) \neq \emptyset$, which implies that $H_i'(\rho) = H_i(\rho)$. Thus, $H_0'(\exists s.C) = H_0(\exists s.C) = H_i(\rho) \otimes H_i(C) = H_i'(\rho) \otimes H_i'(C)$.

Again, the remaining part of Condition (i) can be shown by similar arguments, replacing $\varphi(\exists s.C)$ by an element of $\varphi_s(\mathcal{O})$ and the equality condition by an inequality. The remaining conditions are similarly easy to verify. $\qquad\square$

By Lemma 4, $\mathsf{A}_{\mathcal{O},H_X}$ is empty iff the induced subautomaton $\mathsf{A}_{\mathcal{O},H_X}^S$ is empty. It remains to show that the latter problem can be decided in PSPACE.

Theorem 22. *The construction of $\mathsf{A}_{\mathcal{O},H_X}^S$ from L, $\mathcal{O}$, and H_X is a PSPACE on-the-fly construction.*

Proof. We show that the automata $\mathsf{A}_{\mathcal{O},H_X}^S$ are m-blocking for

$$m := \max\{\mathsf{rd}_{\mathcal{T}}(C) \mid C \in \mathsf{sub}(\mathcal{O})\} + 3.$$

The other conditions of Definition 5 have already been shown above.

By the definition of $\mathsf{A}_{\mathcal{O},H_X}^S$, every transition decreases the maximal role depth of the support of the state. Hence, after at most $\max\{\mathsf{rd}_{\mathcal{T}}(C) \mid C \in \mathsf{sub}(\mathcal{O})\} + 1$ transitions, we must reach a state H that is undefined for all $C \in \mathsf{sub}(\mathcal{O})$, and hence $\mathsf{supp}(H) = \emptyset$. From the next transition on, all states additionally assign $\mathbf{0}$ to ρ. Hence, after at most m transitions, we find two states that are equal. Since m is bounded by a polynomial in the size of $\mathcal{O}$, the automata $\mathsf{A}_{\mathcal{O},H_X}^S$ satisfy Definition 5. $\qquad\square$

Theorem 6 yields the desired PSPACE upper bound for consistency in $L\text{-}\mathcal{ALCO}$. PSPACE-hardness follows from PSPACE-hardness of consistency w.r.t. the empty TBox in classical $\mathcal{ALC}$ [31].

Theorem 23. *In L-$\mathcal{ALCO}$, the problem of deciding consistency w.r.t. acyclic TBoxes is* PSPACE-*complete.*

Using a different faithful family of functions, it was shown in [11] that consistency of L-$\mathcal{ALCI}$ ontologies with acyclic TBoxes is also PSPACE-complete. As for $\mathcal{ALCO}$, this matches the complexity of reasoning in classical DLs.

Notice that the notions of Hintikka functions and Hintikka trees are independent of the operators used. One could use the residual negation $\ominus\ell := \ell \Rightarrow \mathbf{0}$ to interpret the constructor $\neg$, or the Kleene-Dienes implication $\ell_1 \Rightarrow \ell_2 := {\sim}\ell_1 \vee \ell_2$ instead of the residuum. The only restrictions are that the semantics must be truth functional, i.e. the value of a formula must depend only on the values of its direct subformulae, and the underlying operators must be computable in polynomial time from the lattice values. We could also use a slightly different semantics for concept definitions in which $\otimes$ is replaced by the simple meet t-norm $\wedge$.

The algorithm can be modified for reasoning w.r.t. n-witnessed models for $n > 1$. One needs only extend the arity of the Hintikka trees to account for n witnesses for each quantified concept in $\mathsf{sub}(\mathcal{O})$; the arity of $\mathsf{A}_{\mathcal{O},H_X}$ then grows polynomially in n. This does not affect the obtained complexity upper bounds, and hence Theorems 19 and 23 still hold.

6   Conclusions

We have shown that reasoning in L-$\mathcal{ALCOI}$ is not harder than in the underlying crisp DL $\mathcal{ALCOI}$, if L is a finite De Morgan lattice. More precisely, all the standard reasoning problems in this logic are EXPTIME-complete, even if the TBox is assumed to be empty. If we disallow either nominals or inverse roles, obtaining the logics L-$\mathcal{ALCI}$ and L-$\mathcal{ALCO}$, respectively, then reasoning w.r.t. acyclic TBoxes is PSPACE-complete.

These complexity bounds extend previously known results for lattice-based fuzzy DLs [11,13,14,18] and complements the work in [17], in which L-$\mathcal{ALCI}$ is extended to allow transitivity and role inclusion axioms. Thus, tight complexity bounds are also obtained for logics up to L-$\mathcal{SHI}$, under some restrictions in the interpretation of roles. Our methods demonstrate, once again, that automata can show PSPACE results for (fuzzy) DLs [2].

It is reasonable to expect that the construction from [17] for L-$\mathcal{SHI}$ can be combined with the ideas from this paper for handling nominals, to obtain an automata-based algorithm for reasoning in L-$\mathcal{SHOI}$. A missing step is to further generalize these methods, or develop new ones, to prove tight complexity bounds for fuzzy variants of the current standard ontology languages, like $\mathcal{SROIQ}(\mathcal{D})$. We also need to understand the effect of removing the restrictions on roles from [17] to the complexity of reasoning.

Although their run-time behavior is optimal w.r.t. the complexity of the problem, automata-based methods are typically not used in practice since their *best-case* behavior is as bad as in the worst-case. It would thus be desirable to produce reasoning algorithms that preserve the properties of the algorithms used by current classical reasoners [6,28,32]. First steps in this direction have been made in [12,15], where tableau-based algorithms with better run-time behavior are proposed. Those algorithms, however, require a high-level of non-determinism and are thus inappropriate for efficient implementation. Ideas for improvements will be studied.

References

1. Baader, F., Borgwardt, S., Peñaloza, R.: On the decidability status of fuzzy $\mathcal{ALC}$ with general concept inclusions. J. Philos. Logic (2014). doi:10.1007/s10992-014-9329-3
2. Baader, F., Hladik, J., Peñaloza, R.: Automata can show PSPACE results for description logics. Inf. Comput. **206**(9–10), 1045–1056 (2008)
3. Baader, F., Lutz, C., Miličić, M., Sattler, U., Wolter, F.: Integrating description logics and action formalisms for reasoning about Web services. LTCS-Report 05–02, Chair for Automata Theory, TU Dresden, Germany (2005)
4. Baader, F., Peñaloza, R.: Are fuzzy description logics with general concept inclusion axioms decidable? In: Proceedings of the 2011 IEEE International Conference on Fuzzy Systems (FUZZ-IEEE'11), pp. 1735–1742. IEEE Computer Society Press (2011)
5. Baader, F., Peñaloza, R.: On the undecidability of fuzzy description logics with GCIs and product T-norm. In: Tinelli, C., Sofronie-Stokkermans, V. (eds.) FroCoS 2011. LNCS, vol. 6989, pp. 55–70. Springer, Heidelberg (2011)
6. Baader, F., Sattler, U.: An overview of tableau algorithms for description logics. Stud. Logica **69**(1), 5–40 (2001)
7. Bobillo, F., Bou, F., Straccia, U.: On the failure of the finite model property in some fuzzy description logics. Fuzzy Sets Syst. **172**(1), 1–12 (2011)
8. Bobillo, F., Straccia, U.: A fuzzy description logic with product t-norm. In: Proceedings of the 2007 IEEE International Conference on Fuzzy Systems (FUZZ-IEEE'07), pp. 1–6. IEEE Computer Society Press (2007)
9. Bobillo, F., Straccia, U.: Fuzzy description logics with general t-norms and datatypes. Fuzzy Sets Syst. **160**(23), 3382–3402 (2009)
10. Bobillo, F., Straccia, U.: Reasoning with the finitely many-valued Łukasiewicz fuzzy description logic $\mathcal{SROIQ}$. Inf. Sci. **181**(4), 758–778 (2011)
11. Borgwardt, S., Peñaloza, R.: Finite lattices do not make reasoning in $\mathcal{ALCI}$ harder. In: Bobillo, F., et al. (eds.) Proceedings of the 7th International Workshop on Uncertainty Reasoning for the Semantic Web (URSW'11). CEUR Workshop Proceedings, vol. 778, pp. 51–62 (2011)
12. Borgwardt, S., Peñaloza, R.: Consistency reasoning in lattice-based fuzzy description logics. Int. J. Approximate Reasoning **55**(9), 1917–1938 (2014)
13. Borgwardt, S., Peñaloza, R.: Description logics over lattices with multi-valued ontologies. In: Walsh, T. (ed.) Proceedings of the 22nd International Joint Conference on Artificial Intelligence (IJCAI'11), pp. 768–773. AAAI Press (2011)
14. Borgwardt, S., Peñaloza, R.: Fuzzy ontologies over lattices with t-norms. In: Rosati, R., Rudolph, S., Zakharyaschev, M. (eds.) Proceedings of the 2011 International Workshop on Description Logics (DL'11), pp. 70–80. CEUR Workshop Proceedings (2011)
15. Borgwardt, S., Peñaloza, R.: A tableau algorithm for fuzzy description logics over residuated De Morgan lattices. In: Krötzsch, M., Straccia, U. (eds.) RR 2012. LNCS, vol. 7497, pp. 9–24. Springer, Heidelberg (2012)

16. Borgwardt, S., Peñaloza, R.: Undecidability of fuzzy description logics. In: Brewka, G., Eiter, T., McIlraith, S.A. (eds.) Proceedings of the 13th International Conference on Principles of Knowledge Representation and Reasoning (KR'12), pp. 232–242. AAAI Press (2012)
17. Borgwardt, S., Peñaloza, R.: The complexity of lattice-based fuzzy description logics. J. Data Seman. **2**(1), 1–19 (2013)
18. Bou, F., Cerami, M., Esteva, F.: Finite-valued Łukasiewicz modal logic is PSPACE-complete. In: Walsh, T. (ed.) Proceedings of the 22nd International Joint Conference on Artificial Intelligence (IJCAI'11), pp. 774–779. AAAI Press (2011)
19. Cerami, M., García-Cerdaña, À., Esteva, F.: On finitely-valued fuzzy description logics. Int. J. Approximate Reasoning **55**(9), 1890–1916 (2014)
20. Cerami, M., Straccia, U.: On the (un)decidability of fuzzy description logics under Łukasiewicz t-norm. Inf. Sci. **227**, 1–21 (2013)
21. De Cooman, G., Kerre, E.E.: Order norms on bounded partially ordered sets. J. Fuzzy Math. **2**, 281–310 (1993)
22. Galatos, N., Jipsen, P., Kowalski, T., Ono, H.: Residuated Lattices: An Algebraic Glimpse at Substructural Logics. Studies in Logic and the Foundations of Mathematics, vol. 151. Elsevier, Amsterdam (2007)
23. Hájek, P.: Metamathematics of Fuzzy Logic (Trends in Logic). Springer, London (2001)
24. Hájek, P.: Making fuzzy description logic more general. Fuzzy Sets Syst. **154**(1), 1–15 (2005)
25. Horrocks, I., Sattler, U.: A description logic with transitive and inverse roles and role hierarchies. J. Logic Comput. **9**(3), 385–410 (1999)
26. Klement, E.P., Mesiar, R., Pap, E.: Triangular Norms. Trends in Logic, Studia Logica Library. Springer, New York (2000)
27. Lukasiewicz, T., Straccia, U.: Managing uncertainty and vagueness in description logics for the semantic web. J. Web Seman. **6**(4), 291–308 (2008)
28. Motik, B., Shearer, R., Horrocks, I.: Hypertableau reasoning for description logics. J. Artif. Intell. Res. **36**, 165–228 (2009)
29. Schaerf, A.: Reasoning with individuals in concept languages. Data Knowl. Eng. **13**(2), 141–176 (1994)
30. Schild, K.: A correspondence theory for terminological logics: preliminary report. In: Mylopoulos, J., Reiter, R. (eds.) Proceedings of the 12th International Joint Conference on Artificial Intelligence (IJCAI'91), pp. 466–471. Morgan Kaufmann (1991)
31. Schmidt-Schauß, M., Smolka, G.: Attributive concept descriptions with complements. Artif. Intell. **48**(1), 1–26 (1991)
32. Steigmiller, A., Liebig, T., Glimm, B.: Extended caching, backjumping and merging for expressive description logics. In: Gramlich, B., Miller, D., Sattler, U. (eds.) IJCAR 2012. LNCS, vol. 7364, pp. 514–529. Springer, Heidelberg (2012)
33. Straccia, U.: Description logics over lattices. Int. J. Uncertainty Fuzziness Knowl.-Based Syst. **14**(1), 1–16 (2006)
34. Straccia, U.: Foundations of Fuzzy Logic and Semantic Web Languages. Studies in Informatics. CRC Press, Hoboken (2013)
35. Tobies, S.: The complexity of reasoning with cardinality restrictions and nominals in expressive description logics. J. Artif. Intell. Res. **12**, 199–217 (2000)
36. Tobies, S.: Complexity results and practical algorithms for logics in knowledge representation. Ph.D. thesis, RWTH Aachen, Germany (2001)
37. Vardi, M.Y., Wolper, P.: Automata-theoretic techniques for modal logics of programs. J. Comput. Syst. Sci. **32**(2), 183–221 (1986)

Fuzzy and Cross-Lingual Ontology Matching Mediated by Background Knowledge

Konstantin Todorov[1]([⊠]), Celiné Hudelot[2], and Peter Geibel[3]

[1] LIRMM/Université de Montpellier 2, Montpellier, France
konstantin.get@gmail.com
[2] Ecole Centrale Paris, Châtenay-Malabry, France
[3] TU Berlin, Sekr. MAR 4-2, Berlin, Germany

Abstract. The paper proposes an ontology alignment framework with two core features: the use of background knowledge and the ability to handle imprecision in the matching process and the resulting concept alignments. The procedure is based on the use of a generic reference vocabulary, which is used to define an explicit semantic space for the ontologies to be matched. General-purpose background knowledge sources based on Wikipedia, such as Yago, appear to be appropriate choices of reference vocabularies. The outcome of the procedure is a combined fuzzy knowledge body which captures what is common in two source ontologies. The proposed approach allows to discover cross-concept relations of the kind many-to-many. An important application of the method is found in the field of cross-lingual ontology matching.

1 Introduction

Ontologies are used for semantic annotation of resources on the World Wide Web, allowing for integration and interoperability of information systems and are thus integral part of the semantic web and the web of data. However, the creation and choice of different ontologies in an independent manner for describing similar or identical resources has lead to the problem of ontology heterogeneity. This is understood as any difference in syntax, spelling, meaning, intention or extension in the definitions of two cross-ontology entities which refer to the same or highly similar real world objects.

The field of ontology matching (OM) has taken the challenge of proposing solutions to the heterogeneity problem. As a result of more than 15 years of research and practice, the field has reached a significant level of maturity building on grounds form diverse areas, such as computational linguistics, graph theory, machine learning, information theory, relational algebra and other. Many systems have been proposed capable of aligning ontologies with a very high degree of heterogeneity in an efficient manner. The Ontology Alignment Evaluation Initiative[1] provides an evaluation platform for these systems in annual campaigns.

In spite of the considerable advance, many questions remain open, such as large-scale evaluation, matcher selection, tuning and combination, or user

---

[1] http://oaei.ontologymatching.org

© Springer International Publishing Switzerland 2014
F. Bobillo et al. (Eds.): URSW 2011-2013, LNAI 8816, pp. 142–162, 2014.
DOI: 10.1007/978-3-319-13413-0_8

involvement [31]. This works contributes particularly to the challenges of using background knowledge (BK) and taking into account imprecision in the matching process. An additional feature of the proposed approach is its capability to align cross-lingual ontologies, which is also a current problem in the field.

Using reference background knowledge is helpful in order to recreate the missing semantic context in the matching process. The proposed approach relies on Wikipedia or, alternatively, Yago as BK sources. The advantage of using Wikipedia is that it is available, general-purpose, large, and multilingual. We consider two ontologies as an input, referred to as source ontologies. In step one of our approach, every concept of each source ontology will be represented as a fuzzy set of the concepts of the background ontology - the process of *concept fuzzification*. We arrive at this fuzzy set representation by measuring the similarity of every concept of the source ontologies to every concept of the background ontology by using a concept similarity measure. In step two, with the fuzzy set representations at hand, we proceed to interlink the concepts of the input ontologies. We take their union and construct a novel ontology, which contains all fuzzified concepts of the input ontologies. The inherent semantic relations are translated into the novel ontology, as well.

Using a fuzzy set theoretical framework enables us to capture the vagueness in the definitions of concepts, relations and instances. It allows to take into account different aspects of the similarity of concepts by using their fuzzy set definitions. And eventually, this enables the computation of fuzzy relations (such as subsumption and equivalence) between cross-ontology concepts together with the degrees to which they hold. Within this framework, we are able to handle one-to-many or many-to-many mappings, where crisp methods usually fail. This is especially useful on a large scale where the likelihood that a 1:1 mapping is "the best possible" is very low, notably in the case of matching a relatively small-size ontology to a large ontology. Finally, we show an application of the approach to the task of matching cross-lingual ontologies. Indeed, the problem of using different natural languages in the concepts and relations labeling process is yet another source of ontology heterogeneity. Our framework relies on the possibility of using multilingual background knowledge in order to efficiently align cross-lingual ontologies without the use of machine translation tools.

The rest of the paper is organized as follows. Related work on ontology matching with accents on the use of BK, fuzziness, and multilingualism is presented in Sect. 2. Section 3 introduces definitions and notations from fuzzy set theory and logics. We describe our concept fuzzification algorithm in Sect. 4. A procedure for construction of a combined knowledge body is presented in Sect. 5. Several applications of this procedure are discussed. Experimental results and conclusions are presented in Sects. 6 and 7, respectively.

2   Related Work

Ontology matching is understood as the process of establishing relations between the elements of two or more ontologies [16]. This section discusses related work from three sub-fields of OM strongly related to the contribution of this paper.

2.1 Ontology Matching with Background Knowledge from the Web

Certain approaches include the use of background knowledge in the matching pipeline. Shvaiko et al. [31] outline several groups of relevant techniques.

Sabou et al. [29] provide a good motivation for the use of BK: the fact that two ontologies are always inherently different in terms of intention. Background knowledge comes to bridge the inherent semantic gap between them. The originality of the proposal is the use of the web in order to discover (by crawling) automatically appropriate BK sources (instead of using a single fixed source). Thus, the question of the availability and the coverage of the BK is addressed.

Jain et al. [19] propose a framework designed particularly with the task of mapping concepts that describe linked open data on the LOD cloud. The approach is based on the use of Wikipedia as a background mediator. For a given concept, a graph is constructed by using the connectivity properties of all Wikipedia pages that contain the words in the concept name. The mapping of two concepts from two different datasources is given as an equivalence or subsumption relation computed on the basis of the structural similarity of their respective Wikipedia graphs. An enhanced version of this method is provided in [20]. The main improvements include the use of a more elaborate measure of concept relatedness and the use of contextual information.

Zhang et al. [40] and Aleksovski et al. [2] propose straightforward anchoring procedures in which two input domain ontologies are aligned directly to a background ontology. On the basis of the produced direct alignments, an indirect alignment is inferred between the input ontologies. Aleksovski et al. situate their study in the medical field by using FMA as reference ontology.

Discussion. Similarly to [20], we point at the use of Wikipedia (or a knowledge base built upon it) as a motivating feature of our approach. In contrast, our approach does not require structured BK and can potentially work with just a flat vocabulary. Note that in [29], the authors do not consider nor do they compare their approach to the use of already available encyclopedic knowledge sources such as Wikipedia. One is lead to think that this missing knowledge that they are looking for can be also found in such a resource instead of crawling the web for upper ontologies. Contrarily to the propositions in [2,20,40], we do not compute a matching between the domain ontologies and the reference ontology, but simply a similarity matrix (i.e., this procedure is potentially less complex than a full body matching algorithm).

2.2 Fuzzy Ontology Matching

The theory of fuzzy sets and logics provides a suitable framework for handling imprecision in ontologies. A general definition of a fuzzy ontology is given as one *which uses fuzzy logics to provide a natural representation of imprecise and vague knowledge, and eases reasoning over it* [4,6,7,30]. An ontology concept is defined as a fuzzy set on the domain of instances and relations on concepts are defined as fuzzy mappings. Particularly, subsumption is handled by a fuzzy taxonomic relation that expresses the fact that a concept is a specification of another concept up to a certain degree between 0 and 1. In the definition of a fuzzy ontology, which is given later on, we will follow a similar approach.

Work on **fuzzy ontology alignment** can be classified into two families: (1) approaches extending crisp alignment to deal with fuzzy ontologies and (2) approaches addressing imprecision of the matching of (crisp or fuzzy) concepts.

Based on the work on approximate concept mapping by Stuckenschmidt [35] and Akahani et al. [1], Xu et al. [38] suggested a framework for the mapping of fuzzy concepts between fuzzy ontologies. Their approach is based on finding the best approximations in an ontology for all the concepts in another ontology. The approximations (least upper approximation and greatest lower approximation) are defined by using fuzzy concept subsumption and an iterative algorithm is proposed to find a simplified least upper bound. With a similar objective, Bahri et al. [3] proposed a framework to define relations among fuzzy ontology components based on their intentional definitions (i.e. a set of description logics formulas that represent the meaning of a component).

The second family of approaches is characterized by the representation of imprecision of the alignment itself, even with crisp ontologies. For instance, Ferrara et al. [13] propose a fuzzy approach which handles mapping imprecision and provides criteria for its validation. The principle is to interpret and translate each crisp matching result as a set of fuzzy assertions and perform fuzzy reasoning over this set. An ontology matching approach based on fuzzy conceptual graphs and rules is proposed by Buche et al. [5].

Discussion. The alignment framework that we propose does not directly fall into either of the two families outlined above. To our knowledge, none of the existing works on fuzzy alignment is based on the use of background knowledge, which is among the principal motivations of our approach. Of course, in many cases two ontologies can be aligned directly, without taking into consideration any external information. The advantage of using a reference vocabulary is that it allows for taking into account different aspects of the semantics of a concept. In contrast to many other approaches, we are able to make use of this background information, when it is given. This is the case in many situations (we have used Wikipedia) and being able to exploit it in an efficient manner is advantageous. Among the original contributions of this method is the fact that we are able to apply a fuzzy framework to the specific case of instance-populated ontologies.

2.3   Cross-Lingual Ontology Matching

Gracia et al. [17] present a global *vision of a multilingual semantic web* and present several challenges to the multilingual semantic community. According to the authors, multilingualism has to be seen as an extension of the semantic web – a group of techniques which will be added to the existing semantic technologies in order to resolve linguistic heterogeneity where it appears. The semantic web is seen as language-independent, because semantic information is given in formal languages. The main gap is, therefore, between language specific needs of users and the language-independent semantic content. The authors prognosticate that monolingual non-English linked data will increase in years creating "islands" of unconnected monolingual linked data. The challenge is to connect these islands by interconnecting the language-specific information. The authors outline the development of systems for establishing relations between ontology terms or semantic data with labels and instances in different languages as a main direction of future research – a topic which forms the core of our proposal. We proceed to discuss methods that address specific problems related to the multi- and cross-lingual matching task.

The majority of approaches rely on **machine translation (MT)** techniques. Fu et al. [14] follow a standard paradigm of using monolingual matching techniques enhanced with an MT module. As a result of an analysis of the effect of the quality of the MT, the authors propose a noise-minimization method to reduce the flaw in the performance introduced by the translation. Trojahn et al. [9] have implemented an API for multilingual OM applying two strategies: a direct matching by a direct translation of one ontology to the other prior to the matching process and indirect matching, based on a composition of alignments. The latter approach is originally proposed by Jung et al. [21] and it is based on first establishing manual alignment between cross-lingual ontologies and then using these alignments in order to infer new ones. Paulheim et al. [26] apply web-search-based techniques for computing concept similarities by using MT for cross-lingual ontologies.

Spohr et al. [33] rely on a **machine learning approach**. They use a small amount of manually produced cross-lingual alignments in order to learn a matching function for two cross-lingual ontologies. The paper introduces a clear distinction between a multilingual ontology (that which contains annotations given in different languages) and cross-lingual ontologies (two or more monolingual ontologies given in different natural languages).

On the edge of the OM approaches that use **background knowledge**, Rinser et al. [27] propose a method for entity matching by using the info-boxes of Wikipedia. Entities given in different languages are aligned by the help of the explicit relations between Wikipedia pages in different languages. The matching relies mainly on the values of each property, since the actual labels are in different languages (e.g., "population" and "Einwohner" have approximately the same values (3,4M) in the info-boxes of the English and the German Wikipedia pages of Berlin). A very important and useful contribution of this paper is an analysis of the structure of the Wikipedia interlanguage links.

Outside of the context of ontology matching, in the field of **natural language processing**, research has been carried on the topic of measuring semantic distance between cross-lingual terms or concept labels. Mohammad et al. [25] and Eger et al. [11] propose measures of semantic distance between cross-lingual concept labels based on the use of bilingual lexicons. Explicit Semantic Analysis (ESA) applied with Wikipedia has been proposed as a framework for measuring cross-lingual semantic relatedness of terms, first in a paper by Gabrilovich et al. [15] and then in an extended proposal by Hassan et al. [18]. It is suggested to rely on the multiple language versions of Wikipedia in order to measure semantic relatedness between terms. The authors use an ESA framework in order to model a concept as a vector in a space defined by a set of "encyclopedic concepts" in which the concept appears.

Discussion. The methods that have been proposed to deal with multilingualism in ontology matching, with few exceptions, rely on automatic translation of labels to a single target language. However, MT tolerates low precision levels and often external sources are needed in order to achieve good performance. Applying machine learning techniques requires learning corpora that are rarely available in an ontology matching scenario. An inherent problem of translation as such is that there is often a lack of exact one-to-one correspondance between the terms across natural languages. A fuzzy matching approach is able to model this imprecision in a natural way avoiding the use of machine translation. Our approach shares conceptual grounds with the ESA family of approaches, which to the best of our knowledge has not been applied in the ontology matching field.

3   Background

The current section introduces notation and definitions from the field of fuzzy set theory and logics.

3.1   Fuzzy Sets and Logics

Fuzzy set theory emerged as a generalization of classical set theory [39]. A fuzzy set $\mathcal{A}$ is defined on a given domain of objects X by the function

$$\mu_{\mathcal{A}} : X \longrightarrow [0,1]$$

which expresses the degree of membership of every element of X to $\mathcal{A}$ by assigning to each $x \in X$ a value from the interval $[0,1]$. Analogously, fuzzy logics extends two-valued logics by assigning to a proposition a truth value in this interval.

All crisp set and logical operations can be extended to fuzzy sets and logics. Intersection and union are defined, respectively, based on a so-called t-norm function and a t-conorm function. Crisp logical implication is extended to fuzzy logics by the help of a fuzzy implication function. We give definitions by providing examples in terms of Gödel, Łukasiewicz and product semantics, following the introductions found in [10,34]. For the sake of representation within this section, we will denote $a = \mu_{\mathcal{A}}(x)$ and $b = \mu_{\mathcal{B}}(x)$. The *intersection* of two fuzzy sets $\mathcal{A}$ and $\mathcal{B}$ is given by a function $T(a,b)$, referred to as a t-norm. The Gödel t-norm is defined by $T_G(a,b) = min(a,b)$, the Łukasiewicz t-norm is given as $T_L(a,b) = max(a+b-1,0)$ and the product t-norm – by $T_P(a,b) = a \times b$.

The *union* of two fuzzy sets $\mathcal{A}$ and $\mathcal{B}$ is given by $S(a,b)$, where S is a t-conorm. The Gödel definition is given by $S_G(a,b) = max(a,b)$, the Łukasiewicz definition is given by $S_L(a,b) = min(a+b,1)$ and the product t-conorm is defined by $S_P(a,b) = a + b - a \times b$.

Fuzzy *implication* $\mathcal{A} \rightarrow \mathcal{B}$ is defined by $\mu_{\mathcal{A} \rightarrow \mathcal{B}}(x) = i(\mu_{\mathcal{A}}(x), \mu_{\mathcal{B}}(x))$ where i is a function that determines the properties of the implication. Two types of fuzzy implications are commonly used: *S-implications* which extend the proposition $a \rightarrow b = \neg a \vee b$ to fuzzy logics and *R-implications (residuum-based implications)* defined as $\forall a, b \in [0,1]$, $i(a,b) = sup\{c \in [0,1] : T(a,c) \leq b\}$. In terms of the three considered semantics $i(a,b) = 1$ if $a \leq b$. Depending on the particular *t*-norm definition, the case $a > b$ is defined as follows: $i_G(a,b) = b$ (Gödel), $i_L(a,b) = 1 - a + b$ (Łukasiewicz) and $i_P(a,b) = \frac{b}{a}$ (product).

In this study, we have considered the Gödel definitions of intersection, union and implication. As we shall see, this choice is justified by the properties of the Gödel implication which are used to define a fuzzy degree of subsumption preserving the crisp one. This implication is defined for two fuzzy membership functions as

$$\mu_{\mathcal{A} \rightarrow \mathcal{B}}(x) = \begin{cases} 1, & \text{if } \mu_{\mathcal{A}}(x) \leq \mu_{\mathcal{B}}(x), \\ \mu_{\mathcal{B}}(x), & \text{otherwise.} \end{cases} \tag{1}$$

Taking the infimum $\inf_{x \in X} \mu_{\mathcal{A} \rightarrow \mathcal{B}}(x)$ over all $x \in X$ in Eq. (1) gives rise to the definition of a fuzzy subsumption between $\mathcal{A}$ and $\mathcal{B}$. This is used to define a fuzzy version of the ontological *is_a* relation in Sect. 5.

The fuzzy power set of X, denoted by $\mathcal{F}(X,[0,1])$, is the set of all membership functions defined on X.

3.2　Measures of Fuzzy Set Relatedness

Let $\mathcal{A}$ and $\mathcal{B}$ be two fuzzy sets with respective membership functions $\mu_{\mathcal{A}}$ and $\mu_{\mathcal{B}}$. We consider the following measures of fuzzy set relatedness, well-known from the fuzzy ontology literature [8]. These measures are relevant to our approach and will be applied as explained later on in Sects. 5 and 6 for measuring the similarity between fuzzified concepts.

– Base measure:

$$\rho_{base}(\mu_{\mathcal{A}}, \mu_{\mathcal{B}}) = 1 - \max_{x \in X} |\mu_{\mathcal{A}}(x) - \mu_{\mathcal{B}}(x)|. \tag{2}$$

– Euclidean distance–based fuzzy similarity ($\|x\|_2 = \left(\sum_{x \in X} |x|^2\right)^{1/2}$ is the ℓ^2-norm):

$$\rho_{diff}(\mu_{\mathcal{A}}, \mu_{\mathcal{B}}) = 1 - \frac{1}{|X|} \|\mu_{\mathcal{A}} - \mu_{\mathcal{B}}\|_2. \tag{3}$$

– 1-Norm ($\|x\|_1 = \sum_{x \in X} |x|$ is the ℓ^1-norm):

$$\rho_{sum}(\mu_{\mathcal{A}}, \mu_{\mathcal{B}}) = 1 - \frac{1}{|X|} \|\mu_{\mathcal{A}} - \mu_{\mathcal{B}}\|_1. \tag{4}$$

– Zadeh's partial matching index:

$$\rho_{sup-min}(\mu_{\mathcal{A}}, \mu_{\mathcal{B}}) = \sup_{x \in X} T(\mu_{\mathcal{A}}(x), \mu_{\mathcal{B}}(x)). \tag{5}$$

In our experiments, we used a more robust variant of $\rho_{sup-min}$ that applies the average of the k largest values of $T(\mu_{\mathcal{A}}(x), \mu_{\mathcal{B}}(x))$. This measure will be called $\rho_{sup-min(k)}$.

– Jaccard coefficient:

$$\rho_{jacc}(\mu_{\mathcal{A}}, \mu_{\mathcal{B}}) = \frac{\sum_x T(\mu_{\mathcal{A}}(x), \mu_{\mathcal{B}}(x))}{\sum_x S(\mu_{\mathcal{A}}(x), \mu_{\mathcal{B}}(x))} . \tag{6}$$

4 A Hierarchical Algorithm for Concept Fuzzification

An *ontology* consists of a set of semantically related *concepts* which provides in an explicit and formal manner knowledge about a given domain of interest [12]. We are particularly interested in ontologies, whose concepts come equipped with a set of associated instances, defined as it follows.

Definition 1 (Crisp Ontology). *Let C be a finite set of concepts, $is\_a \subseteq C \times C$ a partial order on concepts, R a set of relations on C, I a set of instances, $g : C \to 2^I$ a function that assigns subsets of instances from I to each concept in C. For each considered language L, we assume a function $l_L : C \to 2^{\Sigma_L^*}$ that assigns to each concept a set of labels from a set of labels Σ_L^* coming from some language-specific alphabet Σ_L. With these definitions, the quintuple*

$$O = (C, is\_a, R, I, g, l)$$

forms a crisp ontology.

4.1 Crisp Concept Similarities

Consider the ontologies $O = (C, is\_a, R, I, g, l)$ and $O_{ref} = (X, is\_a_{ref}, R_{ref}, I_{ref}, g_{ref}, l_{ref})$. Determining the similarity $\sigma(A, x)$ of two concepts $A \in C$ and $x \in X$ can be done by comparing their label sets $l_L(A)$ and $l_{ref,L}(x)$ (terminological similarity), by using the structure of the ontologies, or by comparing their instance sets $g(A)$ and $g_{ref}(x)$. Given that we have a similarity measure for each of the three comparisons, we can define a generic measure for concept similarity as

$$\sigma(A, x) = \alpha \sigma_{term} + \beta \sigma_{struct} + \gamma \sigma_{inst}, \tag{7}$$

where σ_{term}, σ_{struct} and σ_{inst} are, respectively, terminological, structural and instance-based similarities, and $\alpha, \beta, \gamma \in [0, 1]$ are coefficients that assign weights to each of the similarity types. We require that σ takes values in the $[0,1]$.

An example of a terminological measure is the well-known normalized Levenshtein similarity. As a structural measure one can use any semantic similarity based on graph path-length (e.g., Wu-Palmer). For computing instance-based concept similarity, we need a similarity measure for instances $\mathbf{i}^A$ and $\mathbf{i}^x$, where $\mathbf{i}^A \in g_1(A)$ and $\mathbf{i}^x \in g_{ref}(x)$. We have used the scalar product and the cosine $s(\mathbf{i}^A, \mathbf{i}^x) = \frac{\langle \mathbf{i}^A, \mathbf{i}^x \rangle}{\|\mathbf{i}^A\| \|\mathbf{i}^x\|}$. Based on this similarity measure for elements, the similarity measure for the sets can be defined by computing the similarity of the mean vectors corresponding to class prototypes [22]:

$$\sigma_{inst(proto)}(A, x) = s\left(\frac{1}{|g(A)|} \sum_{j=1}^{|g(A)|} \mathbf{i}_j^A, \frac{1}{|g_{ref}(x)|} \sum_{k=1}^{|g_{ref}(x)|} \mathbf{i}_k^x \right). \tag{8}$$

Note that other approaches of concept similarity can be employed as well, for instance the variable selection based approach in [37]. As a matter of fact, one can use any similarity measure. In our study, we chose one that worked well in the experiments and had the lowest computational complexity.

4.2 Concept Fuzzification

Let O and O_{ref} be two ontologies as defined in Definition 1. We will call O_{ref} a *reference* ontology whose concepts will be called *reference concepts*, whereas O will be called *source ontology* and its concepts - *source concepts*.

The fuzzification procedure that we propose relies on the idea of modeling every source concept as a function of its similarities to the reference concepts, using the measure (7). Any source concept A is represented by a function of the kind

$$\mu_A(x) = \sigma(A, x), \forall x \in X, \tag{9}$$

where $\sigma(A, x)$ is the similarity between the concept A and a given reference concept x. Since σ takes values between 0 and 1, (9) defines a fuzzy set. We will refer to such a fuzzy set as the *fuzzified* concept A denoted by $\mathcal{A}$. Note that the $\mu_A(x)$ notation is used with slightly different meaning in fuzzy description logics, incorporating the well accepted understanding of a fuzzy set.

In order to fuzzify the concepts of a source ontology O, we propose the following hierarchical algorithm. First, we assign degree-of-membership (*dom*) vectors, i.e., fuzzy membership functions to all leaf-node concepts of O by applying (9). Every non-leaf node, if it does not contain instances of its own, is assigned a *dom* vector as the maximum of the *dom*s of its children for every $x \in X$.

If a non-leaf node has directly assigned instances (not inherited from its children), the node is first assigned *dom*s on the basis of these instances with respect to the reference ontology (following Eq. (9)), and then as the maximum of its children and itself. We redefine (9) for any (leaf or non-leaf) source concept in the following way.

$$\mu_A(x) = max\{max_{A' \in D(A)}\mu_{A'}(x), \sigma(A, x)\}, \forall x \in X, \tag{10}$$

where $D(A)$ is the set of children of A. This reflects the idea that the instances of the descendants of A are also indirect instances of A.

The algorithm is given in Algorithm 1. Note that assigning the *max* of all children to the parent for every x leads to potentially higher values of the membership functions for nodes higher up in the hierarchy. Naturally, the functions of the higher level concepts are expected to be less "specific" than those of the lower level concepts. A concept in a hierarchical structure can be seen as the union of its descendants, which corresponds to taking the *max* (an approach underlying the single link strategy used in clustering).

The hierarchical computation of the similarity vectors has the advantage that it holds that $\mu_{A' \to A}(x) = 1$ for all x and all children A' of A. From computational viewpoint, the procedure has the advantage of scoring only certain nodes. All other information from the source concepts, i.e. their label sets and their relations, are just transferred to their fuzzfizied counterparts without change.

Procedure hierachicalFuzzification(O, O_{ref}, σ)
begin

1. Let C be the list of concepts in O.
2. Let L be a list of nodes, initially empty
3. // compute an ordering L of the ontology concepts from bottom to top:
   Do until C is empty:
   (a) Let L' be the list of nodes in C that have only children in L
   (b) $L = $ append(L, L')
   (c) $C = C - L'$
4. // compute fuzzified vector from bottom to top:
   Iterate over L (first to last), with A being the current element:
   (a) **Let** $D = children(A)$
   (b) **If** $D = \emptyset$ (i.e., leaf concepts)
       i. For all $x \in X$:
          Define $\mu_A(x) = \sigma(A, x)$
       Else If $g^*(A) \neq \emptyset$ (there are children and direct concept instances):
       i. For all $x \in X$:
          Define $\mu_A(x) = max\{max_{A' \in D}\mu_{A'}(x), \sigma(A, x)\}$
       Else (i.e., children and no direct concept instances)
       i. For all $x \in X$:
          Define $\mu_A(x) = max_{A' \in D}\ \mu_{A'}(x)$

return fuzzified concepts defined

Algorithm 1. An algorithm for fuzzification of the source concepts.

5   Heterogeneous Knowledge and Data Integration

The current section describes several applications of the hierarchical fuzzification algorithm presented above. We focus on finding fuzzy relations between cross-ontology concepts in a classical ontology alignment framework, as well as on the discovery of alignments of type many-to-many. We pay some attention to a promising application of the approach for matching cross-lingual ontologies (ontologies, defined each in a different natural language).

An Alignment as a Fuzzy Union of Two Ontologies. The implication $A' \rightarrow A$ holds for any A' and A such that $is\_a(A', A)$. We provide a definition for a fuzzy subsumption of two fuzzified concepts $\mathcal{A}'$ and $\mathcal{A}$ based on the fuzzy implication (1).

Definition 2 (Fuzzy Subsumption). *The subsumption $\mathcal{A}'$ is a $\mathcal{A}$ is defined and denoted in the following manner:*

$$\text{is\_a}(\mathcal{A}', \mathcal{A}) = \inf_{x \in X} \mu_{\mathcal{A}' \rightarrow \mathcal{A}}(x) \tag{11}$$

Equation (11) defines the fuzzy subsumption as a degree between 0 and 1 to which one concept is the subsumer of another. It can be shown that **is_a**, similarly to its crisp version, is reflexive and transitive (i.e. a quasi-order). In addition, the hierarchical procedure for concept fuzzification introduced in the previous section assures that **is_a**$(A', A) = 1.0$ holds for every child-parent concept pair, i.e., the crisp subsumption relation is preserved by the fuzzification process.

We provide a definition of a fuzzy ontology which follows directly from the fuzzification of the source concepts and their **is_a** relations introduced above.

Definition 3 (Fuzzy Ontology). *Let C be a set of (fuzzy) concepts, **is_a** : $C \times C \to [0, 1]$ a fuzzy is_a-relationship (a fuzzy quasi-order), R a set of fuzzy relations on C, i.e., R contains relations $r : C^n \to [0, 1]$, where n is the arity of the relation (for the sake of presentation, we only consider binary relations), X a set of objects, and $\phi : C \to \mathcal{F}(X, [0, 1])$ a function that assigns a membership function to every fuzzy concept in C. For each considered language L, we assume a function $\lambda_L : C \to 2^{\Sigma_L^*}$ that assigns to each concept a set of labels from a set of labels Σ_L^* from some alphabet Σ_L. With these definitions, the quintuple*

$$\mathcal{O} = (C, \text{is\_a}, R, X, \phi, \lambda)$$

forms a fuzzy ontology.

Above, the set X is defined as a set of abstract objects. In our setting, these are the concepts of the reference ontology, i.e. $X = X$. The set C is any subset of C_Ω. In case $C = C_1$, where C_1 is the set of fuzzified concepts of the ontology O_1, $\mathcal{O}$ defines a fuzzy version of the crisp source ontology O_1.

Now let us be given two source ontologies, O_1 and O_2 with their respective sets of fuzzified concepts C_1 and C_2. Looking again at Definition 3, in case $C = C_1 \cup C_2$, $\mathcal{O}$ defines *a common knowledge body* for the two source ontologies. The degree of similarity between any two concepts taken from the union $C_1 \cup C_2$ can be computed inexpensively by applying any of the fuzzy set relatedness measures given in Sect. 3.2.

Quantifying Commonality and Specificity. The union of two fuzzy concepts can be decomposed into three components, each quantifying, respectively, the commonality of both concepts, the specificity of the first compared to the second and the specificity of the second compared to the first expressed in the following manner

$$S(\mathcal{A}, \mathcal{B}) = (\mathcal{A}\mathcal{B}) + (\mathcal{A} - \mathcal{B}) + (\mathcal{B} - \mathcal{A}). \tag{12}$$

Each of these components is defined as follows and, respectively, accounts for:

$$\mathcal{A}\mathcal{B} = T(\mathcal{A}, \mathcal{B}) \text{ // what is common to both concepts;} \tag{13}$$
$$\mathcal{A} - \mathcal{B} = T(\mathcal{A}, \neg\mathcal{B}) \text{ // what is characteristic for A;} \tag{14}$$
$$\mathcal{B} - \mathcal{A} = T(\mathcal{B}, \neg\mathcal{A}) \text{ // what is characteristic for B.} \tag{15}$$

An example is given in Fig. 1. Several heuristics about the concepts of interest can be provided to a potential user with respect to the values of these components. For example, in case (13) is significantly larger than each of (14) and (15) for certain values of $x \in X$, the two concepts can be merged.

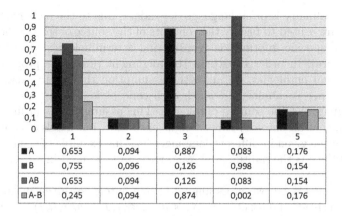

	1	2	3	4	5
■ A	0,653	0,094	0,887	0,083	0,176
■ B	0,755	0,096	0,126	0,998	0,154
■ AB	0,653	0,094	0,126	0,083	0,154
□ A-B	0,245	0,094	0,874	0,002	0,176

Fig. 1. An example of the quantities in Eqs. (13)–(15) for two source concepts A and B, fuzzified by the help of 5 reference concepts.

Mappings of Type Many-to-Many and N-Way Matching. Based on the subsumption relation defined above, we define θ-equivalence of two concepts in the following manner.

Definition 4 (Fuzzy θ-Equivalence). *Fuzzy θ-equivalence between a concept A and a concept B, denoted by $A \leadsto_\theta B$ holds if and only if is_a$(A, B) > \theta$ and is_a$(B, A) > \theta$, where $\theta \in [0, 1]$.*

It is easy to show that the θ-equivalence is an equivalence relation (i.e., reflexive, symmetric and transitive) on a set of concepts. This relation allows us to define classes of equivalence on the set of the union of the concept sets of two (or more) ontologies in the following way. Let C_1 and C_2 be the sets of concepts of ontologies O_1 and O_2, respectively. The equivalence class of a concept $A \in C_1$,

$$[A] = \{B \in C_1 \cup C_2 : A \leadsto_\theta B\}, \tag{16}$$

defines a set of mappings from A to all members of its equivalence class $[A]$. Taking the subset of $[A]$ that contains only concepts from C_2 defines a set of matches of type one-to-many for A to concepts in the ontology O_2.

Generalizing on that, if we have as an input N (N > 2) source ontologies, we construct equivalence classes on the set of the union of their concept sets resulting into alignments of multiple ontologies, or N-way matching.

Cross-Lingual Ontology Matching. An ontology is called *multilingual* if several natural languages are used to label its concepts and properties; it is called *monolingual* otherwise. Two monolingual ontologies are referred to as *cross-lingual* if their respective labels are given in two different natural languages [33]. For simplicity, in what follows we will describe the bi-lingual case (the generalization to the multilingual case follows directly). As shown in the previous section, the fuzzification of a source ontology with respect to a given reference vocabulary is often language dependent, due to the fact that the similarity measure σ used to fuzzify the concepts of the source ontologies often relies on language specific information (considering a terminological or an instance-based measure). In order to extend the alignment procedure proposed above to cross-lingual ontologies, we make the assumption that there exists a multi-lingual reference vocabulary, in which each concept is assigned a label in each of the languages of the sources ontologies. This assumption is in line with the observation that an ontology is not language specific, but labels of concepts and relations can be given in different languages [9]. One example of such a multilingual vocabulary is the multilingual Yago that we have used in our experiments (Sect. 6).

Let O_1 and O_2 be two cross-lingual source ontologies given in languages 1 and 2, respectively, and let $O_{ref} = (X, is\_a_{ref}, R_{ref}, I_{ref}, g_{ref}, l_{ref})$ be a bi-lingual ontology, i.e., there exist two sets of concept labels Σ_1^* and Σ_2^* and l_{ref} comprises two labeling functions $l_{ref,1} : X \rightarrow 2^{\Sigma_1^*}$ and $l_{ref,2} : X \rightarrow 2^{\Sigma_2^*}$ that assign to each concept $x \in X$ a label set from each of the two respective label sets. That means that for a given concept $x \in X$, $l_{ref,1}$ and $l_{ref,2}$ will return its versions in the languages 1 and 2. In the fuzzification process, the labels in Σ_1^* will be used to fuzzify the concepts of O_1 and the labels in Σ_2^* those of O_2.

The rest of the fuzzy alignment procedure remains unchanged. The resulting fuzzy functions for the two sets of source concepts can be compared as described in Sect. 3.2 and submitted to the procedures given in Sects. 4.2 and 5. This enables the construction of a common fuzzy ontology for cross-lingual source ontologies, which reflects the relations and various aspects of the similarities between the concepts of these ontologies, as well as the degrees to which they hold.

6   Experiments and Examples

We describe experiments in matching domain specific ontologies from the multimedia field in the following subsection. An evaluation of the approach is given later on in a multilingual ontology matching scenario produced on the basis of the Multifarm OAEI benchmark.

6.1   Application of the Fuzzy Alignment Procedure

We situate this experiment in the multimedia domain, opposing two complementary heterogeneous ontologies containing annotated pictures. For the alignment task, we have taken an instance-based approach. We chose, on one hand, LSCOM [32] initially built in the framework of TRECVID[2] and populated with the development set of TRECVID 2005. Since this set contains images from broadcast news videos, LSCOM is particularly adapted to annotate this kind of content, thus contains abstract and specific concepts (e.g. SCIENCE_TECHNOLOGY,

---

[2] http://www-nlpir.nist.gov/projects/tv2005/

INTERVIEW_ON_LOCATION). On the other hand, we used WordNet [24] populated with the LabelMe dataset [28], referred to as the LabelMe ontology. Contrarily to LSCOM, this ontology is very general, populated with photographs from daily life and contains concepts such as CAR, COMPUTER, PERSON, etc.

A text document has been generated for every image of the two ontologies, by taking the names of all concepts that an image contains in its annotation, as well as the (textual) definitions of these concepts (the LSCOM definitions for TRECVID images or the WordNet glosses for LabelMe images). Several problems related to this representation are worth noting. The LSCOM keyword descriptions sometimes depend on negation and exclusion which are difficult to handle in a simple bag-of-words approach. Taking the WordNet glosses of the terms in LabelMe introduces problems related to polysemy and synonymy. Additionally, a scene often consists of several objects, which are frequently not related to the object that determines the class of the image. In such cases, the other objects in the image act as noise. A common problem related to polysemy is avoided when applying an instance-based approach because polysemous concepts result in different instance sets, whereas synonymous concepts have similar instance sets.

Table 1. Examples of pairs of matched intra-ontology concepts (above) and cross-ontology concepts (below), column-wise.

Concept $\mathcal{A}$: Concept $\mathcal{B}$:	LSCM:truck vs. LSCM:gr.vehicle	LSCM:sports vs. LSCM:basketball	LM:computer vs. LM:elec. device	LM:animal vs. LM:bird
is_a$(\mathcal{A}, \mathcal{B})$	1	0.007	1	0.004
is_a$(\mathcal{B}, \mathcal{A})$	0.012	1	0.011	1
is_a$^{\text{mean}}(\mathcal{A}, \mathcal{B})$	1	0.052	1	0.062
is_a$^{\text{mean}}(\mathcal{B}, \mathcal{A})$	0.326	1	0.07	1
Base Sim.	0.848	0.959	0.915	0.390
Eucl. Sim.	0.835	0.908	0.854	0.350
SupMin Sim.	0.435	0.545	0.359	0.309
Jacc. Sim.	0.870	0.814	0.733	0.399
Cosine Sim.	0.974	0.994	0.975	0.551

Concept $\mathcal{A}$: Concept $\mathcal{B}$:	LM:gondola vs. LSCM:boat_ship	LSCM:group vs. LM:audience	LSCM:truck vs. LM:vehicle	LSCM:truck vs. LM:conveyance
is_a$(\mathcal{A}, \mathcal{B})$	0.016	0.006	0.022	0.022
is_a$(\mathcal{B}, \mathcal{A})$	0.009	1	0.012	0.012
is_a$^{\text{mean}}(\mathcal{A}, \mathcal{B})$	0.86	0.022	0.748	0.769
is_a$^{\text{mean}}(\mathcal{B}, \mathcal{A})$	0.167	1	0.301	0.281
Base Sim.	0.72	0.78	0.58	0.58
Eucl. Sim.	0.66	0.71	0.40	0.38
SupMin Sim.	0.069	0.082	0.22	0.22
Jacc. Sim.	0.49	0.42	0.54	0.52
Cosine Sim.	0.69	0.82	0.66	0.67

In order to fuzzify the source concepts, we applied the hierarchical algorithm from Sect. 4 independently for each of the source ontologies. As a reference ontology, we have used an extended version of the Wikipedia's so-called main topic classifications, containing more than 30 categories. For each category, we included a set of corresponding documents from the Inex2007 corpus.

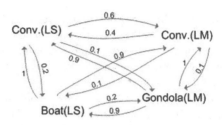

Fig. 2. A fragment of the common fuzzy ontology of LSCOM (LS) and LabelMe (LM).

The new combined knowledge body has been constructed by first taking the union of all fuzzified source concepts. For every pair of concepts, we have computed their Gödel subsumptional relations, as well as the degree of their similarities (applying the measures from Sect. 3 and the standard cosine measure). Apart from the classical Gödel subsumption defined in (11), we consider a version of it which takes the average over all x instead of the smallest value, given as $\text{is\_a}^{\text{mean}}(\mathcal{A}', \mathcal{A}) = \text{avg}_{x \in X} \mu_{\mathcal{A}' \to \mathcal{A}}(x)$. The results for several intra-ontology concepts and several cross-ontology concepts are given in Table 1. Figure 2 shows a fragment of the common fuzzy ontology built for LSCOM and LabelMe. The labels of the edges of the graph correspond to the values of the fuzzy subsumptions between concepts. We see that the two "Conveyance" concepts from the two ontologies have similar degrees of their fuzzy subsumption relations. The hierarchical structure of the concepts is preserved within each ontology (a "Boat" is a "Conveyance" rather than the other way round), and it is also reflected on cross-ontology concepts (a "Gondola" (from LM) is a "Boat" (from LS)).

6.2 Aligning Cross-Lingual Ontologies

To be able to align source ontologies from different languages, we need a multilingual reference ontology, which either contains concept instances from the languages of the source ontologies, or alternatively ontology labels from different languages. In the following, we will use an approach that relies on concept labels and ontology structure.

We assume that the concepts of the source ontologies are labeled with expressions in different languages. Further, often a concept has several alternative labels in a single language. Let us assume that

$$t_L(A) = \{t_1, \ldots, t_m\}$$

denotes the set of m labels $t_i \in \Sigma_L^*$ of a concept A in a language L (e.g., a set of synonyms, abbreviations or syntactic variants). Σ_L is the alphabet of L and Σ_L^* - the set of words on this alphabet.

To solve the matching problem, we assume that there exists a reference ontology whose concepts have labels from the languages that are also used in the source ontologies to be matched. The similarity, σ, of a source and a reference concept is then determined by computing label similarities taking into account the language of the source ontology.

Since each concept can have multiple labels from a specific language, we use the maximum similarity of two cross-ontology concept label pairs to determine the similarity of the concepts in the respective language:

$$\sigma_L(A, x) = \max_{t' \in t_L(A), t'' \in t_L(x)} \tau_L(t', t'') \ .$$

Following our convention, A and x represent a source and a reference concept, respectively. Note that A might only have labels in one language, whereas x is assumed to have labels for all languages. The reference ontology is not assumed to have labels in all languages for each of its concepts. τ_L is the similarity measure used for the respective strings. It may be language dependent, for instance, when a language specific processing of t' and t'' is used for determining their similarity. This typically includes steps like tokenization, spell correction, morphological analysis including lemmatization or stemming, among other things. In the experiment described later in this section, however, we used string-based similarity measures which are defined in a language independent-manner.

To demonstrate the suitability of our approach for multilingual ontology matching, we experimented with the **Multifarm benchmark** described in [23]. The Multifarm data is obtainable from the multifarm homepage http://web. informatik.uni-mannheim.de/multifarm/. The benchmark comprises 5 conference-related source ontologies named CMT (88), CONFERENCE (123), CONFOF (74), IASTED (181), SIGKDD (77). The numbers given in parentheses denote the number of concepts for each ontology. Each ontology has labels in English, Spanish, German, French, Russian, Portuguese, Czech, Dutch, and Chinese (Mandarin).

The concepts in the ontologies include object classes like "Author" and properties such as "hasAuthor". Object classes range from fairly simple classes like "Author", "Paper", and "Title" to semantically relatively complex ones such as "Registration fee for authors who receive the proceedings volume on CD" and "mark conflict of interest".

When looking at the German translations of the ontologies, we found that in some cases the translation is not correct. For instance, the English word "abstract" is translated as "Abstrakt", which, as a noun, is not a word in German. A possible correct translation would be "Zusammenfassung" - many authors would also use the English word. "Submission of abstracts" was translated to "abstrakte Einreichung" which re-translates to "a submission, which is abstract". The French translations also exhibit some flaws. Although these problems do not affect the majority of all translations, it makes the matching task potentially more difficult.

In the experiments, we only considered the so-called type (ii) Multifarm ontology matching task, which consists in the re-identification of concepts for each of the 5 source ontologies. Note that we did not consider the English concept names, which would have made the reidentification of concepts a trivial task since the concept names are the same for all language versions of a source ontology.

We also did not consider structural properties of the concepts (i.e., sub- and superclasses, relations, etc.), but only the concept labels. The only exception is that we mapped classes to each other only, and properties – only to properties.

As reference ontologies, we used parts of Yago2 [36]. In particular, we considered the Wordnet multilingual subontology where the numbers of concepts per language are as follows: 36378 (en), 27933 (de), 20280 (fr), 23620 (cz), 16476 (nl), 14393 (pt), 19372 (es). Obviously the number of reference concepts varies considerably between languages. Russian and Chinese had been excluded from the benchmark because of potential coding problems. However, in our experiments, we did not experience coding problems. The results obtained when including also Russian and Chinese are very similar to the ones reported below.

For matching labels, we considered the following string similarity measures:

Levenshtein (LSH): This similarity measure is based on the Levenshtein distance, which counts the number of operations (replace, insert, delete), which are necessary for transforming one string into another. In order to obtain a similarity measure, we divided the distance by the length of the longer string and subtracted it from 1.0.

Longest Common Substring (LCS): This similarity measure is not based on operations but simply counts the characters in the longest common substring. Note that gaps are allowed.

Longest Continuous Common Substring (LCCS): here, we only consider sequences which do not have gaps in the original strings. This appears more suitable for natural language, although it is not robust to spelling errors and variations.

Since frequently concept labels consist of multiple words, we split these words into tokens and matched these tokens separately, allowing only 1:1 matches between tokens. The optimal match is found using a greedy search strategy recursively starting with the best matching token pair. This two-level approach allows more freedom with respect to word order. However, it is not able to deal properly with compound nouns that occur frequently in German. In order to deal with compound nouns, one has to employ linguistic methods like the morphological analysis of words which allows identifying the semantic "roots" of an expression. However, if the reference ontology is rich enough in concepts, this task might also be solved by the matching process. For instance, a German word like "Konferenzbeitrag" could be similar to two reference concepts, e.g., "Konferenz" and "Beitrag", instead of requiring a directly equivalent reference concept.

For the evaluation, we computed matchings for 5 ontologies in 7 languages, and we considered 7 crisp and 7 fuzzy matchings. All software was programmed in JAVA. On an Intel i7-3770K processor (3.5 GHz), the overall evaluation took 4 h and 20 min. This means that a single match took approximately 1.51 s on average.

The matching results are described in Table 2. According to the definition of the benchmark, we averaged over all ontologies and all language pairs. The best result found in Table 2 corresponds to using LCCS2L together with the Sup-Min(4) similarity measure and a decision threshold of 0.35. LSH2L also performs well. For all crisp similarity measures, matching was improved by using the 2-level approach (marked with the suffix 2L) compared to using the simple version.

Table 2. Results: F-measure averaged over all source ontologies and language pairs. The number in parentheses gives the decision threshold $\theta 1^*$ for the optimal F-measure

	SupMin	SupMin(4)	Jaccard	Base	Cosine	Euclidean	1-Norm
LSH	0.43 (0.5)	0.42 (0.45)	0.23 (0.65)	0.14 (0.25)	0.22 (0.9)	0.17 (0.5)	0.19 (0.5)
LSH2L	0.45 (0.8)	0.49 (0.65)	0.26 (0.5)	0.17 (0.5)	0.21 (0.5)	0.25 (0.5)	0.27 (0.5)
LCS	0.42 (0.5)	0.42 (0.5)	0.25 (0.6)	0.12 (0.35)	0.27 (0.5)	0.21 (0.5)	0.24 (0.5)
LCS2L	0.43 (0.5)	0.43 (0.45)	0.3 (0.5)	0.16 (0.25)	0.31 (0.5)	0.26 (0.5)	0.27 (0.5)
LCCS	0.44 (0.4)	0.46 (0.3)	0.29 (0.5)	0.18 (0.5)	0.33 (0.5)	0.27 (0.5)	0.27 (0.5)
LCCS2L	0.47 (0.4)	**0.5** (0.35)	0.34 (0.5)	0.23 (0.3)	0.35 (0.85)	0.28 (0.5)	0.29 (0.5)

On the Multifarm homepage, the performances of different systems are reported as YAM++ (0.63), AUTOMSv2 (0.27), WeSeE (0.49), CIDER (0.12), MapSSS (0.67), LogMap (0.04), CODI (0.63), MaasMtch (0.23), LogMapLt (0.05), MapPSO (0.05), CSA (0.42), MapEVO (0.02). This means that our approach with an f-measure of 0.5 is among the top 3, which is not bad given that the matching approach relies on labels only and does not use natural language processing. However, we cannot draw strong conclusions from this result since we had the advantage of optimizing the performance with respect to different combinations of crisp and fuzzy similarity measures and decision thresholds.

Other results of the experiments are that the SupMin/SupMin(k) measures appear to perform best, compared to the Jaccard, base, cosine, Euclidean, and 1-norm measure. We assume that the Jaccard, base, cosine, Euclidean, and 1-norm measure perform worse because the final value of the similarity is based on numerical contributions from all reference concepts including irrelevant ones, which leads to high noise. In the case of the base measure, the final results depends on the reference concept with the "worst" combination of concept labels in the two languages. The probability of such bad reference concepts also increases with the number of concepts.

We also considered all language pairs separately. The best-matching language-pairs are English/German (0.75), English/Dutch (0.75), and English/Spanish (0.74). The worst matching results were obtained for Czech/Polish (0.55). The results neither seem to correlate with language families, nor do they correlate only with the number of reference concepts, since there are not that many reference concepts for Dutch either. English/Czech, however, only has an F-Measure of 0.65. We therefore assume that the differences depend on particular properties of the translation of the source ontologies, and on properties of the reference ontology used, which are not yet completely transparent to us.

7 Conclusion and Future Work

The use of background knowledge has shown to be a prominent direction to follow in order to improve and build on top of the current state-of-the-art in the field of ontology matching. In the presented work, we have considered enhancing ontology matching by the use of a generic reference vocabulary from the web of data (the Yago ontology). This setting gives rise to a fuzzy set representation of concepts and concept relations (within and across ontologies), embodying imprecision in the concept definitions and the produced alignments. Our evaluation results showed that the approach is promising for computing alignments between ontologies with cross-lingual entity labels.

In future, we plan to complete the evaluation of the multilingual matching approach by testing it on both tasks of the Multifarm benchmark. In addition, we will explore the application of reasoning techniques to perform fuzzy inference on the matched concepts. Other applications of the proposed method will be considered as well, such as the alignment of ontologies of large scale or of different degrees of expressivity.

References

1. Akahani, J.-I., Hiramatsu, K., Satoh, T.: Approximate query reformulation based on hierarchical ontology mapping. In: Proceedings of International Workshop on SWFAT, pp. 43–46 (2003)
2. Aleksovski, Z., ten Kate, W., van Harmelen, F.: Exploiting the structure of background knowledge used in ontology matching. In: Proceedings of OM, p. 13 (2006)
3. Bahri, A., Bouaziz, R., Gargouri, F.: Dealing with similarity relations in fuzzy ontologies. In: Proceedings of FUZZ-IEEE, pp. 1–6. IEEE (2007)
4. Bobillo, F.: Managing vagueness in ontologies. Ph.D. thesis, University of Granada, Spain (2008)
5. Buche, P., Dibie-Barthélemy, J., Ibanescu, L.: Ontology mapping using fuzzy conceptual graphs and rules. In: ICCS Supplement, pp. 17–24 (2008)
6. Calegari, S., Ciucci, D.: Fuzzy ontology, fuzzy description logics and fuzzy-OWL. In: Masulli, F., Mitra, S., Pasi, G. (eds.) WILF 2007. LNCS (LNAI), vol. 4578, pp. 118–126. Springer, Heidelberg (2007)
7. Calegari, S., Sanchez, E.: A fuzzy ontology-approach to improve semantic information retrieval. In: Bobillo, F., Cesar, P., da Costa, G., d'Amato, C., Fanizzi, N., Fung, F., Lukasiewicz, T., Martin, T., Nickles, M., Peng, Y., Pool, M.l., Smrz, P., Vojtás, P. (eds.) URSW, vol. 327 of CEUR Workshop Proceedings. CEUR-WS.org (2007)
8. Cross, V., Yu, X.: A fuzzy set framework for ontological similarity measures. In: Proceedings of WCCI 2010, FUZZ-IEEE 2010, pp. 1–8. IEEE Computer Society Press (2010)
9. Trojahn dos Santos, C., Quaresma, P., Vieira, R.: An API for multilingual ontology matching. In: Calzolari, N., Choukri, K., Maegaard, B., Mariani, J., Odijk, J., Piperidis, S., Rosner, M., Tapias, D. (eds.) Proceedings of LREC. European Language Resources Association (2010)
10. Dubois, D.: Fuzzy Sets and Systems: Theory and Applications, vol. 144. Academic Press, New York (1980)

11. Eger, S., Sejane, I.: Computing semantic similarity from bilingual dictionaries. In: Proceedings of the 10th International Conference on the Statistical Analysis of Textual Data (JADT-2010), pp. 1217–1225 (2010)
12. Euzenat, J., Shvaiko, P.: Ontology Matching, 2nd edn. Springer, Heidelberg (2013)
13. Ferrara, A., Lorusso, D., Stamou, G., Stoilos, G., Tzouvaras, V., Venetis, T.: Resolution of conflicts among ontology mappings: a fuzzy approach. In: Proceedings of OM at ISWC (2008)
14. Fu, B., Brennan, R., O'Sullivan, D.: Cross-lingual ontology mapping – an investigation of the impact of machine translation. In: Gómez-Pérez, A., Yu, Y., Ding, Y. (eds.) ASWC 2009. LNCS, vol. 5926, pp. 1–15. Springer, Heidelberg (2009)
15. Gabrilovich, E., Markovitch, S.: Computing semantic relatedness using wikipedia-based ESA. IJCAI **7**, 1606–1611 (2007)
16. Gal, A., Shvaiko, P.: Advances in ontology matching. In: Dillon, T.S., Chang, E., Meersman, R., Sycara, K. (eds.) Advances in Web Semantics I. LNCS, vol. 4891, pp. 176–198. Springer, Heidelberg (2008)
17. Gracia, J., Montiel-Ponsoda, E., Cimiano, P., Gomez-Perez, A., Buitelaar, P., McCrae, J.: Challenges to the multilingual web of data. Web Semant. Sci. Serv. Agents World Wide Web **9**(1), 29–51 (2011)
18. Hassan, S., Mihalcea, R.: Cross-lingual semantic relatedness using encyclopedic knowledge. In: Proceedings of the 2009 Conference on Empirical Methods in Natural Language Processing, vol. 3, pp. 1192–1201. Association for Computational Linguistics (2009)
19. Jain, P., Hitzler, P., Sheth, A.P., Verma, K., Yeh, P.Z.: Ontology alignment for linked open data. In: Patel-Schneider, P.F., Pan, Y., Hitzler, P., Mika, P., Zhang, L., Pan, J.Z., Horrocks, I., Glimm, B. (eds.) ISWC 2010, Part I. LNCS, vol. 6496, pp. 402–417. Springer, Heidelberg (2010)
20. Jain, P., Yeh, P.Z., Verma, K., Vasquez, R.G., Damova, M., Hitzler, P., Sheth, A.P.: Contextual ontology alignment of LOD with an upper ontology: a case study with proton. In: Antoniou, G., Grobelnik, M., Simperl, E., Parsia, B., Plexousakis, D., De Leenheer, P., Pan, J. (eds.) ESWC 2011, Part I. LNCS, vol. 6643, pp. 80–92. Springer, Heidelberg (2011)
21. Jung, J.J., Håkansson, A., Hartung, R.: Indirect alignment between multilingual ontologies: a case study of Korean and Swedish ontologies. In: Håkansson, A., Nguyen, N.T., Hartung, R.L., Howlett, R.J., Jain, L.C. (eds.) KES-AMSTA 2009. LNCS, vol. 5559, pp. 233–241. Springer, Heidelberg (2009)
22. Lacher, M.S., Groh, G.: Facilitating the exchange of explicit knowledge through ontology mappings. In: Proceedings of the 14th FLAIRS, pp. 305–309. AAAI Press (2001)
23. Meilicke, C., García-Castro, R., Freitas, F., Van Hage, W.R., Montiel-Ponsoda, E., Ribeiro de Azevedo, R., Stuckenschmidt, H., Šváb-Zamazal, O., Svátek, V., et al.: Multifarm: a benchmark for multilingual ontology matching. Web Semant. Sci. Serv. Agents World Wide Web **15**, 62–68 (2012)
24. Miller, G.A.: WordNet: a lexical database for English. Commun. ACM **38**(11), 39–41 (1995)
25. Mohammad, S., Gurevych, I., Hirst, G., Zesch, T.: Cross-lingual distributional profiles of concepts for measuring semantic distance. In: EMNLP-CoNLL, pp. 571–580 (2007)
26. Paulheim, H., Hertling, S.: Wesee-match results for OAEI 2013. In: Ontology Matching, p. 197 (2013)
27. Rinser, D., Lange, D., Naumann, F.: Cross-lingual entity matching and infobox alignment in wikipedia. Inf. Syst. (2012)

28. Russell, B.C., Torralba, A., Murphy, K.P., Freeman, W.T.: LabelMe: a database and web-based tool for image annotation. IJCV **77**(1), 157–173 (2008)
29. Sabou, M., d'Aquin, M., Motta, E.: Exploring the semantic web as background knowledge for ontology matching. In: Spaccapietra, S., Pan, J.Z., Thiran, P., Halpin, T., Staab, S., Svatek, V., Shvaiko, P., Roddick, J. (eds.) Journal on Data Semantics XI. LNCS, vol. 5383, pp. 156–190. Springer, Heidelberg (2008)
30. Sanchez, E., Yamanoi, T.: Fuzzy ontologies for the semantic web. In: Larsen, H.L., Pasi, G., Ortiz-Arroyo, D., Andreasen, T., Christiansen, H. (eds.) FQAS 2006. LNCS (LNAI), vol. 4027, pp. 691–699. Springer, Heidelberg (2006)
31. Shvaiko, P., Euzenat, J.: Ontology matching: state of the art and future challenges. IEEE Knowl. Data Eng. **25**(1), 158–176 (2013)
32. Smith, J.R., Chang, S.F.: Large-scale concept ontology for multimedia. IEEE Multimedia **13**(3), 86–91 (2006)
33. Spohr, D., Cimiano, P., Hollink, L.: Multilingual and cross-lingual ontology matching and its application to financial accounting standards. In: Proceedings of ISWC (2011)
34. Straccia, U.: Towards a fuzzy description logic for the semantic web (preliminary report). In: Gómez-Pérez, A., Euzenat, J. (eds.) ESWC 2005. LNCS, vol. 3532, pp. 167–181. Springer, Heidelberg (2005)
35. Stuckenschmidt, H.: Approximate information filtering on the semantic web. In: Jarke, M., Koehler, J., Lakemeyer, G. (eds.) KI 2002. LNCS (LNAI), vol. 2479, pp. 114–128. Springer, Heidelberg (2002)
36. Suchanek, F.M., Kasneci, G., Weikum, G.: Yago: a core of semantic knowledge. In: Proceedings of the 16th International Conference on World Wide Web, pp. 697–706. ACM (2007)
37. Todorov, K., Geibel, P., Kühnberger, K.-U.: Mining concept similarities for heterogeneous ontologies. In: Perner, P. (ed.) ICDM 2010. LNCS, vol. 6171, pp. 86–100. Springer, Heidelberg (2010)
38. Xu, B., Kang, D., Lu, J., Li, Y., Jiang, J.: Mapping fuzzy concepts between fuzzy ontologies. In: Khosla, R., Howlett, R.J., Jain, L.C. (eds.) KES 2005. LNCS (LNAI), vol. 3683, pp. 199–205. Springer, Heidelberg (2005)
39. Zadeh, L.A.: Fuzzy sets. Inf. Control **8**(3), 338–353 (1965)
40. Zhang, S., Bodenreider, O.: Alignment of multiple ontologies of anatomy: deriving indirect mappings from direct mappings to a reference. In: AMIA Annual Symposium Proceedings, vol. 2005, p. 864 (2005)

Semantic Knowledge Discovery and Data-Driven Logical Reasoning from Heterogeneous Data Sources

Claudia d'Amato[1], Volha Bryl[2(✉)], and Luciano Serafini[3]

[1] Department of Computer Science, University of Bari, Bari, Italy
claudia.damato@uniba.it
[2] University of Mannheim, Mannheim, Germany
volha@informatik.uni-mannheim.de
[3] Data and Knowledge Management Unit, Fondazione Bruno Kessler, Trento, Italy
serafini@fbk.eu

Abstract. Available domain ontologies are increasing over the time. However there is still a huge amount of data stored and managed with RDBMS. This complementarity could be exploited both for discovering knowledge patterns that are not formalized within the ontology but that are learnable from the data, and for enhancing reasoning on ontologies by relying on the combination of formal domain models and the evidence coming from data. We propose a method for learning association rules from both ontologies and RDBMS in an integrated way. The extracted patterns can be used for enriching the available knowledge (in both format) and for refining existing ontologies. We also propose a method for automated reasoning on grounded knowledge bases (i.e. knowledge bases linked to RDBMS data) based on the standard Tableaux algorithm which combines logical reasoning and statistical inference thus making sense of the heterogeneous data sources.

1 Introduction

From the introduction of the Semantic Web view [3], many domain ontologies have been developed and stored in open access repositories. However, still huge amounts of data are stored in relational databases (DBs) and managed by relational database management systems (RDBMS or simply DBMS). The seamless integration of these two knowledge representation paradigms is becoming a crucial research challenge. Most of the work in this area concerns to what is addressed as *ontology based data access (OBDA)* [4]. In OBDA the ontology "replicates", at a higher conceptual level, the physical schema of the DBMS and provides a "lens" under which the data can be viewed, and possibly adds additional semantic knowledge on the data. The connection between the ontology and the data is represented as conjunctive queries. Roughly speaking, every concept/relation of the ontology is associated to a conjunctive query which retrieves from the DB all and only the instances of such a concept/relation.

© Springer International Publishing Switzerland 2014
F. Bobillo et al. (Eds.): URSW 2011-2013, LNAI 8816, pp. 163–183, 2014.
DOI: 10.1007/978-3-319-13413-0_9

Another common situation is when existing ontologies describe domain aspects that (partially) complement the data in a database. In this case the concepts of the ontologies are linkable to views of the DB and the data within the DB can be exploited for enriching/populating existing domain ontologies. Due to the heterogeneity of the information, a crisp representation of the correspondence between the DB (data) and the classes and relations of the ontologies (such as the one adopted in OBDA) is not possible. A more flexible connection between the two sources of knowledge should be adopted. Additionally, this complementarity could be exploited for (semi-)automatizing the ontology refinement and completion tasks as well as for performing data analysis.

We present a framework for extracting hidden knowledge patterns across ontologies and relational DBMS by building an integrated view of the two and by using some statistical evidence of the correlation of (a subset of) the data stored in the two sources. We exploit then the discovered hidden knowledge patterns for semi-automatizing the ontology population task and for performing an informative form of deductive reasoning.

For giving the intuition of the envisioned idea, let us consider the following scenario.

Example 1. Let us assume that an ontology describing people gender, family status and their interrelations with Italian urban areas[1] is available as well as a demographic DB describing Italian occupations, average salaries, etc. is available. Given a way for analyzing jointly the two sources with the purpose of highlighting some statistical correlation evidence, a pattern like the following one could be discovered, with a certain degree of reliability:

$$\text{``clerks between 35 and 45 years old } \textbf{living in a big city} \text{ are } \textbf{male} \text{ and} \quad (1)$$
$$\text{earn between 40 and 50 } \text{€''} \text{ with a } \textit{confidence} \text{ value of 0.75.}$$

The bold face terms correspond to classes and relations in the ontology, the non bold face terms correspond to attributes/data in the DB.

We call a pattern like this a *hybrid pattern*[2]. The *confidence* value can be interpreted as the reliability or probability that the pattern occurs.

In [7], we introduced an inductive approach for discovering new knowledge, in the form of *association rules* [1], from heterogeneous data sources that is, domain ontologies and relational data bases. We named the discovered hybrid patterns *semantically enriched association rules*. Association rule mining methods are well know in Data Mining [18]. They are generally applied to propositional data representations with the goal of discovering patterns in the data, whilst, to the best of our knowledge, there are very few works concerning the extraction of association rules from hybrid data sources.

---

[1] The concepts *"male"*, *"parent"*, *"big city"*, *"medium-sized town"*, and the relations *"lives_in"* are used in the ontology.

[2] We call patterns like that in (1) a hybrid pattern since it is composed by elements (that is concept names, role names and attributes) coming from different data sources.

In this chapter, we present extensively the revision of the approach introduced in [7] and formalized in [6]. This revision takes into account the *Open
World Assumption* adopted in description logics (DLs), while in [7] association
rules are discovered by adopting an implicit *Closed Word Assumption* which is
not fully compliant with the theory of ontological representation. Additionally,
this revision makes a further step towards the framework for knowledge representation and reasoning by allowing knowledge as represented by a mix of logical
formulas and sets of data, linked together. The notions of *grounded knowledge
base* and *mixed model* are introduced. The latter stands for the integration of
the logical knowledge expressed in terms of a DL language, and a data mining
model that expresses the statistical regularities of the properties associated to a
set of individuals. We also present an approach for (semi-)automatically enriching the existing domain ontology with (some of the) data stored in the DB. The
problem is cast as a classification problem that is solved by exploiting the discovered semantically enriched association rules and algorithms at the state of
the art. Furthermore we argue extensively on a method for automated reasoning
on *grounded knowledge bases* that we proposed in [6]. This method represents
the result of combining logical reasoning and statistical inductive inference. It is
formulated as an extension of the standard Tableaux algorithm. The extension
is grounded on the adoption of an heuristic, exploiting the evidence coming from
the data, to be used when random choices[3] have to be made. For getting the
intuition of this method the following example is briefly illustrated.

Example 2. Assume that for a given individual x, which is known to be a *Person*,
a *high school student*, and has the property of being 15 years old, we need to
decide whether x is instance of the concept *Parent* or not, and there are no
statements in the knowledge base from which it is possible to infer neither x is
a *Parent* nor x is $\neg Parent$. If the following *semantically enriched association
rule* is discovered (with high degree of confidence)

$$IF\ \textsc{Age} = [0,16]\ THEN\ \neg Parent\quad 0.99$$

it can be exploited to conclude that, with high probability, x is not a *Parent*.

The rest of the paper is structured as follows. In Sect. 2, the basics are presented.
In Sect. 3, the framework for learning semantically enriched association rules is
summarized and extended. In Sect. 4 the data-driven Tableaux reasoning algorithm is formalized. In Sect. 5 the approach for enriching a domain ontology with
the data coming from a related DB is shown. A discussion on how to evaluate
the effectiveness of the proposed framework is illustrated in Sect. 6. In Sect. 7
related approaches are discussed while conclusions are drawn in Sect. 8.

---

[3] An example of random choice within Tableaux algorithm occurs when processing
a concepts disjunction, i.e. $C \sqcup D$, that is when it has to be decided whether an
individual x belongs to concept the C or D.

2 Basics

Let $\mathbf{D}$ be a non empty set of objects and $f_1, \ldots, f_n$ be n feature functions defined on every element $\mathbf{d}$ of $\mathbf{D}$, with $f_i : \mathbf{D} \to D_i$. $\mathbf{D}$ is called the set of observed objects and $f_i(\mathbf{d})$ is the value of the i-th feature observed on $\mathbf{d}$. Notationally we use:

- $\mathbf{d}_1, \ldots, \mathbf{d}_n$ to denote the values of $f_1(\mathbf{d}), \ldots, f_n(\mathbf{d})$,
- $A_1, \ldots, A_n$ to denote the labels (also called attribute names) of $f_1(\mathbf{d}), \ldots, f_n(\mathbf{d})$.

Let Σ be a DL alphabet composed of three disjoint sets of symbols: Σ_C, Σ_R and Σ_I, respectively standing for: the set of concepts symbols, the set of role symbols and the set of individual symbols. A knowledge base $\mathcal{K}$ on Σ, is formally defined as a couple $\mathcal{K} = \langle \mathcal{T}, \mathcal{A} \rangle$ where $\mathcal{T}$ contains axioms of the form $X \sqsubseteq Y$ (inclusion axioms), where X and Y are (complex) concepts, and $\mathcal{A}$ contains assertional axioms, namely axioms of the form $X(a)$ where X is a (complex) concept and a is an individual symbol, or $R(a, b)$, and $a = b$, where R is a role symbol and a and b are individual symbols.

An interpretation of a DL alphabet Σ is a pair $\mathcal{I} = \langle \Delta_{\mathcal{I}}, \cdot^{\mathcal{I}} \rangle$ such that $\Delta^{\mathcal{I}}$ is a non empty set, and $\cdot^{\mathcal{I}}$ is a function that assigns to each concept name a subset of $\Delta_{\mathcal{I}}$, to each role name a binary relation on $\Delta_{\mathcal{I}}$, and to each individual an element of $\Delta_{\mathcal{I}}$. The interpretation function can be extended to complex concepts in the usual way [2]. Satisfiability $\models$ of statements is also defined as usual [2]. An interpretation $\mathcal{I}$ satisfies a knowledge base $\mathcal{K}$, (in symbols $\mathcal{I} \models \mathcal{K}$) if $\mathcal{I} \models \alpha$ for every axiom α of $\mathcal{K}$.

The "glue" between a dataset $\mathbf{D}$ and a knowledge base $\mathcal{K}$ is the so called *grounding*, which is a relation that connects (some of) the individuals of the knowledge base with (some of) the objects of the database. More formally: a *grounding* g of Σ on $\mathbf{D}$ is a total[4] function $g : \mathbf{D} \to \Sigma_I$. This implies that for every $\mathbf{d} \in \mathbf{D}$ there is at least an element $a \in \Sigma_I$ such that $g(\mathbf{d}) = a$. Intuitively $g(\mathbf{d}) = a$ represents the fact that the data $\mathbf{d}$ are about/correspond to the individual a of the knowledge base. The grounding g refers to objects that are explicitly mentioned in $\mathbf{D}$ and $\mathcal{K}$ respectively. In our framework (see Sect. 3) we assume that the grounding between $\mathbf{D}$ and $\mathcal{K}$ is already given.

3 Semantically Enriched Association Rules

Association rules (ARs) [1] make it possible to represent in a rule-based form some statistical regularities of the tuples in a relational DB. Roughly speaking, ARs allow one to state conditional probabilities among the values of the attributes (also called features) of the tuples of a DB. In this section we recall the basics for ARs (see Sect. 3.1) and how ARs can be extended to include information coming from an ontological knowledge base (see Sect. 3.2). These rules are called *semantically enriched* ARs [6,7]. Hence we recall the learning algorithm for discovering semantically enriched ARs (see Sect. 3.3).

---

[4] The restriction of g to the subsets of $\mathbf{D}$ and Σ_I can be considered if a mapping for all objects does not exist.

3.1   Association Rules: An Overview

Association rules [1] provide a form of rule patterns for data mining. Association rule mining methods are well know in Data Mining [18]. They are generally applied to propositional data representations with the goal of discovering patterns in the data.

Let $\mathbf{D}$ be a dataset described by a set of attributes $\{A_1, \dots, A_n\}$ with domains D_i for $i \in \{1, \dots, n\}$. The basic components of an AR for $\mathbf{D}$ are called itemsets. An itemset ϕ is a finite set of assignments of the form $A_i = a$ with $a \in D_i$. An itemset $\{A_{i_1} = a_1, \dots, A_{i_m} = a_m\}$ can be also written as: $A_{i_1} = a_1 \wedge \cdots \wedge A_{i_m} = a_m$

An AR has the general form

$$\theta \Rightarrow \varphi \tag{2}$$

where θ and φ are itemsets. The *frequency* of an itemset θ, denoted by $freq(\theta)$, is the number of cases in $\mathbf{D}$ that match θ, i.e.

$$freq(\theta) = |\{\mathbf{d} \in \mathbf{D} \mid \forall (A = a) \in \theta \ : \ f_A(\mathbf{d}) = a\}|$$

where f_A is the feature function for $\mathbf{d}$ with respect to the attribute A (see beginning of Sect. 2).

The *support* of a rule $\theta \Rightarrow \varphi$ is computed as $freq(\theta \wedge \varphi)$. The *confidence* of a rule $\theta \Rightarrow \varphi$ is the fraction of items in $\mathbf{D}$ that match φ among those matching θ:

$$conf(\theta \Rightarrow \varphi) = \frac{freq(\theta \wedge \varphi)}{freq(\theta)}$$

A frequent itemset expresses the attributes and the corresponding values that occur reasonably often together. In terms of conditional probability, the confidence of a rule $\theta \Rightarrow \varphi$, can be seen as the maximum likelihood (frequency-based) estimate of the conditional probability that φ is true given that θ is true [12].

3.2   Semantically Enriched Association Rules

Given $\mathcal{K}$ a knowledge base on Σ, $\mathbf{D}$ a dataset and g a grounding of Σ on $\mathbf{D}$, we call a *semantically enriched itemset* a set of statements of the form: $A_i = a$ and/or $C = tv$ and/or $R = tv$ where, A_i is an attribute of $\mathbf{D}$, a is a value in the domain D_i of A_i, C is a concept name of Σ_C, R is a role name of Σ_R and tv is a truth value in $\{true, false, unknown\}$.

The elements of the *semantically enriched itemset* that are of the form $A_i = a$ are called *data items* whilst the elements of the form $C = tv$ and $R = tv$ are called *knowledge items*.

A *semantically enriched* AR is an association rules made by *semantically enriched itemsets*.

Given the grounding g of Σ on $\mathbf{D}$, the *frequency* of a *semantically enriched itemset* $\theta = \theta_d \wedge \theta_k$, where θ_d stands for the set of data items and θ_k stands for the set of knowledge items, is defined as:

$$freq(\theta_d \wedge \theta_k) = |F|$$

where F is the following set:

$$F = \left\{ \mathbf{d} \in \mathbf{D} \left| \begin{array}{l} \forall (A_i = a) \in \theta_d, \ f_i(\mathbf{d}) = a \\ \forall (C = true) \in \theta_k, \ \mathcal{K} \models C(g(\mathbf{d})) \\ \forall (C = false) \in \theta_k, \ \mathcal{K} \models \neg C(g(\mathbf{d})) \\ \forall (C = unknown) \in \theta_k, \ \mathcal{K} \not\models C(g(\mathbf{d})) \ \& \ \mathcal{K} \not\models \neg C(g(\mathbf{d})) \\ \forall (R = true) \in \theta_k, \ \mathcal{K} \models \exists R.\top(g(\mathbf{d})) \\ \forall (R = false) \in \theta_k, \ \mathcal{K} \models \neg \exists R.\top(g(\mathbf{d})) \\ \forall (R = unknown) \in \theta_k, \ \mathcal{K} \not\models \exists R.\top(g(\mathbf{d})) \ \& \ \mathcal{K} \not\models \neg \exists R.\top(g(\mathbf{d})) \end{array} \right. \right\}$$

Support and *confidence* of a *semantically enriched* AR are defined accordingly. Please note that the notion of frequency of a semantically enriched itemset and the notions of confidence and support of a *semantically enriched AR* are compliant with the *Open World Semantics* adopted in DL.

3.3 Learning Semantically Enriched Association Rules

The approach for learning *semantically enriched* ARs is grounded on the assumption that a dataset $\mathbf{D}$ and an ontological knowledge base $\mathcal{K}$ share (a subset of) common individuals, and a grounding g of Σ on $\mathbf{D}$ is available (see the end of Sect. 2 for details on the grounding function). This assumption is reasonable in practice since, in the real world, there are several cases in which different information aspects concerning the same entities come from different data sources not always sharing the same conceptual model. An example is given by the public administration, where different administrative organizations have information about the same persons but concerning complementary aspects such as: personal data, income data, ownership data. Another example is given by the biological domain where research organizations have their own databases that could be complemented with existing domain ontologies which also allow performing standard deductive inferences on them [2].

In order to learn *semantically enriched* ARs from a dataset $\mathbf{D}$ and an ontological knowledge base $\mathcal{K}$ grounded by g to $\mathbf{D}$, a unique view of the two information sources is necessary. It is obtained by performing a propositionalization step as formalized in the following:

1. choose the primary entity of interest in $\mathbf{D}$ or $\mathcal{K}$ for extracting association rules and set this entity as the first attribute A_1 in the table $\mathbf{T}$ to be built; A_1 will be the primary key of the table; its values will be its corresponding values in $\mathbf{D}$ or $\mathcal{K}$
2. choose (a subset of) the attributes in $\mathbf{D}$ that are of interest for A_1 and set them as additional attributes in $\mathbf{T}$; the corresponding values are obtained as the result of a SQL query involving the selected attributes and the corresponding A_1 values
3. choose (a subset of) concept names $\{C_1, \ldots, C_m\}$ in $\mathcal{K}$ that are of interest for A_1 and set their names as additional attribute names in $\mathbf{T}$

4. for each $C_k \in \{C_1, \ldots, C_m\}$ and for each value a_i of A_1, if $\mathcal{K} \models C_k(a_i)$ then set to *true* the corresponding value of C_k in **T**, else if $\mathcal{K} \models \neg C_k(a_i)$ then set the value to *false*, otherwise set to *unknown* the corresponding value of C_k in **T**
5. choose (a subset of) role names $\{R_1, \ldots, R_t\}$ in $\mathcal{K}$ that are of interest for A_1 and set their names as additional attribute names in **T**
6. for each $R_l \in \{R_1, \ldots, R_t\}$ and for each value a_i of A_1, if $\exists y \in \mathcal{K}$ s.t. $\mathcal{K} \models R_l(a_i, y)$ then set to *true* the value of R_l in **T**, else if $\forall y \in \mathcal{K} : \mathcal{K} \models \neg R_l(a_i, y)$ then set the value of R_l in **T** to *false*, otherwise set the value of R_l in **T** to *unknown*
7. choose (a subset of) datatype property names $\{T_1, \ldots, T_v\}$ in $\mathcal{K}$ that are of interest for A_1 and set their names as additional attribute names in **T**
8. for each $T_j \in \{T_1, \ldots, T_v\}$ and for each value a_i of A_1, if $\mathcal{K} \models T_j(a_i, dataValue_j)$ then set to $dataValue_j$ the corresponding value of T_j in **T**, set *unknown* otherwise.

Clearly, for all but the datatype properties, the Open World assumption is in effect during the propositionalization step. After propositionalization, a post-processing step could be performed for the case of numerical attributes, in order to proceed, as usual in data mining, with data discretization [18] which consists in transforming single numerical values in corresponding ranges of values that could be then treated as categorical data.

An example of a unique tabular representation for the demographic domain depicted in Sect. 1 is reported in Table 1 where *Person*, *Parent*, *Male* and *Female* are concepts of an ontological knowledge base $\mathcal{K}$, and JOB and AGE are attributes of a relational dataset **D**. The numeric attribute AGE has been discretized.

Table 1. Demographic example: the unified view in the table **T**

OBJECT	JOB	AGE	*Person*	*Parent*	*Male*	*Female*
x_1	Engineer	[36, 45]	*true*	*true*	*true*	*false*
x_2	Policeman	[26, 35]	*true*	*false*	*true*	*unknown*
x_3	Student	[16, 25]	*true*	*false*	*true*	*false*
x_4	Student	[16, 25]	*true*	*false*	*false*	*true*
x_5	Housewife	[26, 35]	*true*	*true*	*false*	*true*
x_6	Clerk	[26, 35]	*true*	*false*	*unknown*	*unknown*
x_7	Primary school teacher	[46, 55]	*true*	*unknown*	*unknown*	*unknown*
x_8	Policeman	[16, 25]	*true*	*true*	*unknown*	*unknown*
x_9	Student	[16, 25]	*true*	*unknown*	*unknown*	*unknown*
...	...	...	...	...	...	...

The propositionalization step is exploited for building a uniform view (in the form of a single table) of the two different sources of information. Additionally, since a unique table is obtained, state of the art algorithms could be directly applied for learning semantically enriched ARs. Thus, the process for learning *semantically enriched* ARs, similarly to the one for learning ARs can be basically articulated in two phases. In the first phase, frequent itemsets are discovered. In the second phase, association rules are built from the discovered frequent itemsets.

For the first phase, the well known APRIORI algorithm [1] is exploited. It is grounded on the assumption that a set X of variables can be frequent only if all subsets of X are frequent. The algorithm is recalled in Algorithm 1. The first step consists in discovering all frequent sets L_1 (with respect to a support threshold) consisting of one item. Hence, the candidate sets of two items are built by joining L_1 with itself and by depurating them of the sets having a frequency that is lower than the fixed threshold, thus obtaining the sets L_2 of frequent itemsets of length 2. The process is iterated, incrementing the length of the itemsets at each step, until the set of candidate itemsets is empty. Once the set L of all frequent itemsets is determined, the second phase is performed by building the ARs as shown in Algorithm 2, given a certain confidence threshold which ensures that only significant ARs are considered while the others are discarded. The confidence value of the learnt *semantically enriched* ARs is interpreted as the conditional probability on the values of the items in the consequence of the rule given that the left hand side of the rule is satisfied in (a model of) the available knowledge. Examples of *semantically enriched* ARs that could be learned, given Table 1, are shown in Table 2.

4   Data-Driven Inference

Semantically enriched ARs (see Sect. 3.3) can be exploited when performing deductive reasoning given DL (namely ontological) representations. Since almost all DL inferences can be reduced to concept satisfiability [2], we focus on this inference procedure. For most expressive DL (such as $\mathcal{ALC}$) the Tableaux algorithm is employed. Its goal is to built a possible model, namely an interpretation, for the concept whose satisfiability has to be shown. If building such a model, all clashes (namely contradictions) are found, the model does not exist and the concept is declared to be unsatisfiable.

Algorithm 1. Sketch of Apriori algorithm.

Input: T:data table, *sp-tr*: support threshold;
Output: L frequent itemsets
   $L = \emptyset$
   $L_1 = \{$frequent itemsets of length 1$\}$
   for all (k = 1; $L_k \neq \emptyset$; k++) **do**
      $C_{k+1} =$ candidates generated by joining L_k with itself
      $L_{k+1} =$ candidates in C_{k+1} with frequency equal or greater than *sp-tr*
      $L = L \cup L_{k+1}$
   return L

Table 2. Demographic example: association rules

#	RULE	Confidence
1	$(\text{AGE} = [16, 25]) \wedge (\text{JOB} = Student) \Rightarrow (Parent = false)$	0.98
2	$(\text{JOB} = Policeman) \Rightarrow (Male = true)$	0.75
3	$(\text{AGE} = [16, 25]) \wedge (Parent = true) \Rightarrow (Female = true)$	0.75
4	$(\text{JOB} = Primary\ school\ teacher) \Rightarrow (Female = true)$	0.78
5	$(\text{JOB} = Housewife) \wedge (\text{AGE} = [26, 35]) \Rightarrow (Parent = true) \wedge (Female = true)$	0.85

In this section we set up a modified version of the Tableaux algorithm, representing the result of combining logical reasoning and statistical inductive inference. It is grounded on the adoption of an heuristic exploiting the evidence coming from the data in the form of semantically enriched ARs. The output of such an algorithm, if any, is the **most plausible model**, namely the model that best fits the available data. This means to set up a data driven heuristic that should allow to reduce the computational effort in finding a model for a given concept and should be also able to supply the model that is most coherent with/match the available knowledge. In this way also the *variance due to intended diversity and the variance due to incomplete knowledge*, that is the number of possible models that could be built (see [11] for formal definitions), is reduced.

The inference problem we want to solve is formally defined as follows:

Definition 1 (Inference Problem).

Given: *the following*
  - **D**, $\mathcal{K}$, *the set R of semantically enriched ARs,*
  - *a (possibly complex) concept E of $\mathcal{K}$,*
  - *the individuals $x_1, \ldots, x_k \in \mathcal{K}$ that are instances of E,*
  - *the grounding g of Σ on* **D**

Determine: *the model $\mathcal{I}_r$ for E representing the* **most plausible model** *given the $\mathcal{K}$,* **D**, *g and R.*

Algorithm 2. Building Association Rules given the discovered frequent itemsets.

Input: L: frequent itemsets, *min-conf*: minimum confidence threshold
Output: R learnt association rules
  $R = \emptyset$
  for all $(I \in L)$ **do**
    for all $((S \subset I)\ \text{AND}\ (S \neq \emptyset))$ **do**
      $r := S \Rightarrow (I - S)$ {//build the rule}
      if $((\frac{fr(I)}{fr(S)} \geq min\text{-}conf)$ **then**
        $R := R \cup r$
  return R

Intuitively, the most plausible model $\mathcal{I}_r$ for E is the one on top of the ranking of the possible models $\mathcal{I}_i$ for E. Such a ranking is built according to the degree up to which the models are compliant with the set R of ARs and $\mathcal{K}$. The detailed procedure for building the *most plausible model* is illustrated in the following.

Here we first briefly recall the standard Tableaux algorithm. In order to find (or not) a model, the standard Tableaux algorithm exploits a set of transformation rules that are applied to the considered concept. Specifically, a transformation rule for each constructor of the considered language exists. In the following, the transformation rules for $\mathcal{ALC}$ logic are briefly recalled (see [2] for more details).

⊓-rule: IF the ABox $\mathcal{A}$ contains $(C_1 \sqcap C_2)(x)$, but it does not contain both $C_1(x)$ and $C_2(x)$ THEN $\mathcal{A} = \mathcal{A} \cup \{C_1(x), C_2(x)\}$

⊔-rule: IF $\mathcal{A}$ contains $(C_1 \sqcup C_2)(x)$, but it does not contain neither $C_1(x)$ nor $C_2(x)$ THEN $\mathcal{A}_1 = \mathcal{A} \cup \{C_1(x)\}$, $\mathcal{A}_2 = \mathcal{A} \cup \{C_2(x)\}$

∃-rule: IF $\mathcal{A}$ contains $(\exists R.C)(x)$, but there is no individual name z s.t. $C(z)$ and $R(x, z)$ are in $\mathcal{A}$ THEN $\mathcal{A} = \mathcal{A} \cup \{C(y), R(x, y)\}$ where y is an individual name not occurring in $\mathcal{A}$.

∀-rule: IF $\mathcal{A}$ contains $(\forall R.C)(x)$ and $R(x, y)$, but it does not contain $C(y)$ THEN $\mathcal{A} = \mathcal{A} \cup \{C(y)\}$

To test the satisfiability of a concept E, the algorithm starts with considering the ABox $\mathcal{A} = \{E(x_0)\}$ (with x_0 being a new individual) and applies to the ABox the consistency preserving transformation rules reported above until no more rules apply. The result could be all clashes, which means the concept is unsatisfiable, or an ABox containing a model for the concept E that means the concept is satisfiable.

The transformation rule for the disjunction (⊔-rule) is non-deterministic. Specifically, if a disjunctive concept has to be processed, two (or more depending on the number of disjuncts) different ABoxes have to be considered, each one containing the assertion concerning one of the two disjoint concepts. The original ABox is consistent if and only if one of the new ABoxes is consistent. The choice on one of the two (or more) ABoxes to be processed is absolutely non-deterministic. In order to save the computational complexity, the ideal solution (for the case of a consistent concept) should be to choose directly the ABox containing a model.

Moving from this observation, we propose an alternative version of the Tableaux algorithm. The main differences with respect to the standard Tableaux algorithm are:

1. the starting model for the inference process is given by the set of all attributes (and corresponding values) of the unified tabular representation (see Sect. 3.3) that are related to the individuals $x_1, \ldots, x_k$ that are instances of E, differently from the standard Tableaux algorithm where the initial model is simply given by the assertion concerning the concept of which the satisfiability (or unsatisfiability) has to be shown;

2. a heuristic is adopted for performing the $\sqcup$-rule, differently from the standard case where no heuristic is used and the choice on the new ABox to be considered is arbitrary;
3. the most plausible model for the concept E and the individuals $x_1, \ldots, x_k$ is built with respect to the available knowledge $\mathcal{K}$, $\mathbf{D}$ and R. The obtained model is a *mixed model*, namely a model containing both information from R and $\mathcal{K}$. Differently, in the standard Tableaux algorithm the model that is built refers only to $\mathcal{K}$ and does not take into account the (assertional) available knowledge.

In the following these three characteristics are analyzed and the way to accomplish each of them is illustrated.

Firstly, the process for building the starting model $\mathcal{I}_r$ is illustrated. For each individual $x_i \in \{x_1, \ldots, x_k\}$ that is instance of E, all attribute names A_i in the unified tabular representation $\mathbf{T}$ (see Sect. 3.3) that related to x_i are selected jointly with the corresponding values a_i. The assertions $A_i(a_i)$ are added to $\mathcal{I}_r$. For simplicity and without loss of generality, a single individual x will be considered in the following. The generalization to multiple individuals is straightforward. It is sufficient to applying the same procedure to all individuals that are (or assumed to be) instances of the considered concept.

Once the initial model $\mathcal{I}_r$ is built, all deterministic expansion rules, namely all but $\sqcup$-rule, are applied to $\mathcal{I}_r$ following the standard Tableaux algorithm. For the case of the $\sqcup$-rule, a heuristic is adopted. The goal of such a heuristic is twofold: (a) choosing a new consistent ABox in almost one step, for saving the computational complexity, if $E(x)$ is consistent (see discussion above concerning the $\sqcup$-rule); (b) driving the construction of the most plausible model given $\mathcal{K}$ and R. The heuristic is determined as follows.

Let $C \sqcup D$ be the disjunctive concept to be processed by $\sqcup$-rule. The choice on C rather than D (or vice versa) will be driven by the following procedure[5].

- The ARs in R containing C (resp. D) or its negation in the *knowledge items* of the right hand side of the rules are selected.
- For each selected rule, the left hand side is considered and the degree of match between the left hand side and the model under construction $\mathcal{I}_r$, is computed. Specifically, the degree of match is calculated by counting the number of (both data and semantic) items in the left hand side of a rule that are contained in $\mathcal{I}_r$, and averaging this number with respect to the length of the left hand side of the rule. Items with uncertain (*unknown*) values are not considered for assessing the degree of match. The degree of match for the rules whose (part of the) left hand side is contradictory with respect to the model $\mathcal{I}_r$ is set to 0. Details for the matching procedure are reported in Algorithm 3.
- Rules having 0 as a degree of match are discarded.
- For each of the remaining rules the weighted confidence value is computed as $weightedConf = ruleConfidence * degreeOfMatch$.

---

[5] The case of a negated concept within the disjunction is treated similarily.

- Rules that have the weighted confidence value below a given threshold are discarded.
- The rule having the highest weighted confidence value is selected. In case of multiple rules having the same weighted confidence value, a random choice is performed.
- If the chosen rule contains $C = true$ (resp. $D = true$) in the right hand side, the model under construction $\mathcal{I}_r$ is enriched with $C(x)$ (resp. $D(x)$), where x is the individual under consideration.
- If the chosen rule contains $C = false$ (resp. $D = false$) in the right hand side, the model under construction $\mathcal{I}_r$ is enriched with $D(x)$ (resp. $C(x)$).
- In the general case, the right hand side of the selected AR may contain additional items besides that involving C or D. Assertions concerning such additional items will be also added in $\mathcal{I}_r$ accordingly[6].
- If no rules are available for one of the two concepts, e.g. concept D, the concept for which some evidence, via existing rules, is available, i.e. C, will be chosen for expanding $\mathcal{I}_r$.

If no rule in R contains C (resp. D) or its negation in the right hand side, the following approach may be adopted. Given $\mathcal{I}_r$, a corresponding itemset is created by transforming:

- each assertion $A_i(a_i)$ referring to an attribute in $\mathbf{D}$ into a data item $A_i = a_i$;
- each positive (not negated) concept/role assertion into a knowledge item *concept/role name = true*;
- each negative assertion into a knowledge item *concept/role name = false*.

Let θ be the conventional name of such a built itemset, hence four rules are created, namely: (1) $\theta \Rightarrow (C = true)$, (2) $\theta \Rightarrow (C = false)$, (3) $\theta \Rightarrow (D = true)$ and (4) $\theta \Rightarrow (D = false)$, and their confidence value is computed (see Sect. 3.2). Then, the rule having the highest confidence value (satisfying a given confidence threshold) is selected and the corresponding right hand side will be used as a guideline for expanding $\mathcal{I}_r$.

This solution may yield a rule with a lower reliability (the created items do not necessarily refer to the same individual/object) and it is also computationally expensive. As an alternative, the prior probability of C (resp. D) could be used. It is computed by adopting a frequency-based approach as: $P(C) = |ext(C)|/|\mathcal{A}|$ where $ext(C)$ is the extension of the concept C, namely the number of individuals that are instances (asserted or derived) of C and $|\cdot|$ returns the cardinality of the set extension. $P(D)$ is computed similarly. The concept to be chosen for extending $\mathcal{I}_r$ is the one having the highest prior probability.

In the cases discussed above, the disjunctive expression is assumed to be made by atomic concept names. However, in $\mathcal{ALC}$, complex concepts may occur as part of a disjunctive expression such as: existential concept restrictions

---

[6] If a most conservative behavior of the heuristic has to be considered, only the assertion concerning the disjunct C (resp. D) is added to $\mathcal{I}_r$ while the additional items in the right hand side of the selected rule are not taken into account.

Algorithm 3. Sketch of the Matching Algorithm.

Input: r:ass-rule, $\mathcal{I}_r$:model under construction
Output: degreeOfMatch:real
  let $r := \theta \Rightarrow \varphi$
  matchFound := 0
  lenghtLeft := number of items in θ
  for all (item $(A_i = a_i)$ in θ) **do**
    if the item is a *data item* **then**
      build an assertion $ass_i := A_i(a_i)$
      if $ass_i \in \mathcal{I}_r$ **then**
        matchFound = matchFound + 1
    else if the item is a *knowledge item* **then**
      if (A_i is a concept name) **then**
        if ($a_i = true$) **then**
          if there exists an assertion $A_i(x) \in \mathcal{I}_r$ with x arbitrary individual **then**
            matchFound = matchFound + 1
        else if ($a_i = false$) **then**
          if there exists an assertion $\neg A_i(x) \in \mathcal{I}_r$ with x arbitrary individual **then**
            matchFound = matchFound + 1
      else if (A_i is a role name) **then**
        if ($a_i = true$) **then**
          if there exists an assertion $A_i(x, y) \in \mathcal{I}_r$ with x, y arbitrary individuals
          then
            matchFound = matchFound + 1
        else if ($a_i = false$) **then**
          if there exists an assertion $\neg A_i(x, y) \in \mathcal{I}_r$ with x, y arbitrary individuals
          then
            matchFound = matchFound + 1
      else if (A_i is a data type property name) **then**
        if there exists an assertion $A_i(x, a_i) \in \mathcal{I}_r$ with x arbitrary individual **then**
          matchFound = matchFound + 1
  degreeOfMatch := matchFound / lenghtLeft
  return degreeOfMatch

(i.e. $\exists R.A \sqcup \exists R.B$), universal concept restrictions (i.e. $\forall R.A \sqcup \forall S.B$), nested concept expression (i.e. $\exists R.\exists S.A$ or $\exists R.(A \sqcap B)$). To cope with these cases a straightforward solution could be to create new artificial concept names for naming the complex concepts so that a disjunction of atomic concept names is finally obtained. These new artificial concept names have to be added in the unified tabular representation **T** jointly with their corresponding values (see Sect. 3.3) and the process for discovering ARs has to be run (see Sect. 3.3). This is because potentially useful ARs for treating the disjuncts may be found. It is important to note that the artificial concept names are not used for discovering new knowledge in itself (as illustrated in Sect. 3.3) but only for the reasoning purpose.

In the following we present an example showing how the modified Tableaux algorithm works when random choices occur and namely when processing a $\sqcup$-rule.

Table 3. Demographic example: data given at the inference stage

OBJECT	JOB	AGE	Parent	Male	Female
x_7	Primary school teacher	[46, 55]	unknown	unknown	unknown
x_8	Policeman	[16, 25]	true	unknown	unknown
x_9	Student	[16, 25]	unknown	unknown	unknown

Example 3. Let us consider: (a) the demographic domain introduced in Sect. 1 and Sect. 3.3; (b) the extracted semantically enriched ARs reported in Table 2 and (c) the model $\mathcal{I}_r$ under construction for the inference procedure, as reported in Table 3. Let us assume that the reasoning process needs to evaluate the expansion of $(Male \sqcup Female)(x)$ with respect to the model $\mathcal{I}_r$ under construction reported in Table 3. The heuristic illustrated above is applied as follows.

- As a first step, the rules having *Male* (resp. *Female*) in the right hand side are selected. Looking at Table 2, the selected rules are hence r_2, r_3, r_4 and r_5.
- For each of the selected rule, the degree of match is computed as follows:
  - r_2: matchFound = 1 (because of the item JOB = Policemen in Table 3 (for x_8)) ⇒ degreeOfMatch = 1 (note that lengthLeft = 1 since the left hand side of r_2 is made by a single item, see Algorithm 3 for details)
  - r_3: matchFound = 2 (because of the items AGE = [16, 25] and *Parent* = *True* (for x_8) in Table 3) ⇒ degreeOfMatch = 2 (note that lengthLeft = 2 since the left hand side of r_3 is made by two items)
  - r_4: matchFound = 1 (because of the item JOB = PrimarySchoolTeacher (for x_7) in Table 3) ⇒ degreeOfMatch = 1 (note that lengthLeft = 1 since the left hand side of r_4 is made by one item)
  - r_5: matchFound = 0 (because no item in Table 3 matches the left hand side of r_5 ⇒ degreeOfMatch = 0 (note that lengthLeft = 2 since the left hand side of r_5 is made by two items)
- Rules having a null degree of match are discarded, hence, for the particular case r_5 is descarded
- For each of the remaining rules, the weighted confidence value is computed (for the rule confidence values see Table 2):
  - r_2: $weightedConf = ruleConfidence * degreeOfMatch = 0,75 * 1$
  - r_3: $weightedConf = 0.75 * 1 = 0.75$
  - r_4: $weightedConf = 0.78 * 1 = 0.78$
- Rules having a weighted confidence value lower than a given threshold are discarded. Provided that the threshold for the weighted confidence value is 0.5, none of the above rules is discarded
- The rule having the highest weighted confidence value is selected, for the specific case r_4 is selected
- Since the right hand side of r_4 is the concept *Female*, the model under construction $\mathcal{I}_r$ is enriched with the assertion $Female(x)$ (where x is the individual under consideration) and this enriched model is the one to be considered for the application of the successive expansion rules, until the stopping criterion is met.

Since processing a $\sqcup$-rule, assertions coming from the evidence of the available knowledge are always added, the proposed approach should ensure that the resulting model is the one that is mostly compliant with the statistical regularities learned from data.

5   Data-Driven Ontology Population

Semantically enriched ARs (see Sect. 3.3) can be exploited for (semi-)automatically enriching the considered domain ontology. The problem is cast as a classification problem that is solved by regarding the discovered semantically enriched ARs as classification rules to be used by a rule-based classifier. The problem we want to solve is formally defined as follows:

Definition 2 (Approximate concept membership).

Given: *the following*

- **D**, $\mathcal{K}$, *the set R of semantically enriched ARs,*
- *the grounding g of Σ on* **D**
- *the unified tabular representation* **T**
- *a tuple $t_j \in$* **T**

Determine: *the set of concepts in $\mathcal{K}$ of which (some of) the attribute values $f_i(t_j)$ for $i \in \{1, \ldots, n\}$ is instance of*

This problem can be easily generalized to the case of multiple tuples by repeating the same problem solution for each tuple. The assumption underlying this problem is that the knowledge base $\mathcal{K}$ is assumed to be stable over the time, specifically equality and inclusion axioms are assumed to be stable over the time, while the assertional knowledge is evolving over the time with the addition of new facts. This is what generally happens for DBs, where once the schema is defined, it is generally considered stable while new data, in terms of new tuple are added or deleted from the DB. Here we consider the monotonic scenario where new tuples are basically added.

A rule-based classifier is a technique for classifying tuples/objects using a collection of "if...then" rules. The set of the rules that are used for the classification is called rule set. Rules within the rule set are assumed to be disjoint [18]. The left hand side of a classification rule is made by the conjunction of one or more attribute tests while the right hand side is made by a unique element representing the predicted class.

It is straightforward to notice the similarity between our semantically enriched ARs and the classification rules described above, where our classes are given by (some of) the concepts within the considered ontology. Differently from standard classification rules, our ARs may have more than a single item in the right hand side. We consider all the items in the right hand side of our ARs as multiple classes with respect to which the classification is performed, similarly to a multi-class classifier.

A classification rule r_c is said to *cover* a tuple in a table if the precondition (the left hand side of the rule) matches the attributes of the tuple. r_c is also said to be *fired* or *triggered* whenever it *covers* a given tuple. A rule based classifier classifies a test tuple based on the rule triggered by the tuple. Ideally a rule set should have rules that are: (a) *mutually exclusive*, that is no more than one rule is triggered by a tuple, and (b) *exhaustive*, that is there is a rule for each combination of the attribute values. Together, these properties ensure that every tuple is covered exactly by one rule. However in practice, most of the rule-based classifiers do not have such properties. This is also valid for the case of our classification rules for which these two properties are not necessarily satisfied.

Since our purpose is not classifying individuals/objects with respect to all possible classes, namely concepts in $\mathcal{K}$, rather we are only interested in completing the available knowledge by exploiting the evidence of the data represented by the extracted semantically enriched ARs, we can safely discard the *exhaustive* property of a rule set.

However, if a rule set is not made by mutually exclusive classification rules, it may happen that the same tuple triggers more than one rule, possibly also conflicting. In order to cope with this problem usually two different approaches are adopted. One consists in considering the rule set as an ordered (in decreasing priority order) list (also called decision list) with respect to a given metric (e.g accuracy, coverage, total description length). The other solution consists in applying a majority voting mechanism to the classes assessed by the fired rules. Since the semantically enriched ARs may have more that one item in the rule consequent (making the application of the majority vote not straightforward) we adopt the ordered rule approach. Specifically, our rules are ordered in decreasing order with respect to their confidence value.

By applying a rule-based classifier (whose classification rules are the semantically enriched ARs) to the tuples in the unified tabular representation **T** (built as described in Sect. 3.3), new knowledge can be induced as illustrated in the following example.

Example 4. Rule r_4 in Table 2 is fired by the tuple corresponding to x_7 in Table 1 for which it is possible to induce that x_7 is instance of the concept *Female*.

The induced new knowledge could be possibly validated before actually adding it within the ontology.

Please note that the rule-based classifier could be also used for classifying individuals with respect to the attributes in DB. However, this could led to introduce tuples with several missing values, so we exploit this approach only for enriching the ontology. This specifically implies that, at the current stage, *data items* in the right hand side of a semantically enriched ARs are ignored for the classification step.

The presented approach could be also used for classifying new tuples occurring in the DB that are not known in the considered ontology. In order to do this, firstly the grounding g of Σ on **D** has to be updated in order to include in $\mathcal{K}$ the new objects of **D**. Indeed, the grounding g generally implements a correspondence between two (or more) entities (attribute-concept and/or attribute value-individual) of $\mathbf{D}$ and Σ, consequently the new values of an attribute may be added as instances of the corresponding concept in $\mathcal{K}$. Consequently, new tuples with respect to the new objects in $\mathbf{D}$ are added in the unified tabular representation $\mathbf{T}$ (see Sect. 3.3) and the rule-based classifier is applied for these new tuples. Since the absence of the new objects in the ontology could provoke several missing (unknown) information for the new tuples inserted in $\mathbf{T}$, this could have as a consequence the reduction of the number of rules that can be fired (basically rules containing only data items in the left hand side could be fired). In order to cope with this problem a relaxed matching process could be applied for the classification phase. Specifically, the match can be computed only with respect to data items in the left hand side of the rules, hence a degree of match between the left hand side of the rule and the new tuple is computed, similarly to the computation of the degree of match in Algorithm 3. The ordered decision list is dynamically rebuilt with respect to the computed degree of match.

6   Towards the Method Evaluation

Because on the lack of existing freely available complementary data that are stored on relational DB and ontologies, an actual empirical evaluation of the proposed approach is hard to be performed at this stage. In the following we sketch the procedure that we envision and intend to follow for evaluating the proposed solution, as soon proper data are available[7]. The experimental evaluation should focuses on:

- assessing the validity of the proposed data-driven Tableaux algorithm when compared to the standard Tableaux algorithm
- assessing the validity of the data-driven ontology population procedure that is grounded on a rule-based classifier

As regards the first point, an essential aspect to be considered is the number of ABox expansions that are performed by the data-driven Tableaux algorithm when compared to the standard Tableaux algorithm. Indeed, since the heuristic that is adopted by the data-driven Tableaux algorithm aims to reduce the number of ABox expansions, the experimental evaluation should empirically show that the heuristic is able to significantly decreases the number of ABox expansion when the consistency of a (possibly complex) disjoint concept, that is already known to be consistent, is performed. In order to strengthen the result, a statistical significance test should be performed and the test should be applied on a number of different disjunctive concept descriptions (possibly of increasing complexity). Additionally, since the data-driven Tableaux algorithm requires some additional computations (for instance of computing the degree of match) with respect to the standard Tableaux algorithm, a comparison of the two in terms of execution time has to be performed. How to formally assess that the model computed by the data-driven Tableaux algorithm is actually the one that is mostly

---

[7] We are working for building an heterogeneous dataset for empirically evaluating the proposed approach.

compliant with the statistical regularities that are learnt from the data, namely the most reliable model, is still an open issue.

As regards the evaluation of the data-driven ontology population procedure, different aspects could be evaluated. One is the ability of the classifier to classify instances with respect to one or more concepts[8]. At this regards the usual setting proposed in [5,8] could be adopted, where the performances of an inductive classifier are compared with the performances of a standard deductive reasoner and the mistakes, correct predicted class-memberships and the new induced class-memberships are measured. Particularly important is the ability of the inductive classifier of inducing new knowledge that is not logically derivable. Notably, this new knowledge should be validated by a domain expert in order to assess its actual validity.

Even if other inductive classifiers for assessing the class-membership of individuals in an ontological knowledge base have been proposed in the literature (see [15] for a survey), they cannot be used for comparing the performance our developed data-driven ontology population procedure since, differently from the methods at the state of the art, the method proposed in this chapter exploits, besides of an ontological knowledge base, a relational DB as an additional source of information. Instead, an aspect that could be interesting to evaluate is the ability of our data-driven ontology population procedure to induce new knowledge, when compared with other inductive classifiers at the state of the art, with the final goal to show that exploiting a hybrid source of information actually help to induce a larger (and/or more accurate) amount of new knowledge.

7 Related Works

In this paper we make a step towards the framework for knowledge representation and reasoning in which knowledge is specified by a mix of logical formulas and data linked together. Differently from [16], where federated DBs are considered with the goal of removing structural conflicts automatically while maintaining unchanged the views of the different DBs, we focus on building a new knowledge base that is able to collect the complementary knowledge that is contained in heterogeneous sources of information. We also proposed a method for data-driven logical reasoning, which is the result of combining logical reasoning and data mining methods embedded in a Tableaux algorithm. Differently from [9], where an integrated system ($\mathcal{AL}$-log) for knowledge representation based on DL and the deductive database language Datalog is presented, here purely relational databases are considered. Additionally, while in [9] a method for performing query answering based on constrained resolution is proposed, where the usual deduction procedure defined for Datalog is integrated with a method for reasoning on the structural knowledge, here a more expressive DL is considered and *semantically enriched* ARs are exploited. The reason for which

---

[8] Concept names of the considered ontology could be considered as well as new query concepts that are built starting (by using the constructors of the chosen DL language [2]) from the concept names that are formalized in the ontology.

the extracted association rules are adopted for driving the construction of the model performed by the Tableaux algorithm is that association rules (supported by a certain degree of confidence) represent meaningful and recurrent pattern in the data. Hence association rules containing one of the two disjunctive concepts in the right hand side represent a form of evidence coming from the available data and as such a suitable heuristic for building the most reliable model, namely the model that best fit the available heterogeneous knowledge.

To the best of our knowledge there are very few works concerning the extraction of ARs from hybrid sources of information. The one closest to ours is [14], where a hybrid source of information is considered, specifically an ontology and a constrained DATALOG program. Association rules at different granularity levels (w.r.t. the ontology) are extracted, given a specified query involving both the ontology and the DATALOG program. A similar approach has been adopted in [13] for discovering ARs from expressive ontologies. In our framework, no query is specified. A collection of data is built and all possible patterns (possibly across the DB and the ontology) are learnt. Furthermore, some restrictions are required in [14], i.e. the set of DATALOG predicate symbols has to be disjoint from the set of concept and role symbols appearing in the ontology. In our framework, such restrictions are not required. Additionally, [14] assumes that the alphabet of constants in the DATALOG program coincides with the alphabet of the individuals in the ontology. In our case a partial overlap of the constants would be sufficient.

Other usages of the ARs in the SW context have been presented in [10,17]. In [17] association rules are extracted from RDF-data collections for learning useful axioms to be exploited for building/enriching the related ontology schema. Differently from this work, we cope with heterogeneous sources of information and extract rules to be used for enriching an ontology at the instance level. At this regards, a related approach is the one proposed in [10] where association rules are extracted from RDF data collections for performing instance/resource classification. Differently from this approach, we cope with heterogeneous data sources and besides of exploiting the extracted association rules for classifying individuals we also exploit association rules for an empowered, data driven, logical reasoning.

8   Conclusions

This chapter summarized the proposal of a framework for knowledge representation and reasoning in which knowledge is specified by a mix of logical formulas and data linked together. We focused on building a new knowledge base that is able to collect the complementary knowledge that is contained in heterogeneous sources of information and specifically in a relational DB and a domain ontology.

Eventually, we proposed a method for data-driven logical reasoning, which is the result of combining logical reasoning and data mining methods embedded in a Tableaux algorithm. The method is intended to improve the effectiveness of the Tableaux algorithm when random choices, such as processing a disjointness axiom, have to be processed.

In order to accomplish with this goal, we exploited the *semantically enriched ARs* extracted from the heterogeneous sources of information "packed" in a unique tabular representation. The output of the extended Tableaux algorithm is the so called *most reliable model* that is the model, if any, most close to the evidence coming from the available knowledge.

We also exploited the extracted semantically enriched ARs for enriching the available domain ontology at the instance level. We regarded the problem as a classification problem and exploited the extracted semantically enriched ARs as classification rules to be used by a rule-based classifier.

The proposed framework has to be intended as the backbone of a mixed models representation and reasoning. Several improvements and open issues need still to be focused on. Firstly, in building the unified tabular representation, currently concepts and roles are managed without considering any inclusion relationships potentially existing among them. The explicit treatment of this information could save computational costs and could also avoid the extraction of redundant association rules. Additionally, at the current stage we focused on the instance prediction problem with respect to the concepts of the ontology. It would be of interest to be able to focus also on the membership of the individuals with respect to the roles of the ontology. An aspect that has not been treated extensively concerns with the potential missing values in the original DB and their impact on the presented framework. As regards the classification phase, a decision list has been considered for the classification process. It would be interesting to use an unordered list as an alternative for this phase. An additional improvement could be to apply the algorithm for learning association rules directly on a relational representation, without building an intermediate propositional representation.

References

1. Agrawal, R., Imielinski, T., Swami, A.N.: Mining association rules between sets of items in large databases. In: Buneman, P., Jajodia, S. (eds.) Proceedings of the 1993 ACM SIGMOD International Conference on Management of Data, pp. 207–216. ACM Press (1993)
2. Baader, F., Calvanese, D., McGuinness, D.L., Nardi, D., Patel-Schneider, P.F. (eds.): The Description Logic Handbook: Theory, Implementation, and Applications. Cambridge University Press, New York (2003)
3. Berners-Lee, T., Hendler, J., Lassila, O.: The Semantic Web. Scientific American, Singapore (2001)
4. Calvanese, D., De Giacomo, G., Lembo, D., Lenzerini, M., Poggi, A., Rodriguez-Muro, M., Rosati, R., Ruzzi, M., Savo, D.F.: The mastro system for ontology-based data access. Semant. Web J. **2**(1), 43–53 (2011)
5. d'Amato, C.: Similarity-based learning methods for the semantic web. Ph.D. Thesis (2007)
6. d'Amato, C., Bryl, V., Serafini, L.: Data-driven logical reasoning. In: Bobillo, F., et al. (eds.) Proceedings of International Workshop on Uncertainty Reasoning for the Semantic Web (URSW 2012). CEUR-Workshop Proceedings, pp. 51–62. CEUR (2012)

7. d'Amato, C., Bryl, V., Serafini, L.: Semantic knowledge discovery from heterogeneous data sources. In: ten Teije, A., Völker, J., Handschuh, S., Stuckenschmidt, H., d'Acquin, M., Nikolov, A., Aussenac-Gilles, N., Hernandez, N. (eds.) EKAW 2012. LNCS, vol. 7603, pp. 26–31. Springer, Heidelberg (2012)
8. d'Amato, C., Fanizzi, N., Esposito, F.: Query answering and ontology population: an inductive approach. In: Bechhofer, S., Hauswirth, M., Hoffmann, J., Koubarakis, M. (eds.) ESWC 2008. LNCS, vol. 5021, pp. 288–302. Springer, Heidelberg (2008)
9. Donini, F.M., Lenzerini, M., Nardi, D., Schaerf, A.: $\mathcal{AL}$-log: integrating datalog and description logics. J. Intell. Inf. Syst. **10**, 227–252 (1998)
10. Galárraga, L., Teflioudi, C., Suchanek, F., Hose, K.: Amie: association rule mining under incomplete evidence in ontological knowledge bases. In: Proceedings of the 20th International World Wide Web Conference (WWW 2013). ACM (2013)
11. Grimm, S., Motik, B., Preist, C.: Variance in e-business service discovery. In: Proceedings of the ISWC Workshop on Semantic Web Services (2004)
12. Hand, D.J., Smyth, P., Mannila, H.: Principles of Data Mining. MIT Press, Cambridge (2001)
13. Józefowska, J., Lawrynowicz, A., Lukaszewski, T.: The role of semantics in mining frequent patterns from knowledge bases in description logics with rules. Theory Pract. Logic Program. **10**(3), 251–289 (2010)
14. Lisi, F.A.: Al-quin: an onto-relational learning system for semantic web mining. Int. J. Semant. Web Inf. Syst. **7**, 1–22 (2011)
15. Rettinger, A., Lösch, U., Tresp, V., d'Amato, C., Fanizzi, N.: Mining the semantic web: statistical learning for next generation knowledge bases. Data Min. Knowl. Discov. **24**(3), 613–662 (2012)
16. Spaccapietra, S., Parent, C.: View integration: a step forward in solving structural conflicts. IEEE Trans. Knowl. Data Eng. **6**(2), 258–274 (1994)
17. Völker, J., Niepert, M.: Statistical schema induction. In: Antoniou, G., Grobelnik, M., Simperl, E., Parsia, B., Plexousakis, D., De Leenheer, P., Pan, J. (eds.) ESWC 2011, Part I. LNCS, vol. 6643, pp. 124–138. Springer, Heidelberg (2011)
18. Witten, I., Frank, E., Hall, M.A.: Data Mining: Practical Machine Learning Tools and Techniques, 3rd edn. Morgan Kaufman, San Francisco (2011)

Learning Probabilistic Description Logic Concepts Under Alternative Assumptions on Incompleteness

Pasquale Minervini[(✉)], Claudia d'Amato, Nicola Fanizzi,
and Floriana Esposito

LACAM Laboratory, Dipartimento di Informatica, Università Degli Studi di Bari
Aldo Moro, via E. Orabona, 4, 70125 Bari, Italy
{pasquale.minervini,claudia.damato,nicola.fanizzi,
floriana.esposito}@uniba.it

Abstract. Real-world knowledge often involves various degrees of uncertainty. For such a reason, in the Semantic Web context, difficulties arise when modeling real-world domains using only purely logical formalisms. Alternative approaches almost always assume the availability of probabilistically-enriched knowledge, while this is hardly known in advance. In addition, purely deductive exact inference may be infeasible for Web-scale ontological knowledge bases, and does not exploit statistical regularities in data. Approximate deductive and inductive inferences were proposed to alleviate such problems. This article proposes casting the concept-membership prediction problem (predicting whether an individual in a Description Logic knowledge base is a member of a concept) as estimating a conditional probability distribution which models the posterior probability of the aforementioned individual's concept-membership given the knowledge that can be entailed from the knowledge base regarding the individual. Specifically, we model such posterior probability distribution as a generative, discriminatively structured, Bayesian network, using the individual's concept-membership w.r.t. a set of *feature* concepts standing for the available knowledge about such individual.

1 Introduction

Real-world knowledge often involves various degrees of uncertainty. For such a reason, in the context of Semantic Web (SW), difficulties arise when trying to model real-world domains using purely logical formalisms. For this purpose, the Uncertainty Reasoning for the World Wide Web Incubator Group[1] (URW3-XG) identified the requirements for representing and reasoning with uncertain knowledge in the SW context, and provided a number of use cases showing the clear need for explicitly representing and reasoning in presence of uncertainty [23]. As a consequence, several approaches, particularly focusing on enriching knowledge bases and inference procedures with probabilistic information has been proposed.

---

[1] http://www.w3.org/2005/Incubator/urw3/

© Springer International Publishing Switzerland 2014
F. Bobillo et al. (Eds.): URSW 2011-2013, LNAI 8816, pp. 184–201, 2014.
DOI: 10.1007/978-3-319-13413-0_10

Some approaches extend knowledge representation formalisms actually used in the SW (such as [7]), while others rely on probabilistic enrichment of Description Logics [1] (DLs) or logic programming formalisms (such as [28]).

Uncertainty is pervasive in real-world knowledge, but it is often hard to elicit it on both the logical and the probabilistic side. Machine Learning (ML) methods have been proposed to overcome several potential limitations of purely deductive reasoning and ontology engineering [9,19,34]. These limitations are inherent to (i) the difficulty of engineering knowledge bases in expressive SW formalisms, (ii) taking regularities in data into account, (iii) performing approximate reasoning on Web-scale SW knowledge bases, and (iv) reasoning in presence of incomplete knowledge (because of the Open-World Assumption), noise and uncertainty.

Various ML techniques have been extended to tackle SW representations. These encode regularities emerging from data as statistical models that can later be exploited to perform efficiently a series of useful tasks, bypassing the limitations of deductive reasoning and being able to cope with potential cases of inconsistency.

One of these tasks is the prediction of assertions, which is at the heart of further often more complex tasks such as query answering, clustering, ranking and recommendation. Data-driven forms of assertion prediction could be useful for addressing the cases where, for various reasons related to cases of incompleteness and inconsistency, it is not possible to logically infer the truth value of some statements (i.e. assertions which are not explicitly stated in nor derivable from the knowledge base). An example of such cases is the following:

Example 1. Consider a knowledge base $\mathcal{K}$ modeling familial relationships, where persons (each represented by an individual in the ontology) are characterized by multiple classes (such as Father, Uncle) and relationships (such as hasChild, hasSibling). By relying on purely deductive reasoning, it might not be possible to assess whether a certain property holds for a given person. For example, it might not be possible to assess whether John is an uncle or not. Assuming the property is represented by the concept Uncle and the person by the individual john, this can be formally expressed as:

$$\mathcal{K} \not\models \text{Uncle(john)} \land \mathcal{K} \not\models \neg\text{Uncle(john)},$$

i.e. it is not possible to deductively infer from $\mathcal{K}$ whether the property "uncle" holds for the person John.

Semantic Web knowledge representation languages make the *Open World Assumption*: a failure on deductively infer the truth value of a given fact does not imply that such fact is false, but rather that its truth value cannot be deductively inferred from the KB. This differs from the *Negation As Failure*, commonly used in databases and logic programs. Other issues are related to the distributed nature of the data across the Web: multiple, mutually conflicting pieces of knowledge may lead to contradictory answers or flawed inferences.

Most approaches to circumvent the limitations of incompleteness and inconsistency rely on extensions of the representation languages or of the inference services (e.g. *ontology repairing* [37] and *epistemic* reasoning [13] and *paraconsistent* reasoning [29]).

An alternative solution consists in relying on data-driven approaches to address the problem of missing knowledge. The prediction of the truth value of an assertion can be cast as a *classification* problem, to be solved through *statistical learning* [41]: domain entities described by an ontology can be regarded as *statistical units*, and their properties can be statistically inferred even when they cannot be deduced from the KB. Several methods have been proposed in the SW literature (see [34] for a recent survey). In particular, Statistical Relational Learning [14] (SRL) methods face the problem of learning in domains showing both a complex relational and a rich probabilistic structure. A major issue with the methods proposed so far is that the induced statistical models (as those produced by kernel methods, tensor factorization, etc.) are either difficult to interpret by experts and to integrate in logic-based SW infrastructures, or computationally impractical (see Sect. 5.1).

Contribution. A learning task can be either *generative* or *discriminative* [24], depending on the structure of the target distribution. Generative models describe the joint probability of all random variables in the model (e.g. a joint probability distribution of two sets of variables $\Pr(\mathbf{X}, \mathbf{Y})$). Discriminative models directly represent aspects of the distribution that are important for a specific task (e.g. a conditional probability distribution of a set of variables given another $\Pr(\mathbf{Y} \mid \mathbf{X})$). The main motivation behind the choice of a discriminative model is described by the *main principle* in [41]: "If you possess a restricted amount of information for solving some problem, try to solve the problem directly and never solve a more general problem as an intermediate step. It is possible that the available information is sufficient for a direct solution but is insufficient for solving a more general intermediate problem". Discriminative learning can also be useful for feature selection (e.g. in the context of a mining or ontology engineering task). In [5], authors show that many feature selection methods grounding on information theory ultimately try to optimize some approximation of a conditional likelihood (that is, a quantity proportional to the true posterior class probabilities in a set of instances).

In this article, we propose a method for predicting the concept-membership relation of an arbitrary individual with respect to a given target concept given a set of training individuals (members and non-members) within a DL knowledge base. The proposed method relies on a Bayesian network with generative parameters (which can be computed efficiently) and discriminative structure (which maximize the predictive accuracy of the model). The proposed model can be used also with knowledge bases expressed in expressive DLs used within the SW context, such as $\mathcal{SHOIN}(\mathcal{D})$ and $\mathcal{SROIQ}(\mathcal{D})$.

In particular, the proposed method relies on a committee of features (represented by possibly complex concepts) to define a set of random variables $\{F_1, \dots, F_n\}$. Such variables are then used to model a posterior probability distribution of the target concept-membership, conditioned the membership w.r.t. the aforementioned feature concepts $\Pr(C \mid F_1, \ldots, F_n)$: the value of each F_i depends on the concept-membership w.r.t. the i-th feature concept, and C is a Boolean random variable whose conditional probability distribution depends on the value of the F_i's. The proposed method relies on an inductive process: it incrementally builds a Bayesian classifier through a set of hill-climbing searches in the space of feature concepts using DL refinement operators [26].

This paper is organized as follows: Sect. 2 contains an introduction to the Bayesian network formalism of describing independence relations among a set of variables. In Sect. 3 we describe Terminological Bayesian Classifier models for class-membership prediction in DL knowledge bases, and how such models can be learned from data. In Sect. 4 we provide an empirical evaluation of the discussed model. Finally, in Sect. 6 we summarize the proposed approach and discuss possible research directions.

2   Bayesian Networks and Bayesian Classifiers

Graphical models [20] (GMs) are a popular framework that allows a compact description of the joint probability distribution for a set of random variables, by representing the underlying structure through a series of modular factors. Depending on the underlying semantics, GMs can be grouped into two main classes, i.e. *directed* and *undirected* graphical models, based directed and undirected graphs respectively.

A Bayesian network (BN) is a directed GM which represents the conditional dependencies in a set of random variables by using a directed acyclic graph (DAG) $\mathcal{G}$ augmented with a set of conditional probability distributions $\theta_{\mathcal{G}}$ (also referred to as *parameters*) associated with $\mathcal{G}$'s vertexes.

In a BN, each vertex corresponds to a random variable X_i, and each edge indicates a *direct influence* relation between the two random variables. A BN stipulates a set of *conditional independence assumptions* over its set of random variables: each vertex X_i in the DAG is conditionally independent of any subset $S \subseteq Nd(X_i)$ of vertexes that are not descendants of X_i given a joint state of its parents. More formally: $\forall X_i : X_i \perp\!\!\!\perp S \mid parents(X_i)$, where the function $parents(X_i)$ returns the parent vertexes of X_i in the DAG representing the BN.

The conditional independence assumption allows representing the *joint probability distribution* $\Pr(X_1, \ldots, X_n)$ defined by a Bayesian network over a set of random variables $\{X_1, \ldots, X_n\}$ as a production of the individual probability distributions, conditional on their parent variables:

$$\Pr(X_1, \ldots, X_n) = \prod_{i=1}^{n} \Pr(X_i \mid parents(X_i)).$$

As a result, it is possible to define $\Pr(X_1, \ldots, X_n)$ by only specifying, for each vertex X_i in the graph, the conditional probability distribution $\Pr(X_i \mid parents(X_i))$. Given a BN specifying a joint probability distribution over a set of variables, it is possible to evaluate inference queries by marginalization, like calculating the posterior probability distribution for a set of query variables given some observed event (i.e. assignment of values to the set of evidence variables).

In BNs, common inference tasks (such as calculating the most likely value for some variables, their marginal distribution or their conditional distribution given some evidence) are NP-hard. However, such inference tasks are less complex for particular classes of BNs such tasks, approximate inference algorithms exist to efficiently infer in restricted classes of networks. For example, the *variable elimination* algorithm has linear complexity in the number of vertexes if the BN is a singly connected network [20].

Approximate inference methods for BNs also exist in literature such as *Monte Carlo* algorithms, *belief propagation* or *variational methods* [20]. The compact parametrization in graphical models allows for effective learning both model selection (structural learning) and parameter estimation. In the case of BNs, however, finding a model which is optimal with respect to a given scoring criterion (which measures how well the model fits observed data) may be not trivial: the number of possible BN structures is super-exponential in the number of vertexes, making it generally impractical to perform an exhaustive search through the space of its possible models.

Looking for a trade-off between efficiency and expressiveness, we focus on *Bayesian network classifiers*, where a Bayesian network is used to model the conditional probability distribution of a single variable, representing a concept-membership relation.

For its simplicity, accuracy and low time complexity in both inference and learning, we first focused on a particular subclass of Bayesian network classifiers. *Naïve Bayesian classifier* models the dependencies between a set of random variables $\mathcal{X} = \{X_1, \ldots, X_n\}$, also called *features*, and a random variable C, also called *class*, so that each pair of features are independent of each other given the class, i.e. $\forall X_i, X_j \in \mathcal{X} : i \neq j \Rightarrow (X_i \perp\!\!\!\perp X_j \,|\, C)$. This category of models is especially interesting since it proved to be effective also in contexts in which the underlying independence assumptions do not hold [12], outperforming more recent approaches [6].

However, such strong independence assumptions may not capture correlations between feature concepts properly. Therefore, we also consider employing generic Bayesian network structures and polytree structures among feature variables, while retaining the edges from the class variable to feature variables. We avoid performing an exhaustive search in the space of possible structures (that, in the case of Bayesian classifiers, may be too complex to perform) and take the path also used in [17,33] of performing an hill climbing search, making modifications at the network structures at each step until we get to an (possibly local) optimal solution.

3   Terminological Bayesian Classifiers for Concept-membership Prediction

We propose employing the Bayesian network classifier [20] formalism to represent the statistical relations among a set of concepts in a given knowledge base. In particular, we aim at using such BN to model the conditional probability distribution $\Pr(C \mid F_1, \ldots, F_n)$, representing the probability that a generic individual in a knowledge base is a member of a target concept C given its concept-membership relation w.r.t. a set of *feature* concepts $\{F_1, \ldots, F_n\}$ (the random variables in the network can be considered as indicator functions taking different values depending on the concept-membership relation between the individual and the corresponding concept).

An intuitive method for mapping the values of the random variable to the corresponding concept-membership relation is considering the variable as a Boolean indicator function, assuming value **True** iff the individual is an instance of the concept, **False** iff it is an instance of its complement, and otherwise considering the variable as *non-observable*: this allows to consistently handle the Open World Assumption (OWA) characterizing the semantics of standard DLs, where it is common to have *partial knowledge* about the concept-membership relations of an individual.

However, this setting implies that not knowing the concept-membership relation w.r.t. a feature concept is *uninformative* [36] when predicting the concept-membership relation w.r.t. a given target concept; this is a strong assumption that does not hold in general. We will refer to such kind of networks as *Terminological Bayesian Classifiers* (TBCs). More formally:

Definition 1. *(Terminological Bayesian Classifier) A Terminological Bayesian Classifier (TBC) $\mathcal{N}_{\mathcal{K}}$, with respect to a knowledge base $\mathcal{K}$, is defined as a pair $\langle \mathcal{G}, \Theta_{\mathcal{G}} \rangle$, representing respectively the structure and parameters of a BN, in which:*

- $\mathcal{G} = \langle \mathcal{V}, \mathcal{E} \rangle$ *is an augmented directed acyclic graph, in which:*
  - $\mathcal{V} = \{F_1, \ldots, F_n, C\}$ *(vertexes) is a set of random variables, each linked to a concept defined over $\mathcal{K}$. Each F_i ($i = 1, \ldots, n$) is a Boolean random variable, whose value depends on the membership w.r.t. a feature concept, while C is a Boolean variable which indicates the membership relation to the target concept (we will use the names of variables in $\mathcal{V}$ to represent the corresponding concept for brevity);*
  - $\mathcal{E} \subseteq \mathcal{V} \times \mathcal{V}$ *is a set of edges, which model the (in)dependence relations among the variables in $\mathcal{V}$.*
- $\Theta_{\mathcal{G}}$ *is a set of conditional probability distributions (CPD), one for each variable $V \in \mathcal{V}$, representing the conditional probability distribution of the feature concept given the state of its parents in the graph.*

A very simple but effective structure is naïve Bayesian one (also described in Sect. 2), which relies on the assumption the concept-membership w.r.t. each of the feature concepts are independent given the concept-membership relation w.r.t. the target concept; this results in the edge set $\mathcal{E} = \{\langle C, F_i \rangle \mid i \in \{1, \ldots, n\}\}$.

Example 2. (Example of Terminological Naïve Bayesian Classifier) Given the following set of feature concepts[2]:

$$\mathcal{F} = \{Fe := Female, HC := \exists hasChild.\top, HS := \exists hasSibling.\top\},$$

and a target concept $FWS := FatherWithSibling$, a terminological naïve Bayesian classifier expressing the target concept in terms of the feature concepts is the following:

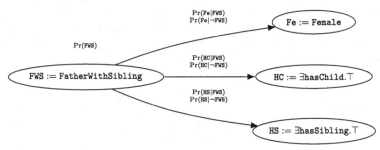

We can also express correlations between feature concepts which may be useful for making the conditional probability distribution more accurate, by relaxing the constraints on the edge set $\mathcal{E}$; we consider allowing for generic (acyclic) graph structures among feature variables, and for polytree (or singly connected tree) graph structures, which allow for exact inference to be calculated in polynomial time [20].

Let $\mathcal{K}$ be a knowledge base and a a generic individual so that $\mathcal{K} \models HC(a)$, and the membership relation between a to the concepts Fe and HS is not known, i.e. $\mathcal{K} \not\models C(a)$ and $\mathcal{K} \not\models \neg C(a)$, where C is either Fe or HS. It is possible to infer, through the given network, the probability that the individual a is a member of the target concept FWS:

$$\Pr(FWS(a) \mid HC(a)) = \frac{\Pr(FWS)\Pr(HC \mid FWS)}{\Pr(HC)},$$

where $\Pr(HC) = \Pr(FWS)\Pr(HC \mid FWS) + \Pr(\neg FWS)\Pr(HC \mid \neg FWS)$. □

In the following, we define the problem of learning a TBC $\mathcal{N}_{\mathcal{K}}$, given a knowledge base $\mathcal{K}$ and a set of positive, negative and neutral training individuals.

The problem consists in finding a TBC $\mathcal{N}_{\mathcal{K}}^*$ maximizing an arbitrary scoring criterion, given a set of training individuals $Ind_C(\mathcal{K})$. Such individuals are organized in positive, negative and neutral examples, accordingly to their concept-membership relation w.r.t. the target concept C in $\mathcal{K}$.

More formally:

Definition 2. *(Terminological Bayesian Classifier Learning Problem)*
   The TBC learning problem can be defined as follows:

---

[2] Here concepts have been aliased for brevity.

Given :

- *a target concept C;*
- *a set of training individuals $Ind_C(\mathcal{K})$ in a knowledge base $\mathcal{K}$ such that:*
  $\forall a \in Ind_C^+(\mathcal{K})$ *positive example:* $\mathcal{K} \models C(a)$,
  $\forall a \in Ind_C^-(\mathcal{K})$ *negative example:* $\mathcal{K} \models \neg C(a)$,
  $\forall a \in Ind_C^0(\mathcal{K})$ *neutral example:* $\mathcal{K} \not\models C(a) \wedge \mathcal{K} \not\models \neg C(a)$;
- *A scoring function specifying a measure of the quality of an induced terminological Bayesian classifier $\mathcal{N}_\mathcal{K}$ w.r.t. the samples in $Ind_C(\mathcal{K})$;*

Find *a network $\mathcal{N}_\mathcal{K}^*$ maximizing a given scoring function Score w.r.t. the samples:*

$$\mathcal{N}_\mathcal{K}^* \leftarrow \arg\max_{\mathcal{N}_\mathcal{K}} Score(\mathcal{N}_\mathcal{K}, Ind_C(\mathcal{K}))).$$

The search space for finding the optimal network $\mathcal{N}_\mathcal{K}^*$ may be too large to be exhaustively explored. For such a reason, the learning approach proposed here works by incrementally building the set of feature concepts, with the aim of obtaining a set of concepts maximizing the score of the induced network. Each feature concept is individually searched by an inner search process, guided by the scoring function itself, and the whole strategy of adding and removing feature concepts follows a forward selection/backward elimination strategy. This approach is motivated by the literature about *selective Bayesian classifiers* [21], where forward selection of attributes generally increases the classifier accuracy. The algorithm proposed here is organized in two nested loops: the inner loop is concerned with exploring the space of possible features (concepts), e.g. by means of DL refinement operators; the outer loop implements the abstract greedy feature selection strategy (such as forward selection [18]). Both procedures are guided by a scoring function defined over the space of TBC models.

Algorithm 1. Scoring function-driven hill climbing search for a new concept to add to the committee of DL concepts used to construct the Terminological Bayesian Network.

---

function $Grow(\mathcal{F}, Ind_C(\mathcal{K}), Start)$

1: $C \leftarrow Start$;
2: {Iteratively refine the concept C until a stopping criterion is met}
3: **repeat**
4: {Let $\mathcal{C}$ be the set of (upward and downward) refinements of the concept C obtained by means of the ρ refinement operator:}
5: $\mathcal{C} \leftarrow \{C' \in \rho_\uparrow(C) \cup \rho_\downarrow(C) \mid |C'| \le min(|C| + depth, maxLength)\}$;
6: {Select the concept in the set of refinements $\mathcal{C}$ providing the highest increase to the score (measured by the *Score* function) to the TBC obtained (using the *ConstructNetwork* function) by adding the selected concept to the set $\mathcal{C}$}
7: $C \leftarrow \arg\max_{C' \in \mathcal{C}} Score(ConstructNetwork(\mathcal{F} \cup \{C'\}, Ind_C(\mathcal{K})), Ind_C(\mathcal{K}))$;
8: **until** Stopping criterion; {E.g. no further improvements in score}
9: **return** C

In the inner loop, outlined in Algorithm 1, the search through the space of concept definitions is performed as a hill climbing search, using the $\rho_\downarrow^{cl}$ refinement operator [26] ($\rho_\downarrow^{cl}(C)$ returns a set of refinements D of C so that $D \sqsubseteq C$, which we consider only up to a given concept length n). For each new complex concept being evaluated, the algorithm creates a new set of concepts $\mathcal{F}'$ and finds an optimal structure, under a given set of constraints (which, in the case of terminological naïve Bayesian classifiers, is already fixed) and parameters (which may vary depending on the assumptions on the nature of the ignorance model). Then, the new network is scored, with respect to a given scoring criterion.

Algorithm 2. Forward Selection Backward Elimination method for the incremental construction of terminological Bayesian classifiers.

---

function $FSBE(\mathcal{K}, Ind_C(\mathcal{K}))$

1: $t \leftarrow 0, \mathcal{F}^t \leftarrow \emptyset$;
2: **repeat**
3:     $t \leftarrow t + 1$;
4:     {A new committee is selected among a set of possible candidates (represented by the set of committees $\hat{\mathcal{F}}$), obtained by either adding of removing a set of concepts to the structure, so as to maximize the score of the corresponding TBC (measured by means of the *Score* function)}
5:     $\hat{\mathcal{F}} = \{Grow(\mathcal{F}^{t-1}, Ind_C(\mathcal{K}), \top), Shrink(\mathcal{F}^{t-1}, Ind_C(\mathcal{K}), max)\}$;
6:     $\mathcal{F}^t \leftarrow \arg\max\limits_{\mathcal{F} \in \hat{\mathcal{F}}} Score(ConstructNetwork(\mathcal{F}, Ind_C(\mathcal{K})), Ind_C(\mathcal{K}))$;
7: **until** Stopping criterion; {E.g. the maximum number of concepts in $\mathcal{F}$ was reached}
8: $\mathcal{N}_\mathcal{K}^t \leftarrow ConstructNetwork(\mathcal{F}^t, Ind_C(\mathcal{K}))$;
9: **return** $\mathcal{N}_\mathcal{K}^t$;

function $Shrink(\mathcal{F}, Ind_C(\mathcal{K}), max)$

1: {Finds the best network that could be obtained by removing at most max feature concepts from the network structure, w.r.t. a given scoring criterion $Score$}
2: $\hat{\mathcal{F}} \leftarrow \{\mathcal{F}' \subseteq \mathcal{F} : |\mathcal{F}| - |\mathcal{F}'| \leq max\}$;
3: $\mathcal{F}^* \leftarrow \arg\max\limits_{\mathcal{F} \in \hat{\mathcal{F}}} Score(ConstructNetwork(\mathcal{F}, Ind_C(\mathcal{K})), Ind_C(\mathcal{K}))$;
4: **return** $\mathcal{F}^*$;

In the outer loop, outlined in Algorithm 2, it is possible to implement a variety of feature selection strategies [18]. In this specific case, we propose a *Forward Selection Backward Elimination* (FSBE) method, which at each iteration considers adding a new concept to the network (by means of the *Grow* function) or removing an existing one (by means of the *Shrink* function).

Different Assumptions on the Ignorance Model. During the learning process, it may happen that the concept membership between a training individual and some of the feature concepts is unknown. The reason of such missingness can be taken into account, when learning the parameters of the statistical model [20]. Formally, the missing data handling method depends on the probability distribution underlying the missingness pattern [36], which in turn can be classified on the basis of its behavior with respect to the variable of interest:

- **Missing Completely At Random** (MCAR) – the variable of interest $\mathbf{X}$ is independent from its observability $O_{\mathbf{X}}$, as any other variable in the probabilistic model. This is the precondition for case deletion to be valid, and missing data does not usually belong to such class [36]:

$$P_{missing} \models (O_{\mathbf{X}} \perp\!\!\!\perp \mathbf{X});$$

- **Missing At Random** (MAR) – happens when the observability of the variable of interest $\mathbf{X}$ depends on the value of some other variable in the probabilistic model:

$$P_{missing} \models (O_{\mathbf{X}} \perp\!\!\!\perp \mathbf{x}^{\mathbf{y}}_{hidden} \mid \mathbf{x}^{\mathbf{y}}_{obs});$$

- **Not Missing At Random/Informatively Missing** (NMAR, IM) – here, the actual value of the variable of interest influences the probability of its observability:

$$P_{missing} \models (O_{\mathbf{X}} \not\!\perp\!\!\!\perp \mathbf{X}).$$

Example 3. (Different Ignorance Models in Terminological Bayesian Classifiers) Consider the network in Example 2: if the probability that the variable Fe is observable is independent on all other variables in the network, then it's missing completely at random; if it only depends, for example, on the value of FWS, then it's missing at random; if it is dependent on the value Fe would have if it was not missing, then it is informatively missing.

Each of the aforementioned assumptions on the missingness pattern implies a different way of learning both network structure and parameters in presence of partially observed data. If **MCAR** holds, *Available Case Analysis* [20] can be used, where maximum likelihood network parameters are estimated using only available knowledge (i.e. ignoring missing data); we are adopting the heuristic used in [17] of setting network parameters to their maximum likelihood value, which is both accurate and efficient. This decision is further motivated by [33], which empirically motivated that generative discriminatively structured Bayesian networks retain both the accuracy of discriminative networks and the efficiency of parameter learning and ability to handle partial evidence typical of generative networks.

As scoring function, similarly to [17], we adopt the conditional log-likelihood on positive and negative training individuals, defined as[3]:

$$CLL(\mathcal{N}_{\mathcal{K}} \mid Ind_C(\mathcal{K})) = \sum_{a \in Ind_C^+(\mathcal{K})} \log \Pr(C(a) \mid \mathcal{N}_{\mathcal{K}})$$

$$+ \sum_{a \in Ind_C^-(\mathcal{K})} \log \Pr(\neg C(a) \mid \mathcal{N}_{\mathcal{K}}).$$

---

[3] When used to score networks, conditional log-likelihoods are calculated ignoring available knowledge about the membership between training individuals and the target concept.

A problem with using simply *CLL* as scoring criterion is that it tends to favor complex structures [20] that overfit the training data. To avoid overfitting, we penalize the conditional log-likelihood through the *Bayesian Information Criterion* (BIC) [20], where the penalty is proportional to the number of independent parameters in a network (according to the minimum description length principle) and is defined as follows:

$$BIC(\mathcal{N}_\mathcal{K} \mid Ind_C(\mathcal{K})) = CLL(\mathcal{N}_\mathcal{K} \mid Ind_C(\mathcal{K})) - \frac{\log N}{2}|\Theta_\mathcal{G}|, \qquad (1)$$

where N is the number of data points and $|\Theta_\mathcal{G}|$ is the number of independent parameters in the network.

Under the naïve Bayes assumption, there is no need to perform a search for finding the optimal network, since the structure is already fixed (each node except the target concept node has only one parent, which is the target concept node).

For relaxing the independence assumptions in naïve Bayes structures, we follow the approach also discussed in [17,33] to perform an hill-climbing search in the space of structures, by looking for the one maximizing the (penalized) CLL. The exploration in the space of possible structures is performed by making atomic modification to the structure between feature variables, and consist in atomic operation of either edge addition, removal or reversal.

When learning network parameters from **MAR** data, a variety of techniques is available, such as Expectation-Maximization (EM) or gradient ascent [20]. In this work, we employ the EM algorithm, as outlined in Algorithm 3: it first initializes network parameters using estimates that ignore missing data; then, it considers individuals whose membership w.r.t. a generic concept D is not known as several fractional individuals belonging, with different weights (corresponding to the posterior probability of their concept membership), to both the components D and $\neg D$. Such fractional individuals are used to recalculate network parameters (obtaining the so-called *expected counts*) and the process is repeated until convergence (e.g. when the improvement in log-likelihood is lower than a specific threshold).

At each iteration, the EM algorithm applies the following two steps:

- **Expectation:** using available data and the current network parameters, infers a distribution over possible completions for the missing knowledge;
- **Maximization:** considering each possible completion as a fully available data case (weighted by its probability), infers next parameters through frequency counting.

When data is **NMAR/IM** it may be harder to model, since we cannot assume that observed and missing values follow the same distributions.

However, it is generally possible to extend the probabilistic model to produce one where the MAR assumption holds; if the value of a variable associated to the feature concept F_i is informatively missing, we can consider its observability as a indicator Boolean variable O_i (such that $O_i = \texttt{False}$ iff $\mathcal{K} \not\models F_i(a)$ and

Algorithm 3. Outline for our implementation of the EM algorithm for parameter learning from MAR data in a terminological Bayesian classifier.

---

function $EM(\mathcal{N}_\mathcal{K}^0, Ind_C(\mathcal{K}))$

1: $\{\mathcal{N}_\mathcal{K}^0$ is initialized with arbitrary heuristic parameters $\Theta_\mathcal{G}^0\}$
2: $\mathcal{N}_\mathcal{K}^0 = \langle \mathcal{G}, \Theta_\mathcal{G}^0 \rangle, \mathcal{G} = \langle \mathcal{V}, \mathcal{E} \rangle; t \leftarrow 0;$
3: **repeat**
4: $\quad \{\bar{n}(x_i, \pi_{x_i})\} \leftarrow ExpCounts(\mathcal{N}_\mathcal{K}, Ind_C(\mathcal{K}));$
5: $\quad \{$Network parameters $\Theta_\mathcal{G}^{t+1}$ are updated according to the inferred expected counts$\}$
6: $\quad$ **for** $X_i \in \mathcal{V}, \langle x_i, \pi_{x_i} \rangle \in vals(X_i, parents(X_i))$ **do**
7: $\qquad \theta_\mathcal{G}^{t+1}(x_i, \pi_{x_i}) \leftarrow \dfrac{\bar{n}(x_i, \pi_{x_i})}{\sum\limits_{x_i' \in vals(X_i)} \bar{n}(x_i', \pi_{x_i})};$
8: $\quad$ **end for**
9: $\quad t \leftarrow t + 1;$
10: $\quad \mathcal{N}_\mathcal{K}^t = \langle \mathcal{G}, \Theta_\mathcal{G}^t \rangle;$
11: $\quad \{$The iterative process stops when improvements in log-likelihood are $\leq$ a threshold$\}$
12: **until** $\mathcal{L}(\mathcal{N}_\mathcal{K}^t \mid Ind_C(\mathcal{K})) - \mathcal{L}(\mathcal{N}_\mathcal{K}^{t-1} \mid Ind_C(\mathcal{K})) \leq \tau;$
13: **return** $\mathcal{N}_\mathcal{K}^t;$

function $ExpCounts(\mathcal{N}_\mathcal{K}, Ind_C(\mathcal{K}))$

1: $\mathcal{N}_\mathcal{K} = \langle \mathcal{G}, \Theta_\mathcal{G} \rangle, \mathcal{G} = \langle \mathcal{V}, \mathcal{E} \rangle;$
2: **for** $X_i \in \mathcal{V}, \langle x_i, \pi_{x_i} \rangle \in vals(X_i, parents(X_i))$ **do**
3: $\quad \bar{n}(x_i, \pi_{x_i}) \leftarrow 0;$
4: **end for**
5: $\{\bar{n}(x_i, \pi_{x_i})$ will contain the expected counts for $(X_i = x_i, parents(X_i) = \pi_{x_i})\}$
6: **for** $a \in Ind_C(\mathcal{K})$ **do**
7: $\quad \{vals(X_i, parents(X_i))$ represents the set of possible values for X_i and its parents$\}$
8: $\quad$ **for** $X_i \in \mathcal{V}, \langle x_i, \pi_{x_i} \rangle \in vals(X_i, parents(X_i))$ **do**
9: $\qquad \bar{n}(x_i, \pi_{x_i}) \leftarrow \bar{n}(x_i, \pi_{x_i}) + \Pr(x_i, \pi_{x_i} \mid \mathcal{N}_\mathcal{K});$
10: $\quad$ **end for**
11: **end for**
12: **return** $\{\bar{n}(x_i, \pi_{x_i})\};$

$\mathcal{K} \not\models \neg F_i(a)$, $O_i = \texttt{True}$ otherwise) and include it in our probabilistic model, so that F_i's ignorance model satisfies the MAR assumption (since the probability of F_i to be observable depends on the always observable indicator variable O_i).

Doing this may however raise some problems, since the induced probabilistic model will be dependent on the specific ignorance model in the training set, and changes in such missingness pattern may impact on the model's effectiveness. However, to empirically evaluate the impact of doing so, we include the observability of a variable in the model by allowing its possible values to be a part of $\{\texttt{True}, \texttt{False}, \texttt{Unknown}\}$ (the best partition is chosen by the search process itself, considering each of the alternatives and choosing the one providing the major increase in the penalized CLL), and compare it with the result obtained allowing variables to vary in $\{\texttt{True}, \texttt{False}\}$ only.

4   Experiments

In this section we empirically evaluate the impact of adopting different missing knowledge handling methods and search strategies, during the process of learning Terminological Bayesian Classifiers from real world ontologies.

Table 1. Ontologies considered in the experiments.

Ontology	Expressivity	#Axioms	#Inds.	#Classes	#ObjProps.
BioPax (Proteomics)	$\mathcal{ALCHN}(\mathcal{D})$	773	49	55	47
Family-Tree	$\mathcal{SROIF}(\mathcal{D})$	2059	368	22	52
MDM0.73	$\mathcal{ALCHOF}(\mathcal{D})$	1098	112	196	22
NTNames	$\mathcal{SHOIN}(\mathcal{D})$	4434	724	49	29
Wine	$\mathcal{SHOIN}(\mathcal{D})$	1046	218	142	21

Starting from a set of real ontologies[4] (outlined in Table 1), we generated a set of 20 random query concepts (each corresponding to a DL complex concept) for each ontology[5], so that the number of individuals belonging to the target query concept C (resp. $\neg C$) was at least of 10 elements and the number of individuals in C and $\neg C$ was in the same order of magnitude. A DL reasoner[6] was employed to decide deductively about the concept-membership of individuals to query concepts.

Experiments consisted in predicting the membership w.r.t. automatically generated concept queries in the form of Terminological Bayesian Classifiers, using different sets of constraints on possible structures (and then obtaining naïve Bayes structures, polytrees or generic Bayesian networks), and on the possible values taken by variables. For predicting the membership w.r.t. the generated query concepts, different constraints on the available values for the variable in networks were empirically evaluated, allowing them to be either {True, False} or to also take a Unknown value, which represents the case in which it is not possible to entail an individual's membership w.r.t. a concept nor to its complement.

During the learning process, we set the *depth* parameter to 3 and *maxLength* to 6 (3 in the case of Family-Tree, for efficiency reasons); for exploring the space of concepts we employed the ψ refinement operator [26], available in the DL-Learner [25] framework, for moving both upwards and downwards in the concept lattice starting from the concept $\top$.

Regarding the feature selection strategy (corresponding to the outer loop in Algorithm 2), two different methods were empirically evaluated, namely Forward Selection (FS) and Forward Selection Backward Elimination (FSBE), where the

---

[4] From TONES Ontology Repository: http://owl.cs.manchester.ac.uk/repository/.
[5] Using the query concept generation method available at http://lacam.di.uniba.it/~nico/research/ontologymining.html.
[6] Pellet v2.3.0 – http://clarkparsia.com/pellet/.

Table 2. Statistics for cross-validated accuracy results on the generated data sets: for each of the ontologies, 20 query concepts were generated, and each was used to obtain a sample of positive/negative individuals, which were then used to evaluate the methods using k-fold cross validation (with $k = 10$) through the *accuracy* (left) and the *area under the precision-recall curve* (right) metrics.

Biopax (Proteomics)	Generic		Polytree		Naïve Bayes	
{ T, F }, FS	0.95 ± 0.1	0.94 ± 0.15	0.96 ± 0.1	0.94 ± 0.15	0.95 ± 0.1	0.94 ± 0.15
Any in {T,U,F}, FS	0.95 ± 0.1	0.93 ± 0.16	0.95 ± 0.1	0.92 ± 0.17	0.95 ± 0.1	0.92 ± 0.17
{ T, F }, FSBE	0.95 ± 0.1	0.94 ± 0.15	0.95 ± 0.1	0.94 ± 0.15	0.95 ± 0.1	0.94 ± 0.15
Any in {T,U,F}, FSBE	0.95 ± 0.1	0.92 ± 0.17	0.95 ± 0.1	0.93 ± 0.17	0.95 ± 0.1	0.93 ± 0.17
Family-Tree	Generic		Polytree		Naïve Bayes	
{ T, F }, FS	1 ± 0	1 ± 0	1 ± 0	1 ± 0	1 ± 0	1 ± 0
Any in {T,U,F}, FS	1 ± 0	1 ± 0	1 ± 0	1 ± 0	1 ± 0	1 ± 0
{ T, F }, FSBE	1 ± 0	1 ± 0	1 ± 0	1 ± 0	1 ± 0	1 ± 0
Any in {T,U,F}, FSBE	1 ± 0	1 ± 0	1 ± 0	1 ± 0	1 ± 0	1 ± 0
MDM0.73	Generic		Polytree		Naïve Bayes	
{ T, F }, FS	0.95 ± 0.08	0.87 ± 0.26	0.95 ± 0.08	0.87 ± 0.26	0.95 ± 0.08	0.87 ± 0.26
Any in {T,U,F}, FS	0.97 ± 0.06	0.9 ± 0.23	0.97 ± 0.06	0.9 ± 0.23	0.97 ± 0.06	0.9 ± 0.23
{ T, F }, FSBE	0.95 ± 0.07	0.87 ± 0.26	0.95 ± 0.08	0.87 ± 0.26	0.95 ± 0.08	0.87 ± 0.26
Any in {T,U,F}, FSBE	0.97 ± 0.06	0.9 ± 0.23	0.97 ± 0.06	0.9 ± 0.23	0.97 ± 0.06	0.9 ± 0.23
NTNames	Generic		Polytree		Naïve Bayes	
{ T, F }, FS	1 ± 0	1 ± 0	1 ± 0	1 ± 0	1 ± 0	1 ± 0
Any in {T,U,F}, FS	1 ± 0	1 ± 0	1 ± 0	1 ± 0	1 ± 0	1 ± 0
{ T, F }, FSBE	1 ± 0	1 ± 0	1 ± 0	1 ± 0	1 ± 0	1 ± 0
Any in {T,U,F}, FSBE	1 ± 0	1 ± 0	1 ± 0	1 ± 0	1 ± 0	1 ± 0
Wine	Generic		Polytree		Naïve Bayes	
{ T, F }, FS	0.92 ± 0.1	0.89 ± 0.18	0.92 ± 0.1	0.89 ± 0.18	0.92 ± 0.1	0.89 ± 0.18
Any in {T,U,F}, FS	0.92 ± 0.12	0.92 ± 0.14	0.92 ± 0.12	0.92 ± 0.14	0.92 ± 0.12	0.92 ± 0.14
{ T, F }, FSBE	0.92 ± 0.1	0.89 ± 0.18	0.92 ± 0.1	0.89 ± 0.18	0.92 ± 0.1	0.89 ± 0.18
Any in {T,U,F}, FSBE	0.92 ± 0.12	0.9 ± 0.16	0.92 ± 0.12	0.9 ± 0.16	0.93 ± 0.11	0.91 ± 0.16

former only adds (at most) one concept and the latter also considers removing one concept from the committee at each iteration.

Results (expressed using the Accuracy and the Area Under the Precision-Recall curve, calculated as proposed in [10]) have been obtained through k-fold cross validation (with $k = 10$); we evaluated the proposed approach in the Concept-membership prediction task, which consisted in predicting the membership w.r.t. automatically generated query concepts, which was also used in [34] and whose results are summarized in Table 2.

From empirical evaluations, it emerged that looking for more complex structures under the penalized CLL did not provide any significant gain over simple naïve Bayesian structures, confirming the simplicity and the accuracy of naïve Bayes network classifiers. There was no statistically significant difference observed adopting different feature selection methods.

On the other hand, it was shown that the missing value handling method impacted on the effective accuracy of the proposed approach: including the *observability* of a concept-membership relation, i.e. whether it can or cannot be proved true or false from the knowledge base, within the probabilistic model, positively impacted on the final accuracy (but making the induced model dependent on the particular ignorance mechanism).

5   Related Works

The problem of managing uncertain knowledge in the SW context has been focused particularly from the knowledge representation perspective. Several approaches, particularly focusing on enriching knowledge and inference procedures with probabilistic information has been proposed. Some of them extend knowledge representation formalisms actually used in the SW. For example: PR-OWL [7] extends the semantics of OWL through the first-order probabilistic logic formalism of Multi-Entity Bayesian Networks [22]. Other approaches rely on probabilistic enrichment of Description Logics [1] (DLs) or logic programming formalisms. Specifically, [15,28] rely on probabilistic lexicographic entailment from probabilistic default reasoning.

Log-Linear DLs [31] and CR$\mathcal{ALC}$ [8] extend DLs by means of probabilistic graphical models [20]. Similarly, in [16] authors propose probabilistic extension of the DL-Lite language based on Bayesian Networks. In [2], authors propose using Binary Decision Diagrams for efficient reasoning over probabilistic ontologies based on distribution semantics.

To handle vagueness, also fuzzy extensions of Description Logics have been proposed in literature (see e.g. [3,38,39]).

5.1   Machine Learning Methods for Knowledge Base Completion

The idea of leveraging Machine Learning methods for handling incomplete and noisy knowledge bases is being explored in SW literature. A variety of methods have been proposed for predicting the truth value of assertions in Web ontologies: those include generative probabilistic models (e.g. [11,32,35]), kernel methods (e.g. [27,42]), matrix and tensor factorization methods (e.g. [30,40]) and energy-based models (e.g. [4]).

An issue with existing methods is that they either rely on a possibly expensive search process, or induce statistical models that are not meaningful to human experts. For example, kernel methods induce models (such as separating hyperplanes) in a high-dimensional feature space implicitly defined by a kernel function. The underlying kernel function itself usually relies on purely syntactic features of the neighborhood graphs of two individual resources (such as their common subtrees [27] or isomorphic subgraphs [42]): in both cases, there is not necessarily an explicit meaning of such syntactic features in terms of domain knowledge.

The Latent variable method in [35], the matrix or tensor factorization methods in [30,40], and the energy-based models in [4], try to explain the observations (assertions) in terms of latent classes or attributes, which also may be not meaningful to the domain experts and knowledge engineers.

The approaches in [11,32] try to overcome this limitation by expressing the induced model using a probabilistic extension of the $\mathcal{ALC}$ Description Logic and Markov Logic, respectively. However, inference in these models is intractable in general: inference in [11,32] reduces to probabilistic inference to the corresponding ground graphical model.

6   Conclusions and Future Work

This article proposes a method based on discriminatively structured Bayesian networks to predict whether an individual is an instance of a given target concept, given the available knowledge about the individual (in the form of its concept-membership relation w.r.t. a set of feature concepts. Instead of modeling a fully fledged joint probability distribution among concepts in the knowledge base, we face the simpler problem directly model the conditional probability distribution of the aforementioned target concept-membership given other, informative and eventually inter-correlated, feature concept-memberships. We then propose a score-based approach to incrementally build the discriminatively structured Bayesian network, using Description Logic refinement operators [26].

References

1. Baader, F., Calvanese, D., McGuinness, D.L., Nardi, D., Patel-Schneider, P.F. (eds.): The Description Logic Handbook. Cambridge University Press, Cambridge (2003)
2. Bellodi, E., Lamma, E., Riguzzi, F., Albani, S.: A distribution semantics for probabilistic ontologies. In: Bobillo, F., Carvalho, R.N., da Costa, P.C.G., d'Amato, C., Fanizzi, N., Laskey, K.B., Lukasiewicz, T., Martin, T., Nickles, M. (eds.) URSW. CEUR Workshop Proceedings, vol. 778, pp. 75–86. CEUR-WS.org (2011)
3. Bobillo, F., Straccia, U.: fuzzydl: an expressive fuzzy description logic reasoner. In: FUZZ-IEEE, pp. 923–930. IEEE (2008)
4. Bordes, A., Glorot, X., Weston, J., Bengio, Y.: A semantic matching energy function for learning with multi-relational data - application to word-sense disambiguation. Mach. Learn. **94**(2), 233–259 (2014)
5. Brown, G., Pocock, A., Zhao, M.J., Luján, M.: Conditional likelihood maximisation: a unifying framework for information theoretic feature selection. J. Mach. Learn. Res. **13**, 27–66 (2012)
6. Caruana, R., Niculescu-Mizil, A.: An empirical comparison of supervised learning algorithms. In: Proceedings of the 23rd international conference on Machine learning, ICML '06, pp. 161–168. ACM, New York, NY, USA (2006)
7. Carvalho, R.N., Laskey, K.B., da Costa, P.C.G.: Pr-owl 2.0 - bridging the gap to owl semantics. In: Bobillo, F., Carvalho, R.N., da Costa, P.C.G., d'Amato, C., Fanizzi, N., Laskey, K.B., Laskey, K.J., Lukasiewicz, T., Martin, T., Nickles, M., Pool, M. (eds.) URSW. CEUR Workshop Proceedings, vol. 654, pp. 73–84. CEUR-WS.org (2010)
8. Cozman, F.G., Polastro, R.B.: Complexity analysis and variational inference for interpretation-based probabilistic description logic. In: Bilmes, J., Ng, A.Y. (eds.) UAI, pp. 117–125. AUAI Press (2009)
9. d'Amato, C., Fanizzi, N., Esposito, F.: Inductive learning for the semantic web: what does it buy? Semant. Web **1**(1–2), 53–59 (2010)
10. Davis, J., Goadrich, M.: The relationship between precision-recall and roc curves. In: ICML 2006, pp. 233–240. ACM, New York, NY, USA (2006)
11. Domingos, P., Lowd, D., Kok, S., Poon, H., Richardson, M., Singla, P.: Just add weights: markov logic for the semantic web. In: da Costa, P.C.G., d'Amato, C., Fanizzi, N., Laskey, K.B., Laskey, K.J., Lukasiewicz, T., Nickles, M., Pool, M. (eds.) URSW 2005 - 2007. LNCS (LNAI), vol. 5327, pp. 1–25. Springer, Heidelberg (2008)

12. Domingos, P., Pazzani, M.J.: On the optimality of the simple bayesian classifier under zero-one loss. Mach. Learn. **29**(2–3), 103–130 (1997)
13. Donini, F.M., Lenzerini, M., Nardi, D., Nutt, W., Schaerf, A.: An epistemic operator for description logics. Artif. Intell. **100**(1–2), 225–274 (1998)
14. Getoor, L., Taskar, B.: Introduction to Statistical Relational Learning (Adaptive Computation and Machine Learning). The MIT Press, Cambridge (2007)
15. Giugno, R., Lukasiewicz, T.: P-$\mathcal{SHOQ}$(**D**): a probabilistic extension of $\mathcal{SHOQ}$(**D**) for probabilistic ontologies in the semantic web. In: Flesca, S., Greco, S., Ianni, G., Leone, N. (eds.) JELIA 2002. LNCS (LNAI), vol. 2424, pp. 86–97. Springer, Heidelberg (2002)
16. Chomicki, J.: Consistent query answering: the first ten years. In: Greco, S., Lukasiewicz, T. (eds.) SUM 2008. LNCS (LNAI), vol. 5291, pp. 1–3. Springer, Heidelberg (2008)
17. Grossman, D., Domingos, P.: Learning bayesian network classifiers by maximizing conditional likelihood. In: Brodley, C.E. (ed.) ICML, vol. 69 (2004)
18. Guyon, I., Gunn, S., Nikravesh, M., Zadeh, L. (eds.): Feature Extraction, Foundations and Applications. Springer, Heidelberg (2006)
19. Hitzler, P., van Harmelen, F.: A reasonable semantic web. Semant. Web **1**(1–2), 39–44 (2010)
20. Koller, D., Friedman, N.: Probabilistic Graphical Models: Principles and Techniques. MIT Press, Cambridge (2009)
21. Langley, P., Sage, S.: Induction of selective bayesian classifiers. In: de Mántaras, R.L., Poole, D. (eds.) UAI, pp. 399–406. Morgan Kaufmann (1994)
22. Laskey, K.B.: Mebn: A language for first-order bayesian knowledge bases. Artif. Intell. **172**(2–3), 140–178 (2008)
23. Laskey, K.J., Laskey, K.B.: Uncertainty reasoning for the world wide web: report on the urw3-xg incubator group. In: URSW2008 (2008)
24. Lasserre, J., Bishop, C.M.: Generative or discriminative? getting the best of both worlds. Bayesian Stat. **8**, 3–24 (2007)
25. Lehmann, J.: Dl-learner: learning concepts in description logics. J. Mach. Learn. Res. **10**, 2639–2642 (2009)
26. Lehmann, J., et al.: Concept learning in description logics using refinement operators. Mach. Learn. **78**, 203–250
27. Lösch, U., Bloehdorn, S., Rettinger, A.: Graph kernels for RDF data. In: Simperl, E., Cimiano, P., Polleres, A., Corcho, O., Presutti, V. (eds.) ESWC 2012. LNCS, vol. 7295, pp. 134–148. Springer, Heidelberg (2012)
28. Lukasiewicz, T.: Expressive probabilistic description logics. Artif. Intell. **172**(6–7), 852–883 (2008)
29. Maier, F., Ma, Y., Hitzler, P.: Paraconsistent OWL and related logics. Semant. Web **4**(4), 395–427 (2013)
30. Nickel, M., Tresp, V., Kriegel, H.P.: A three-way model for collective learning on multi-relational data. In: Getoor, L., et al. (eds.) Proceedings of ICML'11, pp. 809–816. Omnipress (2011)
31. Niepert, M., Noessner, J., Stuckenschmidt, H.: Log-linear description logics. In: Walsh, T. (ed.) IJCAI, IJCAI/AAAI, pp. 2153–2158 (2011)
32. Ochoa-Luna, J.E., Cozman, F.G.: An algorithm for learning with probabilistic description logics. In: Bobillo, F., da Costa, P.C.G., d'Amato, C., Fanizzi, N., Laskey, K.B., Laskey, K.J., Lukasiewicz, T., Martin, T., Nickles, M., Pool, M., Smrz, P. (eds.) URSW, pp. 63–74 (2009)
33. Pernkopf, F., Bilmes, J.A.: Efficient heuristics for discriminative structure learning of bayesian network classifiers. J. Mach. Learn. Res. **11**, 2323–2360 (2010)

34. Rettinger, A., Lösch, U., Tresp, V., d'Amato, C., Fanizzi, N.: Mining the semantic web - statistical learning for next generation knowledge bases. Data mining and knowledge discovery - special issue on web mining (2012)

35. Rettinger, A., Nickles, M., Tresp, V.: Statistical relational learning with formal ontologies. In: Buntine, W., Grobelnik, M., Mladenić, D., Shawe-Taylor, J. (eds.) ECML PKDD 2009, Part II. LNCS, vol. 5782, pp. 286–301. Springer, Heidelberg (2009)

36. Rubin, D.B.: Inference and missing data. Biometrika **63**(3), 581–592 (1976)

37. Stoilos, G., Cuenca Grau, B., Motik, B., Horrocks, I.: Repairing ontologies for incomplete reasoners. In: Aroyo, L., Welty, C., Alani, H., Taylor, J., Bernstein, A., Kagal, L., Noy, N., Blomqvist, E. (eds.) ISWC 2011, Part I. LNCS, vol. 7031, pp. 681–696. Springer, Heidelberg (2011)

38. Straccia, U.: A fuzzy description logic. In: Mostow, J., Rich, C. (eds.) AAAI/IAAI, pp. 594–599. AAAI Press / The MIT Press (1998)

39. Straccia, U.: Towards a fuzzy description logic for the semantic web (preliminary report). In: Gómez-Pérez, A., Euzenat, J. (eds.) ESWC 2005. LNCS, vol. 3532, pp. 167–181. Springer, Heidelberg (2005)

40. Tresp, V., Huang, Y., Bundschus, M., Rettinger, A.: Materializing and querying learned knowledge. In: Proceedings of the First ESWC Workshop on Inductive Reasoning and Machine Learning on the Semantic Web (IRMLeS 2009) (2009)

41. Vapnik, V.N.: Statistical learning theory, 1st edn. Wiley, New York (1998)

42. de Vries, G.K.D.: A fast approximation of the Weisfeiler-Lehman graph kernel for RDF data. In: Blockeel, H., Kersting, K., Nijssen, S., Železný, F. (eds.) ECML PKDD 2013, Part I. LNCS, vol. 8188, pp. 606–621. Springer, Heidelberg (2013)

Graph-Based Regularization for Transductive Class-Membership Prediction

Pasquale Minervini[✉], Claudia d'Amato, Nicola Fanizzi, and Floriana Esposito

LACAM Laboratory – Dipartimento di Informatica, Università degli Studi di Bari
Aldo Moro, via E. Orabona, 4, 70125 Bari, Italy
{pasquale.minervini,claudia.damato,nicola.fanizzi,
floriana.esposito}@uniba.it

Abstract. Considering the increasing availability of structured machine processable knowledge in the context of the Semantic Web, only relying on purely deductive inference may be limiting. This work proposes a new method for similarity-based class-membership prediction in Description Logic knowledge bases. The underlying idea is based on the concept of *propagating* class-membership information among similar individuals; it is non-parametric in nature and characterized by interesting complexity properties, making it a potential candidate for large-scale transductive inference. We also evaluate its effectiveness with respect to other approaches based on inductive inference in SW literature.

1 Introduction

Standard Semantic Web (SW) reasoning services rely on purely deductive inference. However, this may be limiting, e.g. due to the complexity of reasoning tasks, availability and correctness of structured knowledge. Approximate deductive and inductive inference were discussed as a possible approach to try to overcome such limitations [25]. Various proposals to extend inductive inference methods towards SW formalisms have been discussed in SW literature: inductive methods can perform some sort of approximate and uncertain reasoning and derive conclusions which are not derivable or refutable from the knowledge base [25].

This work proposes a novel method for transductive inference on Description Logic (DL) representations. In the class-membership prediction task, discriminative methods proposed so far ignore unlabeled problem instances (individuals for which the value of such class-membership is unknown); however, accounting for unlabeled instances during learning can provide more accurate results if some conditions are met [6,35]. Generative methods, on the other hand, try to model a joint probability distribution on both instances and labels, thus facing a possibly harder learning problem than only predicting the most probable label for any given instance.

In Sect. 2 we will shortly survey related works, and introduce a variant to the classic class-membership prediction problem. In Sect. 3 we will introduce the proposed method: the assumptions it relies on, and how it can be used

F. Bobillo et al. (Eds.): URSW 2011-2013, LNAI 8816, pp. 202–218, 2014.
DOI: 10.1007/978-3-319-13413-0_11

for class-membership prediction on large and Web scale ontological knowledge bases. In Sect. 4 we will provide empirical evidence for the effectiveness of the proposed method with respect to other methods in SW literature. In Sect. 5 we summarize the proposed approach, outline its limitations and discuss possible future research directions.

2 Preliminaries

A variety of approaches have been proposed in the literature for solving the class-membership prediction problem, either *discriminative* or *generative* [21]. Assuming instances are i.i.d. samples from a distribution P ranging over a space $X \times Y$ (where X is the space of instances and Y a set of labels), *generative* prediction methods first build an estimate $\hat{P}$ of the joint probability distribution $P(X, Y)$, and then use it to infer $\hat{P}(Y \mid x) = \hat{P}(Y, x)/\hat{P}(x)$ for a given, unlabeled instance $x \in X$. On the other hand, *discriminative* methods simply aim at estimating when $P(y \mid x) \geq 0.5$, for any given $(x, y) \in X \times Y$ (thus facing a possibly easier problem than estimating a joint probability distribution over $X \times Y$). The following shortly surveys class-membership prediction methods proposed so far.

2.1 Discriminative Methods

Some of the approaches proposed for solving the class-membership prediction problem are similarity-based. For instance, methods relying on the k-Nearest Neighbors (k-NN) algorithm are discussed in [8,25]. A variety of (dis-)similarity measures between either individuals or concepts have been proposed: according to [5], they can be based on *features* (where objects are characterized by a set of features, such as in [16]), on the *semantic-network* structure (where background information is provided in the form of a semantic network, such as in [10,17]) or on the *information content* (where both the semantic network structure and population are considered, such as in [9]).

Kernel-based algorithms [27] have been proposed for various learning tasks from DL-based representations. This is made possible by the existence of a variety of kernel functions, either for concepts or individuals (such as [4,11,13]).

By (implicitly) projecting instances into an high-dimensional feature space, kernel functions allow to adapt a multitude of machine learning algorithms to structured representations. SW literature includes methods for inducing robust classifiers [12] or learning to rank [14] from DL knowledge bases using kernel methods.

2.2 Generative Methods

A generative model for learning from formal ontologies is proposed in [26]: each individual is associated to a *latent variable* (similar to a *cluster indicator*) which influences its attributes and the relations it participates in. It proposes using a

Nonparametric Bayesian model for automatically selecting the number of possible values for such latent variables, with an inference method based on Markov Chain Monte Carlo where posterior sampling is constrained by a predefined set of DL axioms.

A different generative model is proposed in [23]: it focuses on learning theories in a probabilistic extension of the $\mathcal{ALC}$ DL named CR$\mathcal{ALC}$, and using DL refinement operators to efficiently explore the space of concepts. It is inspired by literature on Bayesian Logic Programs [19].

2.3 Semi-supervised and Transductive Learning

Classic discriminative learning methods tend to ignore unlabeled instances. However, real life scenarios are usually characterized by an abundance of unlabeled instances and a few labeled ones [6,35]. This may also be the case for class-membership prediction from formal ontologies: class-membership relations may be difficult to obtain during ontology engineering tasks (e.g. due to availability of domain experts) and inference (e.g. since deciding instance-membership may have an intractable time complexity in some languages).

Using unlabeled instances during learning is generally known in the machine learning community as *Semi-Supervised Learning* [6,35] (SSL). A variant to this setting is known as *Transductive Learning* [31] and refers to finding a labeling only to unlabeled instances provided in the training phase, without necessarily generalizing to unseen instances (and thus resulting into a possibly *simpler* learning problem). If the marginal distribution of instances P_X is informative with respect to the conditional probability distribution $P(Y \mid x)$, accounting for unlabeled instances during learning can provide more accurate results [6,35].

A possible approach is including terms dependent from P_X into the objective function. This results in the two fundamental assumptions [6]:

- **Cluster assumption** – The joint probability distribution $P(X, Y)$ is structured in such a way that points in the same *cluster* are likely to have the same label.
- **Manifold assumption** – Assume that the distribution P_X is supported on a low-dimensional manifold: then, $P(Y \mid x)$ *varies smoothly*, as a function of x, with respect to the underlying structure of the manifold.

In the following sections, we discuss a similarity-based, non-parametric and computationally efficient method for predicting missing class-membership relations. This method is discriminative in nature, but also accounts for unknown class-membership during learning.

We will face a slightly different version of the classic class-membership prediction problem, namely *transductive class-membership prediction*. It is inspired to the *Main Principle* in [31]: "If you possess a restricted amount of information for solving some problem, try to solve the problem directly and never solve a more general problem as an intermediate step. It is possible that the available information is sufficient for a direct solution but is insufficient for solving a more general intermediate problem". In this setting, the learning algorithm only aims at estimating the class-membership relation of interest for a given training set of individuals, without necessarily being able to generalize to individuals outside such set.

In this work, we formalize the transductive class-membership prediction problem as a cost minimization problem: given a set of training individuals $\text{Ind}_C(\mathcal{K})$ whose class-membership relation to a target concept C is either known or unknown, find a function $f^* : \text{Ind}_C(\mathcal{K}) \rightarrow \{+1, -1\}$ defined over training individuals and returning a value $+1$ (resp. -1) if the individual likely to be a member of C (resp. $\neg C$), minimizing a given cost function. More formally:

Definition 1 *(Transductive Class-Membership Prediction). The Transductive Class-Membership Prediction problem can be formalized as follows:*

- **Given:**
  - *a target concept C;*
  - *a set of training individuals $\text{Ind}_C(\mathcal{K})$ in a knowledge base $\mathcal{K}$ partitioned in positive, negative and neutral examples or, more formally, such that:*
    $$\text{Ind}_C^+(\mathcal{K}) = \{a \in \text{Ind}_C(\mathcal{K}) \mid \mathcal{K} \models C(a)\} \text{ positive examples,}$$
    $$\text{Ind}_C^-(\mathcal{K}) = \{a \in \text{Ind}_C(\mathcal{K}) \mid \mathcal{K} \models \neg C(a)\} \text{ negative examples,}$$
    $$\text{Ind}_C^0(\mathcal{K}) = \{a \in \text{Ind}_C(\mathcal{K}) \mid \mathcal{K} \not\models C(a) \wedge \mathcal{K} \not\models \neg C(a)\} \text{ neutral examples;}$$
  - *A cost function $cost(\cdot) : \mathcal{F} \mapsto \mathbb{R}$, specifying the cost associated to a set of class-membership relations assigned to training individuals by $f \in \mathcal{F}$, where $\mathcal{F}$ is a space of labeling functions of the form $f : \text{Ind}_C(\mathcal{K}) \mapsto \{+1, -1\}$;*
- **Find** *a labeling function $f^* \in \mathcal{F}$ minimizing the given cost function with respect to training individuals $\text{Ind}_C(\mathcal{K})$:*

$$f^* \leftarrow \arg\min_{f \in \mathcal{F}} cost(f).$$

The function f^* can then be used to estimate the class-membership relation with respect to the target concept C for all training individuals $a \in \text{Ind}_C(\mathcal{K})$: it will return $+1$ (resp. -1) if an individual is likely to be a member of C (resp. $\neg C$). Note that the function is defined on the whole set of training individuals; therefore it can possibly contradict already known class-membership relations (thus being able to handle noisy knowledge). If $\text{Ind}_C(\mathcal{K})$ is finite, the space of labeling functions $\mathcal{F}$ is also finite, and each function $f \in \mathcal{F}$ can be equivalently expressed as a vector in $\{-1, +1\}^n$, where $n = |\text{Ind}_C(\mathcal{K})|$.

3 Propagating Class-Membership Information Among Individuals

This section discusses a *graph-based semi-supervised* [35] method for class-membership prediction from DL representations. The proposed method relies on a weighted *semantic similarity graph*, where nodes represent positive, negative and neutral examples of the transductive class-membership prediction problem, and weighted edges define similarity relations among such individuals.

More formally, let $\mathcal{K}$ be a knowledge base, $\text{Ind}_C(\mathcal{K})$ a set of training individuals with respect to a target concept C in $\mathcal{K}$, and $Y = \{-1, +1\}$ a space of labels each corresponding to a type of class-membership relation with respect to C. Each training individual $a \in \text{Ind}_C(\mathcal{K})$ is associated to a label, which will be $+1$ (resp. -1) if $\mathcal{K} \models C(a)$ (resp. $\mathcal{K} \models \neg C(a)$), and will be unknown otherwise, thus representing an unlabeled instance. For defining a cost over functions $f \in \mathcal{F}$, the proposed method relies on *regularization by graph*: the learning process aims at finding a labeling function that is both consistent with given labels, and changes smoothly between similar instances (where similarity relations are encoded in the semantic similarity graph). This can be formalized through a *regularization framework*, using a measure of the consistency to the given labels as a loss function, and a measure of smoothness among the similarity graph as a regularizer.

Several cost functions have been proposed in SSL literature. An appealing class of functions, from the side of computational cost, relies on the *quadratic cost criterion* framework [6, ch. 11]: for this class of functions, a closed form solution to the cost minimization problem can be found efficiently (Subsect. 3.2).

3.1   Semantic Similarity Graph

A similarity graph can be represented with a weight matrix $\mathbf{W}$, where the value of $\mathbf{W}_{ij}$ represents the strength of the similarity relation between two training examples x_i and x_j. In graph-based SSL literature, $\mathbf{W}$ is often obtained either as a Nearest Neighbor (NN) graph (where each instance is connected to the k most similar instances in the graph, or to those with a distance under a radius ϵ); or by means of a kernel function, such as the Gaussian kernel.

Finding the best way to construct $\mathbf{W}$ is an active area of research. In [6, ch. 20] authors discuss a method to combine multiple similarity measures in the context of protein function prediction, while [1,18,32] propose different methods for data-driven similarity graph construction.

When empirically evaluating the proposed method, we employ the family of dissimilarity measures between individuals in a DL knowledge base defined in [25], since it does not constrain to any particular family of DLs. We refer to the resulting similarity graph among individuals in a formal ontology as the *semantic similarity graph*. Given a set of concept descriptions $F = \{F_1, \ldots, F_n\}$ and a weight vector $\mathbf{w}$, such family of dissimilarity measures $d_p^F : Ind_C(\mathcal{K}) \times Ind_C(\mathcal{K}) \mapsto [0, 1]$ is defined as:

$$\delta_i(x, y) = \begin{cases} 0 \text{ if } (\mathcal{K} \models F_i(x) \wedge \mathcal{K} \models F_i(y)) \vee (\mathcal{K} \models \neg F_i(x) \wedge \mathcal{K} \models \neg F_i(y)) \\ 1 \text{ if } (\mathcal{K} \models F_i(x) \wedge \mathcal{K} \models \neg F_i(y)) \vee (\mathcal{K} \models \neg F_i(x) \wedge \mathcal{K} \models F_i(y)) \\ u_i \text{ otherwise} \end{cases} \quad (1)$$

where $x, y \in Ind_C(\mathcal{K})$ and $p > 0$.

Two examples of (k-NN) semantic similarity graphs among all individuals in the ontologies BioPax (Proteomics) and Leo, obtained using the aforementioned dissimilarity measure, are provided in Fig. 1.

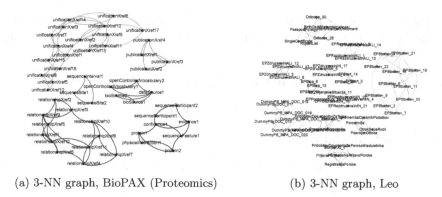

(a) 3-NN graph, BioPAX (Proteomics)   (b) 3-NN graph, Leo

Fig. 1. k-Nearest Neighbor Semantic Similarity graphs for individuals BioPAX (Proteomics) ontology (left) and for the Leo ontology (right), obtained using the dissimilarity measure in [25]: F was defined as the set of atomic concepts in the ontology (each weighted with its normalized entropy [25]) and $p = 2$.

3.2 Quadratic Cost Criteria

In quadratic cost criteria [6, ch. 11], the original label space $\{-1, +1\}$ (binary classification case) is relaxed to $[-1, +1]$. This allows expressing the confidence associated to a labeling (and possibly provide an indicator of $P(Y \mid x)$). For such a reason, in the proposed method, elements of the functions space $\mathcal{F}$ can be relaxed to the form $f : \text{Ind}_C(\mathcal{K}) \mapsto [-1, +1]$.

As in Subsect. 2.3, labeling functions can be equivalently represented as vectors $\mathbf{y} \in [-1, +1]^n$. Let $\hat{\mathbf{y}} \in [-1, +1]^n$ be a possible labeling for a set of n instances. We can see $\hat{\mathbf{y}}$ as a $(l + u) = n$ dimensional vector, where the first l indexes refer to already labeled instances, and the last u to unlabeled instances: $\hat{\mathbf{y}} = [\hat{\mathbf{y}}_l, \hat{\mathbf{y}}_u]$.

Consistency of $\hat{\mathbf{y}}$ with respect to original labels can be formulated in the form of a quadratic cost: $\sum_{i=1}^{l} (\hat{y}_i - y_i)^2 = ||\hat{\mathbf{y}}_l - \mathbf{y}_l||^2$.

To regularize the labellings with respect to the graph structure, the *graph Laplacian* [6] can be used. Let $\mathbf{W}$ be the adjacency (weight) matrix corresponding to the similarity graph G and let $\mathbf{D}$ be the diagonal matrix obtained from $\mathbf{W}$ as $\mathbf{D}_{ii} = \sum_{j=1}^{n} \mathbf{W}_{ij}$ (i.e. by summing the elements in each column of $\mathbf{W}$).

Hence, two alternative definitions for the graph Laplacian can be considered [6]:

Unnormalized graph Laplacian: $\mathbf{L} = \mathbf{D} - \mathbf{W}$;
Normalized graph Laplacian: $\mathcal{L} = \mathbf{D}^{-\frac{1}{2}} \mathbf{L} \mathbf{D}^{-\frac{1}{2}} = \mathbf{I} - \mathbf{D}^{-\frac{1}{2}} \mathbf{W} \mathbf{D}^{-\frac{1}{2}}$.

Another regularization term in the form of $||\hat{\mathbf{y}}||^2$ (or $||\hat{\mathbf{y}}_u||^2$, as in [33]) can be added to the final cost function to prefer smaller values in $\hat{\mathbf{y}}$. This is useful e.g. to prevent arbitrary labellings in a connected component of the semantic similarity graph containing no labeled instances (Fig. 2).

Putting the pieces together, we obtain two quadratic cost criteria discussed in the literature, namely Regression on Graph [2] (RG) and the Consistency Method [33] (CM):

Regression on Graph where the cost function can be written as:

$$cost(\hat{\mathbf{y}}) = ||\hat{\mathbf{y}}_l - \mathbf{y}_l||^2 + \mu \hat{\mathbf{y}}^T \mathbf{L} \hat{\mathbf{y}} + \mu \epsilon ||\hat{\mathbf{y}}||^2; \tag{2}$$

Consistency Method where the cost function can be written as:

$$cost(\hat{\mathbf{y}}) = ||\hat{\mathbf{y}}_l - \mathbf{y}_l||^2 + \mu \hat{\mathbf{y}}^T \mathcal{L} \hat{\mathbf{y}} + ||\hat{\mathbf{y}}_u||^2. \tag{3}$$

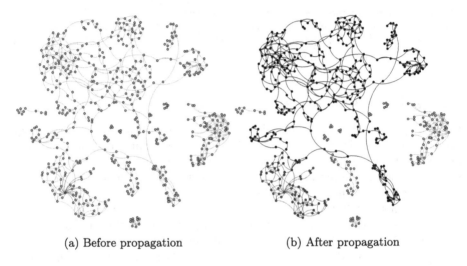

(a) Before propagation   (b) After propagation

Fig. 2. Example of information propagation, from a single individual, to nearby individuals in a sample similarity graph.

We will now derive a closed form solution for the problem of finding a (global) minimum for the quadratic cost criterion in RG; a similar process is also valid in the case of CM. Its first order derivative is defined as follows:

$$\frac{1}{2} \frac{\partial cost(\hat{\mathbf{y}})}{\partial \hat{\mathbf{y}}} = (\mathbf{S} + \mu \mathbf{L} + \mu \epsilon \mathbf{I}) \hat{\mathbf{y}} - \mathbf{S} \mathbf{y},$$

where $\mathbf{S} = diag(\mathbf{s}_1, \ldots, \mathbf{s}_n)$, with $\mathbf{s}_i = 1$ iff $i \leq l$ and 0 otherwise. Its second order derivative is a positive definite matrix if $\epsilon > 0$, since $\mathbf{L}$ is positive semi-definite. Therefore, setting the first order derivative to 0 leads to a global minimum:

$$\hat{\mathbf{y}} = (\mathbf{S} + \mu \mathbf{L} + \mu \epsilon \mathbf{I})^{-1} \mathbf{S} \mathbf{y}, \tag{4}$$

showing that $\hat{\mathbf{y}}$ can be obtained either by matrix inversion or by solving a (possibly sparse) linear system.

Complexity of Inference. The linear system in Eq. 4 can be computed efficiently, with a nearly-linear time complexity in the number of non-zero elements in the coefficient matrix. Indeed, computing $\hat{\mathbf{y}}$ can be reduced to solving a linear system in the form $\mathbf{A}\mathbf{x} = \mathbf{b}$, with $\mathbf{A} = (\mathbf{S} + \mu\mathbf{L} + \mu\epsilon\mathbf{I})$, $\mathbf{b} = \mathbf{S}\mathbf{y}$ and $\mathbf{x} = \hat{\mathbf{y}}$. A linear system $\mathbf{A}\mathbf{x} = \mathbf{b}$ with $\mathbf{A} \in \mathbb{R}^{n \times n}$ can be solved in nearly linear time if the coefficient matrix $\mathbf{A}$ is *symmetric diagonally dominant*[1] (SDD). An algorithm for solving SDD linear systems is proposed in [7]: its time-complexity is $\approx O(m \log^{1/2} n)$, where m is the number of non-zero entries in $\mathbf{A}$ and n is the number of variables in the system of linear equations. This result applies to the calculation in Eq. 4, since the graph Laplacian $\mathbf{L}$ is SDD [30], and thus the coefficient matrix $\mathbf{A}$ is SDD. An efficient parallel solver for SDD linear systems is proposed in [24].

Interpretation as a Probabilistic Graphical Model. The terms enforcing similar labels among nearby individuals and the regularizer in the cost functions in Eqs. 2 and 3 can be seen as *energy functions* [20] over $\hat{\mathbf{y}}$ in the form:

$$E(\hat{\mathbf{y}}) = \hat{\mathbf{y}}^T \tilde{\mathbf{L}} \hat{\mathbf{y}}, \text{ with } \hat{\mathbf{y}} \in \mathbb{R}^n, \tag{5}$$

where $\tilde{\mathbf{L}} = \mu(\mathbf{L} + \epsilon\mathbf{I})$ in Eq. 2 and $\tilde{\mathbf{L}} = \mu\mathcal{L} + \mathbf{I}$ in Eq. 3. The energy function in Eq. 5 corresponds to a *Gaussian Random Field* [20] (GRF):

$$p(\hat{\mathbf{y}}) = \frac{1}{Z} \exp\left[-\beta E(\hat{\mathbf{y}})\right] = \frac{1}{Z} \exp\left[-\beta \hat{\mathbf{y}}^T \tilde{\mathbf{L}} \hat{\mathbf{y}}\right], \tag{6}$$

where Z is a normalization factor and β is an "inverse temperature parameter". The GRF in Eq. 6 defines a multivariate Gaussian distribution $\mathcal{N}(\mathbf{0}, \mathbf{\Sigma})$ on the continuous labellings $\hat{\mathbf{y}}$, where $\mathbf{\Omega} = (2\beta\tilde{\mathbf{L}})$ and $\mathbf{\Sigma} = \mathbf{\Omega}^{-1}$ represent respectively its *information* (or *precision*) and *covariance* matrix. Such matrices encode the independence relations among variables in the multivariate Gaussian distribution (Fig. 3).

Given that $\hat{\mathbf{y}} \sim \mathcal{N}(\boldsymbol{\mu}, \mathbf{\Sigma})$, $\hat{\mathbf{y}}_i$ and $\hat{\mathbf{y}}_j$ are independent iff $\mathbf{\Sigma}_{ij} = 0$ (i.e. $\hat{\mathbf{y}}_i \perp \!\!\! \perp \hat{\mathbf{y}}_j$ iff $\mathbf{\Sigma}_{ij} = 0$), while $\hat{\mathbf{y}}_i$ and $\hat{\mathbf{y}}_j$ are independent conditioned on all the other variables iff $\mathbf{\Omega}_{ij} = 0$ (i.e. $\hat{\mathbf{y}}_i \perp \!\!\! \perp \hat{\mathbf{y}}_j \mid \hat{\mathbf{y}} - \{\hat{\mathbf{y}}_i, \hat{\mathbf{y}}_j\}$ iff $\mathbf{\Omega}_{ij} = 0$).

It is interesting to note that the information matrix $\mathbf{\Omega}$ (and hence the graph Laplacian of the similarity matrix) directly defines a minimal I-map Gaussian Markov random field (GMRF) for the distribution p [20], where non-zero entries in the matrix can be directly translated to edges in the GMRF.

Summary. This work leverages quadratic cost criteria to efficiently solve the transductive class-membership prediction problem. Finding a minimum $\hat{\mathbf{y}}$ for a predefined cost criterion is equivalent to finding a labeling function f^* in the form $f^* : \mathrm{Ind}_C(\mathcal{K}) \mapsto [-1, +1]$, where the labeling returned for a generic training individual $a \in \mathrm{Ind}_C(\mathcal{K})$ correspond to the value in $\hat{\mathbf{y}}$ in the position mapped to a. This can be done by representing the set of training individuals $\mathrm{Ind}_C(\mathcal{K})$

---

[1] A matrix $\mathbf{A}$ is SDD iff $\mathbf{A}$ is symmetric (i.e. $\mathbf{A} = \mathbf{A}^T$) and $\forall i : \mathbf{A}_{ii} \geq \sum_{i \neq j} |\mathbf{A}_{ij}|$.

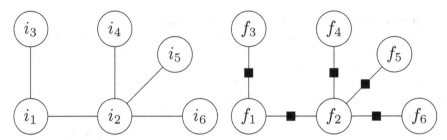

(a) Sample Similarity Graph among a set of 6 individuals in a Knowledge Base

(b) Corresponding GMRF defined over the soft-labels of the individuals

Fig. 3. (a) Sample Similarity Graph among a set of 6 individuals in a Knowledge Base. (b) Corresponding GMRF defined over the soft-labels of the individuals.

as a partially labeled vector **y** of length $|Ind_C(\mathcal{K})| = n$, such that the first l (resp. last u) components correspond to positive and negative (resp. neutral) examples in $Ind_C(\mathcal{K})$. Such **y** can be then used to measure the consistency with original labels in a quadratic cost criterion; while the semantic similarity graph can be employed to enforce smoothness in class-membership predictions among similar training individuals. An advantage of quadratic cost criteria is that their minimization ultimately reduces to solving a large sparse linear system with a SDD coefficient matrix. For large-scale datasets, a subset selection method is discussed in [6, ch. 18], which allows to greatly reduce the size of the original linear system.

4   Empirical Evaluations

In this section, we evaluate several (inductive and transductive) methods for class-membership prediction, with the aim of comparing the methods discussed in Sect. 3 with respect to other methods in SW literature. We are reporting evaluations for the Regularization on Graph [2] (RG) and the Consistency Method [33] (CM); Label Propagation [34] (LP); three kinds of Support Vector Machines [27] (SVM), namely Hard-Margin SVM (HM-SVM), Soft-Margin SVM with L_1 norm (SM-SVM) and Laplacian SVM [3] (LapSVM); and $\sqrt{l}$-Nearest Neighbors for class-membership prediction [25].

4.1   Description of Evaluated Methods

LP is a graph-based SSL algorithm relying on the idea of propagating labeling information among similar instances through an iterative process involving matrix operations. It can be equivalently formulated under the quadratic criterion framework [6, ch. 11]. More formally it associates, to each unlabeled instance in the graph, the probability of performing a random walk until a positively (resp. negatively) example is found.

We also evaluated Support Vector Machines (SVM), which have been proposed for inducing robust classifiers from ontological knowledge bases [13,25]. SVM classifiers come in different flavors: the classic HM-SVM binary classifier aims at finding the hyperplane in the feature space separating the instances belonging to different classes, which maximizes the *geometric margin* between the hyperplane and nearest training points. The SM-SVM classifier is a relaxation of HM-SVM, which allows for some misclassification in training instances (by relaxing the need of having perfectly linearly separable training instances in the feature space). LapSVM is a semi-supervised extension of the SM-SVM classifier: given a set of labeled instances and a set of unlabeled instances, it aims at finding an hyperplane that is also smooth with respect to the (estimated) geometry of instances. More formally, let $(\mathbf{x}_l, \mathbf{y}_l)$ (resp. $\mathbf{x}_u$) be a set of labeled (resp. unlabeled) instances. LapSVM finds a function f in a space of functions $\mathcal{H}_K$ determined by the kernel K (called *Reproducing Kernel Hilbert Space* [27]) minimizing $\frac{1}{l}\sum_{i=1}^{l} V(x_i, y_i, f) + \gamma_L \|f\|^2_{\mathcal{H}_K} + \gamma_{\mathcal{M}}\|f\|^2_{\mathcal{M}}$, where V represents a costs function of errors committed by f on labeled samples (typically the hinge loss function $\max\{0, 1 - y_i f(x_i)\}$), $\|\cdot\|_{\mathcal{H}_K}$ imposes smoothness conditions on possible solutions [27] and $\|\cdot\|^2_{\mathcal{M}}$, intuitively, penalizes rapid changes in the classification function between close instances in the similarity graph. It generalizes HM-SVM ($\gamma_L \to 0, \gamma_{\mathcal{M}} = 0$) and SM-SVM ($\gamma_{\mathcal{M}} = 0$). Our implementation of LapSVM follows the algorithm proposed in [3]; for HM-SVM, SM-SVM and LapSVM, we solve the underlying convex optimization problems using the Gurobi optimizer [15].

RG, CM, LP and LapSVM all rely on a semantic similarity graph $\mathbf{W}$ as a representation of the geometry of instances. We first calculate distances employing the dissimilarity measure defined in [25] and outlined in Eq. 1, with $p = 2$; then we obtain $\mathbf{W}$ by building a k-Nearest Neighbor graph using such distances (since sparsity in $\mathbf{W}$ influences the scalability of quadratic cost criteria, as written in Subsect. 3.2). When building the neighborhood of a node, we handled the cases in which nodes had the same distance by introducing a random ordering between such nodes. The Kernel function used for Hard-Margin SVM, Soft-Margin SVM and Laplacian SVM are also defined in [25], and directly correlated with the aforementioned dissimilarity measure in Eq. 1 (given a committee of concepts F and the parameters $\mathbf{w}$ and p, the dissimilarity was originally obtained as

Table 1. Ontologies considered in the experiments.

Ontology	Expressivity	#Axioms	#Inds.	#Classes	#ObjProps.
BioPax (Proteomics)	$\mathcal{ALCHN}(\mathcal{D})$	773	49	55	47
Family-Tree	$\mathcal{SROIF}(\mathcal{D})$	2059	368	22	52
Leo	$\mathcal{ALCHIF}(\mathcal{D})$	430	61	32	26
MDM0.73	$\mathcal{ALCHOF}(\mathcal{D})$	1098	112	196	22
Wine	$\mathcal{SHOIN}(\mathcal{D})$	1046	218	142	21

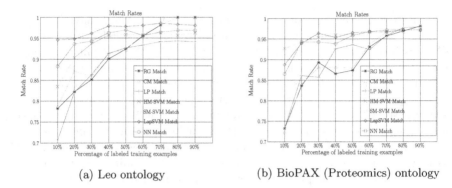

(a) Leo ontology   (b) BioPAX (Proteomics) ontology

Fig. 4. Variation of average Match Rates with respect to the number of folds used in the training step, during a k-Fold Cross Validation (with $k = 10$).

$1 - k(a, b)$, where $k(a, b)$ is the value of the kernel function on a pair of individuals (a, b) in the knowledge base). We also provide a first evaluation for the k-NN algorithm (with $k = \sqrt{l}$, where l is the number of labeled instances, as discussed in [25]): we simply choose the majority class among the $\sqrt{l}$ most similar individuals to label each unlabeled instance.

4.2 Evaluations

Starting from a set of real ontologies[2] (outlined in Table 1), we generated a set of 20 random query concepts for each ontology[3], so that the number of individuals belonging to the target query concept C (resp. $\neg C$) was at least of 10 elements and the number of individuals in C and $\neg C$ was in the same order of magnitude. A DL reasoner[4] was employed to decide on the theoretical concept-membership of individuals to query concepts. We employ the evaluation metrics in [8], which take into account the peculiarities deriving by the presence of missing knowledge:

Match Case of an individual that got the same label by the reasoner and the inductive classifier.

Omission Error Case of an individual for which the inductive method could not determine whether it was relevant to the query concept or not while it was found relevant by the reasoner.

Commission Error Case of an individual found to be relevant to the query concept while it logically belongs to its negation or vice-versa.

Induction Case of an individual found to be relevant to the query concept or to its negation, while either case is not logically derivable from the knowledge base.

---

[2] From TONES Repository: http://owl.cs.manchester.ac.uk/repository/.
[3] Using the methods available at http://lacam.di.uniba.it/~nico/research/ontologymining.html.
[4] Pellet v2.3.0 – http://clarkparsia.com/pellet/.

Before evaluating on the test set, parameter tuning was performed for each of the methods via a k-Fold Cross Validation ($k = 10$) within the training set, for finding the parameters with lower classification error in cross-validation. For LapSVM, the $(\gamma_L, \gamma_\mathcal{M})$ parameters were varied in $\{10^{-4}, 10^{-3}, \ldots, 10^4\}$, while for SM-SVM, which follows the implementation in [27, pg. 223], the C parameter was allowed to vary in $\{10^{-4}, 10^{-3}, \ldots, 10^4\}$. Similarly, the (μ, ϵ) parameters in RG and CM where varied in $\{10^{-4}, 10^{-3}, \ldots, 10^4\}$. The parameter k for building

Table 2. Match, Omission, Commission and Induction [25] results for a k-Fold Cross Validation ($k = 10$) on 20 randomly generated queries. For each experiment, the best parameters within the training were found using a k-Fold Cross Validation ($k = 10$).

Leo	Match	Omission	Commission	Induction
RG	1 ± 0	0 ± 0	0 ± 0	0 ± 0
CM	1 ± 0	0 ± 0	0 ± 0	0 ± 0
LP	0.942 ± 0.099	0.007 ± 0.047	0.052 ± 0.091	0 ± 0
SM-SVM	0.963 ± 0.1	0 ± 0	0.037 ± 0.1	0 ± 0
LapSVM	0.978 ± 0.068	0 ± 0	0.022 ± 0.068	0 ± 0
$\sqrt{l}$-NN	0.971 ± 0.063	0 ± 0	0.029 ± 0.063	0 ± 0
BioPAX (Proteomics)	Match	Omission	Commission	Induction
RG	0.986 ± 0.051	0.004 ± 0.028	0.008 ± 0.039	0.002 ± 0.02
CM	0.986 ± 0.051	0.002 ± 0.02	0.01 ± 0.044	0.002 ± 0.02
LP	0.982 ± 0.058	0.002 ± 0.02	0.014 ± 0.051	0.002 ± 0.02
SM-SVM	0.972 ± 0.075	0 ± 0	0.026 ± 0.068	0.002 ± 0.02
LapSVM	0.972 ± 0.075	0 ± 0	0.026 ± 0.068	0.002 ± 0.02
$\sqrt{l}$-NN	0.972 ± 0.075	0 ± 0	0.026 ± 0.068	0.002 ± 0.02
MDM0.73	Match	Omission	Commission	Induction
RG	0.953 ± 0.063	0.003 ± 0.016	0.011 ± 0.032	0.015 ± 0.039
CM	0.953 ± 0.063	0.001 ± 0.009	0.013 ± 0.036	0.018 ± 0.04
LP	0.942 ± 0.065	0 ± 0	0.026 ± 0.046	0.033 ± 0.054
SM-SVM	0.793 ± 0.252	0 ± 0	0.174 ± 0.255	0.033 ± 0.054
LapSVM	0.915 ± 0.086	0 ± 0	0.052 ± 0.065	0.033 ± 0.054
$\sqrt{l}$-NN	0.944 ± 0.069	0 ± 0	0.023 ± 0.051	0.033 ± 0.054
Wine	Match	Omission	Commission	Induction
RG	0.24 ± 0.03	0 ± 0.005	0.007 ± 0.017	0.5 ± 0.176
CM	0.242 ± 0.028	0 ± 0.005	0.005 ± 0.015	0.326 ± 0.121
LP	0.239 ± 0.035	0 ± 0.005	0.008 ± 0.021	0.656 ± 0.142
SM-SVM	0.235 ± 0.036	0 ± 0	0.012 ± 0.024	0.753 ± 0.024
LapSVM	0.238 ± 0.033	0 ± 0	0.009 ± 0.021	0.753 ± 0.024
$\sqrt{l}$-NN	0.241 ± 0.031	0 ± 0	0.006 ± 0.018	0.753 ± 0.024

the k-NN semantic similarity graph, used by LapSVM, RG, CM and LP, was varied in $\{2, 4, 8, 16\}$. We did not carefully choose the concept committee F defining the dissimilarity measure: we simply used the set of atomic concepts in the ontology, thus ignoring any prior knowledge about the structure of the target concept C or the presence of statistical correlations in the knowledge base. Each concept in the committee F was weighted with its normalized entropy [25]. RG, CM and LP give an indication of the uncertainty associated to a specific labeling by associating values in the set $[-1, +1]$ to each node. A labeling $x \approx 0$ (specifically, when the label was in the set $[-10^{-4}, 10^{-4}]$ we decided to leave the node unlabeled, so to try to provide more robust estimates of labels (and thus a possibly lower commission error and match rates and higher omission error rates). This may happen e.g. when there are no labeled examples within a connected component of the semantic similarity graph.

In Table 2 we report average index rates and standard deviations for each of the ontologies in Table 1; the only exceptions is for the FAMILY-TREE ontology, which provided 0.76 ± 0.13 match rates and 0.24 ± 0.13 induction rates for all methods (except for LP, where the induction rates were 0.21 ± 0.14. In general, LapSVM outperformed the other two non-SSL SVM classification methods. This happened with varying quantities of unlabeled data; this is shown for example in the behavior of match rates in Fig. 4a, where results obtained in a k-Fold Cross Validation using a varying quantity of labeled instances. However, standard SVM training is $O(m^3)$ in general, where m is the number of training instances; therefore, some extra effort may be necessary to make SVM methods scale on SW knowledge bases. Such results may provide some empirical evidence that inductive methods for formal ontologies may take benefit from also accounting for unlabeled instances during learning.

4.3 Limitations

A fundamental problem in graph-based SSL methods relies in the construction of the similarity graph [6,35], which is known to have a strong impact on the effectiveness of SSL methods. In this work, we identified similarity relations among individuals using a measure defined in [25] together with a set of atomic concepts defined in the ontology. However, this might not always be effective (consider e.g. a shallow ontology, where only a few properties of each individual are described by means of atomic concepts).

A possible approach to leverage the relational graph structure between individuals in a DL ontology, would be the use of *graph* and *RDF kernels*, such as the one defined in [4,22,28]. By implicitly mapping individuals into a feature space, a kernel function $k(\cdot, \cdot)$ naturally induces a Euclidean distance in the kernel feature space [27]:

$$||\phi(x_i) - \phi(x_j)||^2 = k(x_i, x_i) + k(x_j, x_j) - 2k(x_i, x_j),$$

where ϕ is a function mapping each instance to some feature space. However, this might not always be effective: Fig. 5 shows a Semantic Similarity Graph constructed

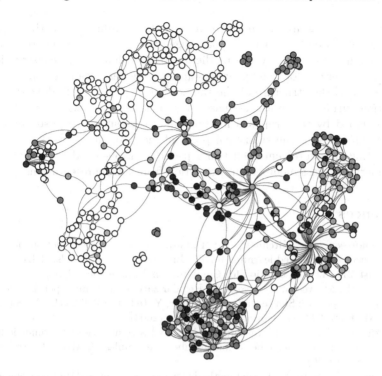

Fig. 5. Semantic Similarity Graph for the persons in the AIFB PORTAL ontology, after removing predicates encoding research group affiliations: each color corresponds to a distinct research group affiliation (white was used when no affiliation was available).

among persons in the AIFB PORTAL Ontology[5], using the RDF kernel described in [22] (ignoring research group affiliations), where each different colors corresponds to a distinct research group. Similarity relations among individuals inferred by such a kernel do not accurately reflect similarities in research group affiliations, suggesting that the choice of a kernel could be task dependent in some contexts.

5   Conclusion and Future Works

This work proposes a method for transductive class-membership prediction based on graph-based regularization from DL representations. It leverages neutral examples by propagating class-membership information among similar individuals in the training set. The proposed method relies on quadratic cost criteria, whose optimization can be reduced to solving a (possibly sparse) symmetric and diagonally dominant linear system. This is a well-known problem in the literature, with a nearly linear time complexity in the number of non-zero entries in the coefficient matrix.

---

[5] http://www.aifb.kit.edu/web/Wissensmanagement/Portal, as of 21 Feb. 2012.

The similarity graph is known to have a strong influence on the effectiveness of graph-based SSL methods [35], suggesting that the graph construction process might be guided by the prediction task at hand. The construction of the similarity graph for class-membership learning tasks can be influenced by factors such as the structure of the target concept C, or by finding statistical correlation within the knowledge base. Also, it is not clear whether continuous labels assigned by the proposed methods may correspond to posterior probability estimates from the statistical point of view. In future work, we aim at investigating the aforementioned two aspects of graph-based transductive and semi-supervised class-membership prediction from DL representations.

References

1. Alexandrescu, A., Kirchhoff, K.: Data-driven graph construction for semi-supervised graph-based learning in NLP. In: Sidner, C., et al. (eds.) HLT-NAACL, pp. 204–211. The Association for Computational Linguistics (2007)
2. Belkin, M., Matveeva, I., Niyogi, P.: Regularization and semi-supervised learning on large graphs. In: Shawe-Taylor, J., Singer, Y. (eds.) COLT 2004. LNCS (LNAI), vol. 3120, pp. 624–638. Springer, Heidelberg (2004)
3. Belkin, M., Niyogi, P., Sindhwani, V.: Manifold regularization: a geometric framework for learning from labeled and unlabeled examples. J. Mach. Learn. Res. **7**, 2399–2434 (2006)
4. Bloehdorn, S., Sure, Y.: Kernel methods for mining instance data in ontologies. In: Aberer, K., et al. (eds.) ASWC 2007 and ISWC 2007. LNCS, vol. 4825, pp. 58–71. Springer, Heidelberg (2007)
5. Borgida, A., Walsh, T., Hirsh, H.: Towards measuring similarity in description logics. In: Horrocks, I., et al. (eds.) Description Logics. CEUR Workshop Proceedings, vol. 147. CEUR-WS.org (2005)
6. Chapelle, O., Schölkopf, B., Zien, A. (eds.): Semi-Supervised Learning. MIT Press, Cambridge (2006)
7. Cohen, M.B., Kyng, R., Miller, G.L., Pachocki, J.W., Peng, R., Rao, A., Xu, S.C.: Solving sdd linear systems in nearly $m\log^{1/2}n$ time. In: Shmoys [29], pp. 343–352
8. d'Amato, C., Fanizzi, N., Esposito, F.: Query answering and ontology population: an inductive approach. In: Bechhofer, S., Hauswirth, M., Hoffmann, J., Koubarakis, M. (eds.) ESWC 2008. LNCS, vol. 5021, pp. 288–302. Springer, Heidelberg (2008)
9. d'Amato, C., Fanizzi, N., Esposito, F.: A semantic similarity measure for expressive description logics. CoRR abs/0911.5043 (2009)
10. d'Amato, C., Staab, S., Fanizzi, N.: On the influence of description logics ontologies on conceptual similarity. In: Euzenat, J., Gangemi, A. (eds.) EKAW 2008. LNCS (LNAI), vol. 5268, pp. 48–63. Springer, Heidelberg (2008)
11. Fanizzi, N., d'Amato, C.: Inductive concept retrieval and query answering with semantic knowledge bases through kernel methods. In: Apolloni, B., Howlett, R.J., Jain, L. (eds.) KES 2007, Part I. LNCS (LNAI), vol. 4692, pp. 148–155. Springer, Heidelberg (2007)
12. Fanizzi, N., d'Amato, C., Esposito, F.: REDUCE: a reduced coulomb energy network method for approximate classification. In: Aroyo, L., et al. (eds.) ESWC 2009. LNCS, vol. 5554, pp. 323–337. Springer, Heidelberg (2009)

13. Fanizzi, N., d'Amato, C., Esposito, F.: Statistical learning for inductive query answering on OWL ontologies. In: Sheth, A.P., Staab, S., Dean, M., Paolucci, M., Maynard, D., Finin, T., Thirunarayan, K. (eds.) ISWC 2008. LNCS, vol. 5318, pp. 195–212. Springer, Heidelberg (2008)
14. Fanizzi, N., d'Amato, C., Esposito, F.: Towards learning to rank in description logics. In: Coelho, H., et al. (eds.) ECAI. Frontiers in Artificial Intelligence and Applications, vol. 215, pp. 985–986. IOS Press (2010)
15. Gurobi Optimization, Inc. Gurobi optimizer reference manual (2012)
16. Hu, B., Dasmahapatra, S., Lewis, P.: Semantic metrics. Int. J. Metadata Semant. Ontologies 2(4), 242–258 (2007)
17. Janowicz, K., Wilkes, M.: SIM-DL$_A$: a novel semantic similarity measure for description logics reducing inter-concept to inter-instance similarity. In: Aroyo, L., et al. (eds.) ESWC 2009. LNCS, vol. 5554, pp. 353–367. Springer, Heidelberg (2009)
18. Kapoor, A., Qi, Y.A., Ahn, H., Picard, R.W.: Hyperparameter and kernel learning for graph based semi-supervised classification. In: Advances in Neural Information Processing Systems 18, Neural Information Processing Systems, NIPS 2005, Vancouver, British Columbia, Canada, 5–8 December (2005)
19. Kersting, K., Raedt, L.D.: Bayesian logic programming: theory and tool. In: Getoor, L., Taskar, B. (eds.) An Introduction to Statistical Relational Learning. MIT Press, Cambridge (2007)
20. Koller, D., Friedman, N.: Probabilistic Graphical Models: Principles and Techniques. MIT Press, Cambridge (2009)
21. Lasserre, J., Bishop, C.M.: Generative or discriminative? Getting the best of both worlds. Bayesian Stat. 8, 3–24 (2007)
22. Lösch, U., Bloehdorn, S., Rettinger, A.: Graph kernels for RDF data. In: Simperl, E., Cimiano, P., Polleres, A., Corcho, O., Presutti, V. (eds.) ESWC 2012. LNCS, vol. 7295, pp. 134–148. Springer, Heidelberg (2012)
23. Ochoa-Luna, J.E., Cozman, F.G.: An algorithm for learning with probabilistic description logics. In: Bobillo, F., et al. (eds.) URSW, pp. 63–74 (2009)
24. Peng, R., Spielman, D.A.: An efficient parallel solver for sdd linear systems. In: Shmoys [29], pp. 333–342
25. Rettinger, A., Lösch, U., Tresp, V., d'Amato, C., Fanizzi, N.: Mining the semantic web - statistical learning for next generation knowledge bases. Data Mining and Knowledge Discovery - Special Issue on Web Mining (2012)
26. Rettinger, A., Nickles, M., Tresp, V.: Statistical relational learning with formal ontologies. In: Buntine, W., Grobelnik, M., Mladenić, D., Shawe-Taylor, J. (eds.) ECML PKDD 2009, Part II. LNCS, vol. 5782, pp. 286–301. Springer, Heidelberg (2009)
27. Shawe-Taylor, J., Cristianini, N.: Kernel Methods for Pattern Analysis. Cambridge University Press, New York (2004)
28. Shervashidze, N., Schweitzer, P., van Leeuwen, E.J., Mehlhorn, K., Borgwardt, K.M.: Weisfeiler-Lehman graph kernels. J. Mach. Learn. Res. 12, 2539–2561 (2011)
29. Shmoys, D.B. (ed.): Symposium on Theory of Computing, STOC 2014. ACM, New York, 31 May–03 June 2014
30. Spielman, D.A.: Algorithms, graph theory, and linear equations in Laplacian matrices. In: Proceedings of the International Congress of Mathematicians 2010 (ICM 2010), pp. 2698–2722 (2010)
31. Vapnik, V.N.: Statistical Learning Theory, 1st edn. Wiley, New York (1998)

32. Zhang, X., Lee, W.S.: Hyperparameter learning for graph based semi-supervised learning algorithms. In: Schölkopf, B., et al. (eds.) Advances in Neural Information Processing Systems 19, Proceedings of the Twentieth Annual Conference on Neural Information Processing Systems, Vancouver, British Columbia, Canada, December 4–7, 2006, pp. 1585–1592. MIT Press (2006)
33. Zhou, D., Bousquet, O., Lal, T.N., Weston, J., Schölkopf, B.: Learning with local and global consistency. In: Thrun, S., et al. (eds.) Advances in Neural Information Processing Systems 16, Neural Information Processing Systems, NIPS 2003, Vancouver and Whistler, British Columbia, Canada, 8–13 December 2003. MIT Press (2003)
34. Zhu, X., Ghahramani, Z.: Learning from labeled and unlabeled data with label propagation. Technical report, CMU CALD tech report CMU-CALD-02 (2002)
35. Zhu, X.: Semi-supervised learning literature survey. Technical report, 1530, Computer Sciences, University of Wisconsin-Madison (2005)

Analyzing User Demographics and User Behavior for Trust Assessment

Davide Ceolin[1]([☒]), Paul Groth[1], Archana Nottamkandath[1], Wan Fokkink[1], and Willem Robert van Hage[2]

[1] VU University, Amsterdam, The Netherlands
{d.ceolin,p.t.groth,a.nottamkandath,w.j.fokkink}@vu.nl
[2] Synerscope B.V., Eindhoven, The Netherlands
willem.van.hage@synerscope.com

Abstract. In many systems, the determination of trust is reduced to reputation estimation. However, reputation is just one way of determining trust. The estimation of trust can be tackled from a variety of other perspectives. In this chapter, we model trust relying on user reputation, user demographics and from provenance. We then explore the effects of combining trust computed through these different methods. Concretely, the first contribution of this chapter is a study of the correlations of demographics with trust. This study helps us to understand which categories of users are better candidates for annotation tasks in the cultural heritage domain. Secondly, we detail a procedure for computing reputation-based trust assessments. The user reputation is modeled in subjective logic based on the user's performance in the system evaluated (*Waisda?* in the case of the work presented here). The third contribution is a procedure for computing trust values based on provenance information, represented using the W3C PROV model. We show how merging the results of these procedures can be beneficial for the reliability of the estimated trust value. We evaluate the proposed procedures and their merger by estimating and verifying the trustworthiness of the tags created within the *Waisda?* video tagging game from the Netherlands Institute for Sound and Vision. Through a quantitative analysis of the results, we demonstrate that using provenance and demographic information is beneficial for the accuracy of trust assessments.

Keywords: Trust · Provenance · Subjective logic · Machine learning · Uncertainty reasoning · Tags

1 Introduction

From deciding the next book to read to selecting the best movie review, we often use the reputation of the author to ascertain the trust in the thing itself. Reputation is an important mechanism in our set of strategies to place trust. In fact, trusting (or placing trust) is an action that we decide or not to perform after having evaluated specific indicators (as specified by O'Hara [25] and also by

© Springer International Publishing Switzerland 2014
F. Bobillo et al. (Eds.): URSW 2011-2013, LNAI 8816, pp. 219–241, 2014.
DOI: 10.1007/978-3-319-13413-0_12

Castelfranchi and Falcone, in their theory reprised by Sabater and Sierra [31]), and reputation, i.e. the quantification of a user trustworthiness, is one of these. However, we may base our trust assessment on a variety of other factors as well, including prior performance, a guarantee, or knowledge of how something was produced. Nevertheless, many systems, especially on the Web, choose to reduce trust to reputation estimation and analysis alone. In this work, we take a multi-faceted approach. We look at trust assessment of Web data based on user reputation, provenance (i.e., how data has been produced), and the combination of the two. We also determine the trust on the user based on user profile stereotypes, that are user groups created on the basis of their demographic information. We try to determine correlations between the demographics and the quality of information provided by the users. We use the term "trust" for the trust in information resources and "reputation" for the trust in agents (see the work of Artz and Gil [1] for complete definitions).

We know that over the Web "anyone can say anything about any topic" [35], and this constitutes one of the strengths of the Semantic Web (and of the Web in general), since it brings democracy to it (everybody has the same right to contribute) and does not prevent a priori any possible useful contribution. This makes the Semantic Web a suitable environment for building crowdsourcing platforms. These platforms are useful to collect data (e.g., annotations) from a variety of users, for instance to help cultural heritage and other institutions to classify their collections. However, this brings along trust concerns, since the variety of the contributors can affect both the quality and the trustworthiness of the data. One mechanism for addressing these concerns is to leverage the reputation of the users and the provenance of data.

We perform a series of analyses to demonstrate the existence of correlation between user demographics and identity and the trustworthiness of the data they provide. On the bases of such results, we first propose a procedure for computing reputation that uses basic evidential reasoning principles and is implemented by means of subjective logic opinions [19]. Secondly, we propose a procedure for computing trust assessments based on provenance information represented in the W3C PROV model [34]. Such a procedure is important because it is not always possible to have complete user demographic information. Here, PROV plays a key role, both because of the availability of provenance data over the Web captured using this standard, and because of its role of interchange format: having modeled our procedure on PROV, any other different input format can be easily treated after having been mapped to PROV. We implement this procedure by discretizing the trust values and applying support vector machine classification. Finally, we combine these two procedures in order to maximize the benefit of both. The procedures are evaluated on data provided by the *Waisda?* [24] tagging game[1], where users challenge each other in tagging videos. We show how to use the FOAF ontology to represent the user information provided in their profiles, and we provide a small extension of it to represent user stereotypes.

---

[1] A zip file containing the R and Python procedures used, together with the dataset, is retrievable at http://trustingwebdata.org/books/URSW_III/Waisda.zip.

A stereotype is an abstraction of user demographics. We then provide a procedure to compute the user trustworthiness based on stereotypes from information in user profiles. Through our experiments, we try to determine correlations between the trust of the users and the stereotype of their profile.

We show that a reputation-based prediction is not significantly different from a provenance-based prediction and, by combining the two, we obtain a small but statistically significant improvement in our predictions. We also show that reputation-based and provenance-based assessments correlate and that there is a correlation between the user profile stereotypes and the trust in a user.

This chapter is based on preliminary results published on the paper "Trust Evaluation through User Reputation and Provenance Analysis" [7], presented at the 8th Uncertainty Reasoning for the Semantic Web Workshop at the 11th International Semantic Web Conference 2012. We have revised these results and added an analysis of the correlation between demographics and trustworthiness.

The rest of the chapter is organized as follows: Sect. 2 describes related work, Sect. 3 describes the dataset used for evaluation, Sects. 5–7 introduce respectively the trust assessment procedures based on reputation, provenance and their combination, including example associated experiments. Section 8 concludes.

2 Related Work

Trust is a widely explored topic within a variety of computer science areas. Here, we focus on those works directly touching upon the intersection of trust, provenance, Semantic Web and Web. We refer the reader to the work of Sabater and Sierra [31], Artz and Gil [1], and Golbeck [16] for comprehensive reviews about trust in respectively artificial intelligence, Semantic Web and Web. The first part of our work focuses on reputation estimation and is inspired by the works collected by Masum and Tovey [23]. Pantola et al. [26] present reputation systems that measure the overall reputation of the authors based on the quality of their contribution and the "seriousness" of their ratings; Javanmardi et al. [18] measure reputation based on user edit patterns and statistics. Their approaches are similar to ours, but they are particularly tailored to wiki-based environments. The second part of our work focuses on the usage of provenance information for estimating trust assessments. In their works, Bizer and Cyganiak [2], Hartig and Zhao [17] and Zaihrayeu et al. [38], use provenance and background information expressed as annotated or named graphs [4] to produce trust values. We do not make use of annotated or named graph, but we use provenance graphs as features for classifying the trustworthiness of artifacts. The same difference also applies to the two works of Rajbhandari et al. [29,30], where they quantify the trustworthiness of scientific workflows and they evaluate it by means of probabilistic and fuzzy models. The use of provenance information for computing trust assessments has also been investigated in a previous work of ours [6] where we determined the trustworthiness of event descriptions based on provenance information by applying subjective logic [19] to provenance traces of event descriptions. In the current chapter, we still represent trust values by means of subjective opinions, but trust assessments are made by means of support vector machines, eventually combined with reputations, again represented by means of subjective opinions. The impact of user information such as age, gender, education and demographics in crowd sourcing tasks have been explored in the works of [20]. In their paper, they explore the relationship between worker characteristics and the quality of their work. Their work has been applied to the crowdsourcing domain and has proven that both the demographics and personality profiles of the workers are strongly linked to the resulting label quality. We apply our algorithm not on a labelling task on a crowdsourcing platform, but on a video annotation task.

Another work by Venanzi et al. [33] addresses the issue of having too few labels from a user to determine their quality by using a community based Bayesian label aggregation model which assumes that crowd workers conform to a few different types, where each type represents a group of workers with similar confusion matrices. We use a similar approach to build stereotypes of users behaviour based on information provided by the users, but not for crowdsourcing systems. Their work is performed on the labeling task while ours is done on annotations of videos. In general, much work has been done in crowdsourcing platforms to determine the effect of a user profile on user accuracy and reputation (see [20,33]). However, these works focus mainly on labeling crowdsourced data where ground truth data is already available. The main difference between our work on determining correlation of user profiles on their quality with the above mentioned work is that we do not have a ground truth. For the labeling tasks on the crowdsourcing platforms, there is ground truth available for both the works. In our case, we lack such information and thus rely on partial evidence, which is that we trust a tag provided by a user more if there are other users who provided the same tag into the system. Also, the procedure introduced in Sect. 5 is a generalization of the procedure that we implemented in a few preceding works [8–10], where we evaluated the trustworthiness of tags of the Steve.Museum [32] artifact collection.

Lastly, the use of stereotyping as a bootstrapping method has already been investigated by Liu et al. [22] and Burnett et al. [5]. There exist relevant similarities between these works and ours, like, for example, the use of subjective logic to represent trust (this probabilistic logic makes use of Beta and Dirichlet distributions to model trust statistically) and the fact that users can be grouped in stereotypes to obtain useful informations to assess unknown users. Nevertheless, there exist also relevant differences. In fact, both these papers take an agent-approach and their final goal is to determine whether we can trust an agent or not. Our goal, instead, is to determine the agent's (user's) trustworthiness to be able to use it to determine the trustworthiness of the artifact that he or she produces. Also, Burnett et al. propose that agents can learn a stereotyping function, and also Liu et al. propose that stereotyping is based on a function, although they do not investigate it. In our work, we propose to create stereotypes based on user characteristics (and hence, implicitly, on a function of these characteristics), although we do not explicitly characterize this function.

3   The *Waisda?* Dataset

Waisda? [24] is a video tagging gaming platform launched by the Netherlands Institute for Sound and Vision in collaboration with the public Dutch broadcaster KRO. The game's logic is simple: users watch video and tag the content. Whenever two or more players insert the same tag about the same video in the same time frame (10 s, relative to the video), they are both rewarded. The number of matches for a tag is used as an estimate of its trustworthiness. When a tag which is not matched by others, it is not necessarily considered to be untrustworthy, because, for instance, it can refer to an element of the video unnoticed by other users, or it can belong to a niche vocabulary. Thus, tags that have no matches are not necessarily wrong. In the game, when counting matching tags, typos or synonymity are not taken into consideration.

We validate our procedures by using tag matching to estimate the trustworthiness of tag entries produced within the game. Our total corpus contains 37850 tag entries corresponding to 115 tags randomly chosen. These tag entries correspond to about 9 % of the total population. We have checked their representativity with respect to the entire dataset. First, we compared the distribution of each relevant feature that we will use in Sect. 6 in our sample with the distribution of the same feature in the entire dataset. A 95 % confidence level Chi-squared test [28] confirmed that the hour of the day and the day of the week distribute similarly in our sample and in the entire dataset. The typing duration distributions (i.e., distributions of the time employed by users to insert tags) instead, are significantly different according to a 95 % confidence level Wilcoxon signed-rank test [37]. However, the mode of the two distributions are the same, and the mean differs only 0.1 s which, according to the KLM-GOMS model [3], corresponds, at most, to a keystroke. So we conclude that the used sample is representative for the entire data set. A second analysis showed that, by randomly selecting other sets of 115 tags, the corresponding tag entries are not statistically different from the sample that we used. We used 26495 tag entries (70 %) as a training set, and the remaining 11355 (30 %) as a test set.

In order to determine the correlation of user profile information with user reputation, we used the data from 17 users who provided information about themselves in their user profiles. The remaining users did not provide their data or chose to remain anonymous. Initially, we tried to cluster the users based on their features such as age, number of contributions etc., and tried to draw conclusions about certain stereotypes. However since we had too few users to draw conclusions based on this approach, we opted, instead, to use standard correlation metrics on our data. We used Pearson correlation for the continuous data such as number of tags provided, number of tags provided which were matched with others and their age. For categorical variables such as gender, we used the point biserial correlation metric.

4   Analysis of Correlation Between User Demographics and Data Trustworthiness

Demographics is the set of quantifiable statistics about a population. A user profile is a collection of personal information about a given user. In this work, we assume that information collected by aggregation of user profiles represents the demographics of the population.

Here, we try to determine if there is a correlation between the user reputation and demographics in the *Waisda?* system. We use the user reputation as a proxy for data trustworthiness.

Our analysis is performed by grouping users based on their demographics and by identifying a correlation between user groups and the trustworthiness of the artifacts they produced. The drawback of our approach is that the users need to provide their details to the system. Since *Waisda?* is an online game, many users chose to participate as anonymous. We realised that the users who actively returned back to the game are mostly the ones who provided their profile information. This is a good indication of which users will actively participate in the system for a longer time. Another thing to note is that, in general, the users may not provide accurate information about themselves in their profile. However, for the sake of this work, we do not take this possibility into account because the users that provided their personal information in the game are known, and hence their information trusted. Moreover, since we take a statistical approach, information inaccuracies, if any, are compensated. The reason why we investigate the correlation between demographics and data trustworthiness is that we hypothesize that certain categories of users may be better performing than others. For instance, younger users may be more attentive or older users may be more accurate. If that is the case, then the stereotype that we define should help us in identifying groups of users whose performance are higher or lower than others.

4.1   User Profiles and Their Representation

The information in the user profile and other quantitative information derived about a user can help to estimate user reputation. Although different systems gather different types of information from a user, there is an overlap between the most common features such as the age, gender, education, etc. Such information provided by the user can be represented using the FOAF ontology [13]. FOAF provides a representation of the individual user along with his details. Apart from the user provided details, we also derive information such as the number of tags contributed by the user, percentage of tags matched with other users, etc. For representing data that are specific to the tagging environment and system, we do not adopt a standard and we use an ad-hoc representation (the property *ex:contributed_tags* for the number of user contributed tags, and the property *ex:matched_tags* for the number of matched tags for a given user).

In our procedure, we also build groups (or stereotypes) of users who share similar characteristics. In order to form groups of users, we use percentiles for each characteristic in their profile and derived characteristics. Percentiles help in obtaining an even distribution of the users across different profile characteristics and grouping them in stereotypes. One example of a stereotype can be users who are at least 30 years old and female. In order to represent these groups or stereotypes, we utilize the *group* class of FOAF. The groups are formed based on the information in the individual FOAF profile. Figure 1 depicts an example of users Alice and Mary who are both females above 30 years of age and belong to the same stereotype. In Fig. 1, the stereotype is represented by an entity of type stereo:stereotype, that is a subclass of the foaf:group class. We propose such a subclass to represent user stereotypes. The fact that we use FOAF and a small extension of it is important, because it eases interoperability with the systems that use this widely adopted ontology.

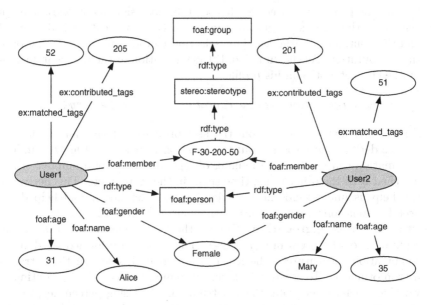

Fig. 1. Graph representation of the users and groups. The group name F-30-200-50 is formed by female users that are older than 30, provided more than 200 tags, and more than 50 of these are matched.

In the next section, we explain a procedure for predicting the reputation of a user based on the aggregation of the reputations of users within the same stereotype.

4.2   Procedure for Analyzing the Correlation Between User Demographics and Reputation

In order to evaluate the correlation between user demographics and the trustworthiness of the artifacts that they produce, we developed a procedure that groups users in stereotypes according to their personal information, and we check the existence of correlations between the fact that a given user belongs to a certain stereotype and their reputation.

The procedure is as follows:

proc reputation_profile_prediction(*user, reputation, user_profile*) ≡
    attribute_set := attribute_selection(user_profile,)
    attributes := attribute_extraction(attribute_profile)
    reputation_levels_aggregation
    classified_testset := classify(testset, trainingset)

The subprocedures used are described below:

attribute_selection. Among all the profile information provided by the user, the first step of our procedure chooses the most significant ones: age and gender. In this process we also distinguish between the categorical variables and the continuous variables. This selection can lead to an optimisation of the computation. As shown in Eq. 1, the reputation of the user is influenced by the characteristics in his profile.

$$user\_reputation = age \otimes education \otimes gender \otimes salary \otimes \ldots \qquad (1)$$

attribute_extraction. Apart from the user provided information in the profile, we derive information about the user contributions in the system. This information can be the total number of tags provided, total number of tags matched with the other users, time spent in the system, etc. This derivation can help us understand the behaviour of the user better and help derive useful correlations about the user behaviour and reputation.

reputation_levels_aggregation. To ease the learning process, we aggregate reputation of the users into n classes. The classes are formed by different combinations of features. The features are created based on the extracted user information. To create a feature, we compute percentiles for continuous variables such as age, total tags contributed, etc. Using percentiles, we discretize the continuous variables into four values per feature with each value representing a quarter of the data. However for categorical variables such as gender, education, etc., we use each of the categories available. Once the classes are formed, we consider them as stereotypes of the users. We assign each user to a particular stereotype.

classification. Machine learning algorithms (or any other kind of classification algorithm) can be adopted at this stage. The choice can be constrained either from the data or by other limitations (e.g., computational power at our disposal). In this subprocedure, we try to predict information about the reputation of a new user belonging to a certain stereotype based on the reputation of other users belonging to that particular stereotype. This prediction helps to give an "a priori value" of reputation for new users in the system based on information in their profiles.

4.3 Application Evaluation

We apply the procedure to the tag entries from the *Waisda?* game as follows.

attribute selection and extraction. In the *Waisda?* dataset, we have 17
users who provided in their profile personal information such as e-mail, id,
age and gender. The remaining users of the tagging game participated as
anonymous. We extract the age and gender from the profiles and derive
information such as total number of tags contributed by each user and, for
each user, the total number of tags matched with the others. We also compute
the reputation of the users using the partial evidence extension of subjective
logic that we introduced in a previous work [11] and is summarized as:

$$b = \frac{1}{l+2}\Sigma_{i=1}^{l}\frac{p_i+1}{p_i+2} \quad d = \frac{1}{l+2}\Sigma_{i=1}^{l}\frac{1}{p_i+2} \quad u = \frac{2}{l+2} \tag{2}$$

where b is the belief, d is the disbelief and u is the uncertainty of a subjective opinion. p is a vector of positive observations about distinct facts
(e.g., number of matches for different tags provided by the same user). l is
the length of p. Each entry in p has a prior probability that is set to the
default non-informative value $\frac{1}{2}$. The value of the reputation corresponds to
the expected value of the opinion we computed, and is determined as follows.

$$E = b + \frac{1}{2}\cdot u \tag{3}$$

reputation stereotypes computation. We discretize the continuous variables such as age, total number of tags contributed, total number of tags
matched into four values using percentiles. Each value represents a quarter
of the total data. We use this approach to ensure equal distribution of the
data for each feature. Categorical variables such as gender represent features
that take two values (male, female). Once the features are formed, we aggregate them in different combinations to form the stereotypes of the users.
In our case for the *Waisda?* dataset we have 7 stereotypes. We compute a
reputation per stereotype based on the evidence at our disposal for the users
that belong to it.

regression/classification algorithm. We used a regression algorithm to predict the trustworthiness of the users belonging to a stereotype. Once we have
sufficient evidence (e.g. at least five or ten users belonging to a stereotype),
we can predict the trustworthiness of new users in the system who belong
to the same stereotype. This prediction can help us to give an idea about
the user trustworthiness in the system and also in the future help to recruit
users with certain characteristics for the system.

4.4 Results

Table 1 shows the results of our analysis about the user reputation per stereotype.
Here the user reputation is computed by using the formulas presented in Eq. (2)

and following Subsect. 5.2: for each user in each stereotype we compute the frequency of matched tags that he or she contributed, weighed on the sample size. Here we want to check if stereotypes are able to discriminate users on their reputations. Hence we compute user reputations based on the evidence at our disposal.

Table 1. Stereotypes of user profiles and their reputation.

Stereotype	# users	User reputations
Stereotype 1	2	[0.96, 0.90]
Stereotype 2	2	[0.97, 0.95]
Stereotype 3	2	[0.91, 0.94]
Stereotype 4	2	[0.97, 0.96]
Stereotype 5	5	[0.97, 0.97, 0.97, 0.98, 0.98]
Stereotype 6	1	[0.94]
Stereotype 7	3	[0.95, 0.93, 0.95]

From Table 1, we observe that the maximum variance and maximum standard deviation between the reputations within a stereotype are 0.001 and 0.03 respectively. This shows that there is not much difference between the reputation values of the users belonging to the same stereotype. Also, the difference between stereotypes is very small. However, this small difference may be due to the fact that these stereotypes do not correlate with users reputations. We will investigate in the future the use of stereotypes based only on demographics features that correlate with user reputations to discriminate users on their reputations. Also, in this specific use case, the variance of the user reputation is quite low, so it may be hard to group users based on their reputation. So, instead of checking the correlation between the user stereotype and the user reputation, we evaluate the correlation between user demographics and user reputation, so we decompose the information that determines the user stereotype and we analyze them independently. For data which is normally distributed, we use the Pearson correlation. For categorical data such as age, we use point biserial correlation [15]. The results of our analysis are shown in Table 2.

Table 2. Results of correlation analysis on *Waisda?* dataset.

X	Y	Correlation method	Corr(X,Y)	p-value
# of tags	Reputation	Pearson	0.53	0.02
# of matched tags	Reputation	Pearson	0.61	0.008
Age	Reputation	Pearson	−0.55	0.02
Gender	Reputation	Point biserial	0.46	0.06

From Table 2 we can see that there is linear positive correlation between the number of tags and the number of tags matched with other tags provided by a user with their reputation. However, there is a negative correlation with the user age and their reputation. The point biserial correlation method shows that there is a positive correlation between the gender of the users and their reputation.

Thus, from our experiments, it can be seen that there is a correlation between the information provided by the user and their reputation, at least in the *Waisda?* dataset. For instance, the age correlation indicates that the youngest users perform best, perhaps because they are more reactive and attentive. Also, users that contributed more tags tend to have a higher reputation. This is probably because they developed a better tagging skill over time. Users that contributed a higher number of matched tags also tend to be more precise (it is not given that to a higher number of matched tags corresponds to a higher reputation, since the matched tags could be accompanied by a lot of unmatched ones; this is not the case here). The gender correlation is not significant, since it is even lower than the probability to guess the correct reputation of a user based on his or her gender. These correlations can help us to predict the reputation of new users based on reputations computed from users with similar characteristics. For the moment, these results hold only for this case study, but in the future, we aim to test these features also on additional use cases and to enrich them (both derived from and provided in the profile) to understand how the user characteristics impact the user reputation, and we aim at identifying a corpus of characteristics (shared among use cases) from which we can infer the user reputation. This information may be useful for expert finding, since once we learn which stereotypes of users perform a certain task well, we can recruit more users of that stereotype into the system.

5   Computing Reputation-Based Trust

In the previous section, we analyze some of the assumptions that underpin the use of user reputations for making trust assessments. We find that there exists a moderate correlation between the user demographics and the trustworthiness of the data that the population produces. This leads us to conclude that by virtue of the correlation between user reputation and demographics, demographics can be used as a foundation for trust prediction, although particular countermeasures need to be taken to compensate the fact that the existing correlation is only moderate.

Here, we provide a generic procedure that allows to build a reputation for a user, based on a set of evaluated artifacts (e.g., annotations), and to use it for assessing trust of other artifacts created by him. We build the reputation based on a set of evaluated tags contributed by the user and not on user demographics because we have such evaluations at our disposal, and this allows tailoring the reputation to the specific user. Still, the analysis presented before lays the foundations for the use of user reputation for trust prediction.

5.1   Procedure

We present a generic procedure for computing the reputation of a user with respect to a given artifact produced by him.

proc reputation(*user*, *artifact*) ≡
  evidence := *evidence_selection*(*user*, *artifact*)
  weighted_evidence := *weigh_evidence*(*user*, *artifact*, *evidence*)
  reputation := *aggregate_evidence*(*weighted_evidence*)

Evidence_election. Reputation is based on historical evidence, hence the first step is to gather all pieces of evidence regarding a given user and select those relevant for trust computation. Typical constraints include temporal (evidence is only considered within a particular time-frame) or semantics (evidence is only considered when is semantically related to the given artifact). *evidence* is the set of all evidence regarding *user* about *artifact*.

proc evidence_selection(*user*, *artifact*) ≡
  for *i* :=1 to *length*(*observations*) do
    if *observations*[*i*].*user* = *user* then *evidence*.*add*(*observation*[*i*]) fi

Evidence_Weighing. Given the set of evidence considered, we can decide if and how to weigh its elements, that is, whether to count all the pieces of evidence as equally important, or whether to consider some of them as more relevant. This step might be considered as overlapping with the previous one since they are both about weighing evidence: evidence selection gives a boolean weight, while here a fuzzy or probabilistic weight is given. However, keeping this division produces an efficiency gain, since it allows computation to be performed only on relevant items.

proc weigh_evidence(*user*, *artifact*, *evidence*) ≡
  for *i* := 1 to *length*(*evidence*) do
    weighted_evidence.*add*(*weigh*(*evidence*[*i*], *artifact*))

Aggregate_evidence. Once the pieces of evidence have been selected and weighed, these are aggregated to provide a value for the user reputation that can be used for evaluation. We can apply several different aggregation functions, depending on the domain. Typical functions are: *count*, *sum*, *average*. Subjective logic [19], a probabilistic logic that we use in the application of this procedure, aggregates the observations in subjective opinions about artifacts being trustworthy based on the reputation of their authors are represented as follows:

$$\omega(b, d, u) \tag{4}$$

where

$$b = \frac{p}{p+n+2} \quad d = \frac{n}{p+n+2} \quad u = \frac{2}{p+n+2} \tag{5}$$

where b, d and u indicate respectively how much we believe that the artifact is trustworthy, non-trustworthy, and how uncertain our opinion is. p and n

are the amounts of positive and negative evidence respectively. Subjective opinions are equivalent to Beta probability distributions (Fig. 2), which range over the trust levels interval $[0 \ldots 1]$ and are shaped by the available evidence.

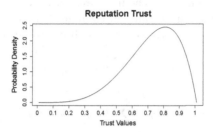

Fig. 2. Example of a Beta probability distribution aggregating 4 positive and 1 negative piece of evidence. The most likely trust value is 0.8 (which is the ratio among the evidence). The variance of the distribution represents the uncertainty about the evaluation.

5.2   Application Evaluation

First, we convert the number of matches that each tag entry has into trust values. We obtain an opinion for a given tag entry by aggregating all the evidence (in form of match or non-match) from the other tag entries. For brevity, we report the details about the computation of p and n (i.e. of the positive and negative evidence counts). The corresponding subjective opinion is always computed as in Eq. (5).

tag selection. For each tag inserted by the user, we select all the matching tags belonging to the same video. In other contexts, the number of matching tags can be substituted by the number of "likes", "retweets", etc.

tag entries weighing. For each matching entry, we weigh it on the time distance between the evaluated entry and the matched entry. The weight is determined from an exponential probability distribution, which is a "memoryless" probability distribution used to describe the time between events. If two entries are close in time, we consider it highly likely that they match. If they match but appear in distant temporal moments, then we presume they refer to different elements of the same video. Instead of choosing a threshold, we give a probabilistic weight to the matching entry.

Evidence that $tagentry_i$ contributes to the determination of the trustworthiness of $tagentry$ is represented as $tagentry_{tagentry_i}$. The *timestamp* of $tagentry$ is represented as $t(tagentry)$.

$$p_{tagentry_{tagentry_i}} = exp(y \cdot (t(tagentry) - t(tagentry_i)))$$

$$n_{tagentry_{tagentry_i}} = 1 - p_{tagentry,tagentry_i}$$

where y is a weighing parameter that allows obtaining that 85 % of probability mass is assigned to tags inserted in a 10 s range (in our case, $y = \frac{1}{5000}$ ms).

tag entries aggregation. In this step, we determine the trustworthiness of every tag. We aggregate the weighed evidence in a subjective opinion about the tag trustworthiness. We have at our disposal only positive evidence (the number of matching entries). The more evidence we have at disposal for the same tag entry, the less uncertain our estimate of its trustworthiness will be. Non-matched tag entries have equal probability to be correct or not.

$$p_{tagentry} = \sum_i p_{tagentry,tagentry_i}$$

$$n_{tagentry} = \sum_i n_{tagentry,tagentry_i}$$

We repeat the procedure above for each tag entry created by the user to compute his reputation.

user tag entries selection. Select all the tag entries inserted by *user*. We denote a *tagentry* inserted by a *user* as $tagentry_{user}$.

user tag entries weighing. Tag entries are weighed by the corresponding trust value previously computed. If an entry is not matched, it is considered as a half positive (tag trust value 0.5) and half negative $(1 - 0.5 = 0.5)$ item of evidence (it has 50 % probability to be incorrect), as computed by means of subjective opinions. The other entries are also weighed according to their trust value. So, user reputation can either rise or decrease as we collect evidence.

$$p_{tagentry_{user}} = E(\omega_{tagentry})$$

$$n_{tagentry_{user}} = 1 - E(\omega_{tagentry})$$

In the future, we plan to use the reputation the user belongs to as a priori value. In that case, if no items of evidence are available for a user, then his reputation coincides with that of the stereotype he belongs to.

user tag entries aggregation. In turn, to compute the reputation of a user with respect to a given tag, we use all the previously computed evidence to build a subjective opinion about the user. This opinion represents the user reputation and can be summarized even more by the corresponding expected value.

$$p_{user} = \sum_{tagentry_{user}} p_{tagentry_{user}}$$

$$n_{user} = \sum_{tagentry_{user}} n_{tagentry_{user}}$$

5.3   Results

We implement the abstract procedure for reputation computation and we evaluate its performance by measuring its ability to make use of the available evidence to compute the best possible trust assessment. Our evaluation does not focus on the ability to predict the exact trust value of the artifact by computing the user reputation, because these two values belong to a continuous space, and they are computed on a different basis. What we expect is that these two values hint at trustworthiness in a similar fashion. We suppose that the trust evaluation system is implemented in such a manner that tags are "accepted" as trustworthy when their trust value is higher than a particular value (also called threshold). So, if the user reputation is a good indicator of trustworthiness, the reputation of a user should be higher than the threshold when the trust values of the artifacts created by him pass the threshold, and vice-versa. The validation, then, depends upon the choice of the threshold which, in turn, depends on constraints imposed by each specific use case. For instance, as we explain below, in the case study we tackle, "false negatives" are preferred over "false positives", and this makes the threshold more likely to be set high (e.g., at least 85 % or 90 %).

We run the procedure with different thresholds as presented in Fig. 4. Low thresholds correspond to low accuracy in our predictions. However, as the threshold increases, the accuracy of the prediction rises. Moreover, we should consider that: (1) it is preferable to obtain "false negatives" (reject correct tags) rather than "false positives" (accept wrong tags), so high thresholds are more likely to be chosen (e.g., see [14]), in order to reduce risks. Rejecting correct tags means rejecting useful information and therefore wasting part of the effort spent in crowdsourcing tags. Accepting wrong tags means to introduce in the system wrong information and therefore, the tasks that rely on these crowdsourced tags may be affected by this (e.g. if we run an information retrieval task using these tags, then we may retrieve wrong items). Hence we prefer the first situation in place of the latter; (2) a Wilcoxon signed-rank test at 95 % confidence level proved that the reputation-based estimates outperform blind guess estimates (having average probability of accuracy 50 %). The average improvement is 8 %, the maximum is 49 %.

We previously adopted this procedure to compute the trustworthiness of tags on the Steve.Museum artifacts [8]. By adapting the procedure to the *Waisda?* case, we were able to formulate the general procedure above.

6   Computing Provenance-Based Trust

User demographics and, in general, user identities are not always available when estimating the trustworthiness of artifacts. Hence, we provide a procedure for estimating the trustworthiness of artifacts based on "how" they were produced rather then on "whom" produced them. Thus, we focus on the "how" part of provenance, i.e., the steps or activities performed in the production of an artifact. (For simplicity, in the rest of the chapter, we will use the word "provenance" to refer to the "how" part). We learn the relationships between PROV and trust values through machine learning algorithms. This procedure allows to process PROV data and, on the basis of previous trust evaluations, predict the trust level of artifacts.

6.1 Procedure

We present the procedure for computing trust estimates based on provenance.

proc provenance_prediction($artifact\_provenance, artifact$) $\equiv$
    $attribute\_set := attribute\_selection(artifact\_provenance)$
    $attributes := attribute\_extraction(attribute\_set)$
    $trust\_levels\_aggregation$
    $classified\_testset := classify(testset, trainingset)$

attribute_selection. Among all the provenance information, the first step of our procedure chooses the most significant ones: agent, processes, temporal annotations and input artifacts can all hint at the trustworthiness of the output artifact. This selection can lead to an optimization of the computation.

attribute_extraction. Some attributes need to be manipulated to be used for our classifications, e.g., temporal attributes may be useful for our estimates because one particular date may be particularly prolific for the trustworthiness of artifacts. However, to ease the recognition of patterns within these provenance data, we extract the day of the week or the hour of the day of production, rather than the precise timestamp. In this way we can distinguish, e.g., between day and night hours (when the user might be less reliable). Similarly, we might refer to process types or patterns instead of specific process instances.

trust_level_aggregation. To ease the learning process, we aggregate trust levels in n classes. Our results will show that this classification process does not affect accuracy significantly.

classification. Machine learning algorithms (or any other kind of classification algorithm) can be adopted at this stage. The choice can be constrained either from the data or by other limitations.

6.2 Application Evaluation

We apply the procedure to the tag entries from the *Waisda?* game as follows.

attribute selection and extraction. The provenance information available in *Waisda?* is represented in Fig. 3, using the W3C PROV ontology. First, for each tag entry we extract: *typing duration, day of the week, hour of the day, game_id (to which the tag entry belongs), video_id.* This is the "how" provenance information at our disposal. Here we want to determine the trustworthiness of a tag given the modality with which it was produced, rather than the author reputation. Some videos may be easier to annotate than others, or, as we mentioned earlier, user reliability can decrease during the night. For similar reasons we use all the other available features.

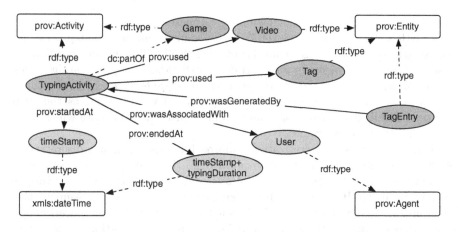

Fig. 3. Graph representation of the provenance information about each tag entry.

trust level classes computation. In our procedure, we are not interested in predicting the exact trust value of a tag entry. Rather we want to predict the range of trust values that hold for an entry. Given the range of trust values $[0 \ldots 1]$, we split it into 20 classes of length 0.5: from $[0, 0.05]$ to $[0.95, 1.0]$. This allows us to increase the accuracy of our classification algorithm without compromising the accuracy of the predicted value or the computation cost. The values in each class were approximated by the middle value of the class itself. For instance, the class $[0.5 \ldots 0.55]$ are approximated as 0.525.

regression/classification algorithm. We use a regression algorithm to predict the trustworthiness of the tags. Having at our disposal five different features (in principle, we might have more), and given that we are not interested in predicting the "right" trust value, but the class of trustworthiness, we adopt the "regression-by-discretization" approach [21], that allows us to use the Support Vector Machines algorithm (SVM) [12] to classify our data after having discretized the continuous ones. The training set is composed by 70 % of our data, and then we predict the trust level of the test set. We used the SVM version implemented in the e1071 R library [36]. In the future, we will consider alternative learning techniques.

6.3   Results

The accuracy of our predictions depends, again, on the choice of a threshold. If we look at the ability to predict the right (class of) trust values, then the accuracy is about 32 % (which still is twice as much as the average result that we would have with a blind guess), but it is more relevant to focus on the ability to predict the trustworthiness of tags within some range, rather than the exact trust value. Depending on the choice of the threshold, the accuracy in this case varies in the range of 40 %–90 %, as we can see in Fig. 4. For thresholds higher than 0.85 (the most likely choices), the accuracy is at least 70 %.

We also compared the provenance-based estimates with the reputation-based ones, with a 95 % confidence level Wilcoxon signed-rank test that proved that the estimates of the two algorithms is not statistically different. *For the* Waisda? *case study, reputation- and provenance-based estimates are equivalent: when reputation is not available or it is not possible to compute it, we can substitute it with provenance-based estimates.* This is particularly important, as the availability of PROV data grows, one can compute trust values for data which is not associated with a trust value.

The "regression-by-discretization" approach consists in first a discretization of the continuous features at our disposal (e.g., timestamps) and a subsequent computation of regression by means of a classification algorithm (e.g., Support Vector Machines). If we apply it for making provenance-based assessments, then we approximate our trust values. This is not necessary with the reputation approach. Had we applied the same approximation to the reputations as well, then provenance-based trust would have performed better, as proven with a 95 % confidence level Wilcoxon signed-ranked test, because reputation can rely only on evidence regarding the user, while provenance-based models can rely on larger data sets. Anyway, we have no need to discretize the reputation and, in general, we prefer it because of its lightweight computational overhead.

7   Combining Reputation and Provenance-Based Trust

Lastly, we provide a procedure for combining reputation- and provenance-based estimates to improve our predictions. If a certain user has been reliable so far, we can reasonably expect him/her to behave similarly in the near future. So we use reputation and we also constantly update it, to reduce the risk of relying on over-optimistic assumptions (if a user that showed to be reliable once, will maintain his/her status forever). However, reputation has an important limitation. To be reliable, a reputation has to be based on a large amount of evidence, which is not always possible. So, both in case the reputation is uncertain, or in case the user is anonymous, other sources of information should be used in order to correctly predict a trust value. The trust estimate based on provenance information, as described in Sect. 6, is based on behavioral patterns which have a high probability to be shared among several users. Hence, if a reputation is not reliable enough, we substitute it with the provenance-based prediction.

7.1   Procedure

The algorithm is as follows:

proc provenance_prediction(*user*, *artifact*) $\equiv$
    $q\_ev = evaluate\_user\_evidence(user, artifact)$
    if $q\_ev > min\_evidence$ then *predict_reputation* else *predict_provenance* fi

evaluate_user_evidence. This function quantifies the evidence. Some implementation examples: (1) *count*; (2) compute a subjective opinion and check if the uncertainty is low enough. As future work we plan to investigate how to automatically determine q_ev and *evaluate_user_evidence*.

7.2 Application Evaluation

We adopt the predictions obtained with each of the two previous procedures. The results are combined as follows: if the reputation is based on a minimum number of observations, then we use it, otherwise we substitute it with the prediction based on provenance. We run this procedure with different values for both the threshold and the minimum number of observations per reputation. We instantiate the *evaluate_user_evidence(user, artifact)* function as a *count* function of the evidence of *user* with respect to a given *tag*.

7.3 Results

The performance of this algorithm depends both on the choice of the threshold for the decision and on the number of pieces of evidence that make a reputation reliable, so we ran the algorithm with several combinations of these two parameters (Fig. 4). The results converge immediately, after having set the minimum number of observations at two. We compared these results with those obtained before. Two Wilcoxon signed-rank tests (at 90 % and 95 % confidence level with respect to respectively reputation- and provenance-based assessments) showed that *the procedure which combines reputation and provenance evaluations in this case performs better than each of them applied alone*. The improvement is, on average, about 5 %. Despite the fact that most of the improvement regards the lower thresholds, which are less likely to be chosen (as we saw in Sect. 5), even at 0.85 threshold there is a 0.5 % improvement. Moreover, we would like to stress how the combination of the two procedures performs better than (in a few cases, equal to) each of them applied alone, regardless of the threshold chosen.

Combining the two procedures allows us to go beyond the limitation of reputation-based approaches. Substituting estimates based on unreliable reputations with provenance-based ones improves our results without significantly increasing risks, since we have previously proven that the two estimates are (on average) equivalent. Hence, when a user is new in a system (and so his/her history is limited) or anonymous, we can refer to the provenance-based estimate to determine the trustworthiness of his/her work, without running a higher risk of poor trust prediction. This improvement is at least partly due to the existing correlation between the reputation and provenance-based trust assessments. A small positive correlation (0.16) has been shown by a Pearson correlation test [27] with a confidence level of 99 %. Thanks to this, we can substitute uncertain reputations with the corresponding provenance-based assessments. This explains also the similarity among the results shown in Fig. 4.

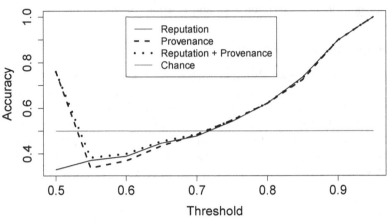

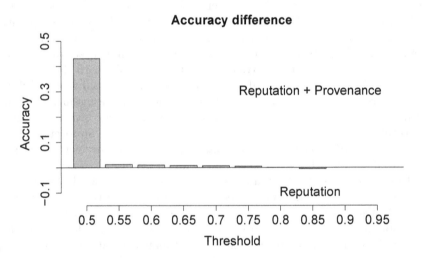

Fig. 4. Absolute and relative (Reputation+Provenance vs. Reputation) accuracy. The gap between the prediction (provenance-based) and the real value of some items explains the shape between 0.5 and 0.55: only very low or high thresholds cover it.

8   Conclusion

In this chapter, we first explored the correlation between user demographics and user reputations and showed the existence of such a correlation in the *Waisda?* tagging dataset. Moreover, we showed how it is possible to use demographics extracted from user profiles to create user stereotypes (user abstractions based on demographics) and to possibly use them as a basis for trust estimation. However, in the *Waisda?* dataset user stereotypes were not useful to discriminate user reputation, although we found a correlation between single demographics (age, gender, etc.) and user reputation. Moreover, we showed how to use the FOAF ontology to both represent user profiles and stereotypes.

Additionally, we proposed and evaluated procedures for computing trust assessments based on reputation, for computing trust assessments based on provenance information, and for combining these two types of assessments. We show that using reputation for trust assessment is simple, computationally light and accurate. We also show the potential of provenance-based trust assessments: these can be at least as accurate as reputation-based methods and can be used to overcome the limitations of a reputation-based approaches (at least within a tagging environment). In *Waisda?* the combination of the two methods was more powerful than each of the two alone. In the future, we will investigate the possibility of automatically extracting provenance patterns usable for trust assessment, to automate, optimize and adapt the process to other case studies and domains. We will also focus on the use of trust assessments as a basis for information retrieval.

Acknowledgements. We thank the Netherlands Institute for Sound and Vision for launching and guiding the *Waisda?* project, and our colleagues Michiel, Riste and Valentina for their support. This research was partially supported by the PrestoPRIME project, in the EC ICT FP7 program, and by the Data2Semantics and SEALINC Media projects in the Dutch national program COMMIT.

References

1. Artz, D., Gil, Y.: A survey of trust in computer science and the semantic web. J. Semant. Web **5**(2), 131–197 (2007)
2. Bizer, C., Cyganiak, R.: Quality-driven information filtering using the WIQA policy framework. J. Web Semant. **7**(1), 1–10 (2009)
3. Card, S., Moran, T.P., Newell, A.: The Psychology of Human Computer Interaction. Lawrence Erlbaum Associates, Hillsdale (1983)
4. Carroll, J., Bizer, C., Hayes, P., Stickler, P.: Named graphs, provenance and trust. In: WWW '05, pp. 613–622. ACM (2005)
5. Burnett, C., Norman, T.J., Sycara, K.: Bootstrapping trust evaluations through stereotypes. In: AAMAS, IFAAMAS, pp. 241–248 (2010)
6. Ceolin, D., Groth, P., Hage, W.R.V.: Calculating the trust of event descriptions using provenance. In: SWPM 2010, pp. 7–12. CEUR-WS (2010)
7. Ceolin, D., Groth, P., van Hage, W.R., Nottamkandath, A., Fokkink, W.: Trust Evaluation through user reputation and provenance analysis. In: URSW, pp. 15–26. CEUR-WS.org (2012)
8. Ceolin, D., Nottamkandath, A., Fokkink, W.: Automated evaluation of annotators for museum collections using subjective logic. In: Dimitrakos, T., Moona, R., Patel, D., McKnight, D.H. (eds.) IFIPTM 2012. IFIP AICT, vol. 374, pp. 232–239. Springer, Heidelberg (2012)
9. Ceolin, D., Nottamkandath, A., Fokkink, W.: Semi-automated assessment of annotation trustworthiness. In: PST 2013, pp. 325–332. IEEE Computer Society, July 2013
10. Ceolin, D., Nottamkandath, A., Fokkink, W.: Efficient semi-automated assessment of annotation trustworthiness. J. Trust Manag. **1**, 1–31 (2014)

11. Ceolin, D., Nottamkandath, A., Fokkink, W.J.: Subjective logic extensions for the semantic web. In: URSW, pp. 27–38. CEUR-WS.org (2012)
12. Cortes, C., Vapnik, V.: Support-vector networks. M. Learn. **20**, 273–297 (1995)
13. Dan Brickley, L.M.: FOAF, Jan 2014. http://xmlns.com/foaf/spec/
14. Gambetta, D.: Can We Trust Trust?. Basil Blackwell, New York (1988)
15. Glass, G.V., Hopkins, K.D.: Statistical Methods in Education and Psychology. Allyn & Bacon, Boston (1995)
16. Golbeck, J.: Trust on the world wide web: a survey. Found. Trends Web Sci. **1**(2), 131–197 (2006)
17. Hartig, O., Zhao, J.: Using web data provenance for quality assessment. In: SWPM 2009, pp. 26–31. CEUR-WS (2009)
18. Javanmardi, S., Lopes, C., Baldi, P.: Modeling user reputation in wikis. Stat. Anal. Data Min. **3**(2), 126–139 (2010)
19. Jøsang, A.: A logic for uncertain probabilities. Int. J. Uncertainty Fuzziness Knowl. Based Syst. **9**(3), 279–311 (2001)
20. Kazai, G., Kamps, J., Milic-Frayling, N.: The face of quality in crowdsourcing relevance labels: demographics, personality and labeling accuracy. In: CIKM, pp. 2583–2586. ACM (2012)
21. Kononenko, I.: Naive bayesian classifier and continuous attributes. Informatica **16**(1), 1–8 (1992)
22. Liu, X., Datta, A., Rzadca, K., Lim, E.-P.: StereoTrust: a group based personalized trust model. In: CIKM, pp. 7–16. ACM (2009)
23. Masum, H., Tovey, M. (eds.): The Reputation Society. MIT Press, Cambridge (2012)
24. Netherlands Inst. for Sound and Vision. Waisda? June 2012. http://waisda.nl
25. O'Hara, K.: A General Definition of Trust. Technical report, University of Southampton (2012)
26. Pantola, A.V., Pancho-Festin, S., Salvador, F.: Rating the raters: a reputation system for wiki-like domains. In: SIN '10, pp. 71–80. ACM (2010)
27. Pearson, K.: Mathematical contributions to the theory of evolution. In: Proceedings of the Royal Society of London, pp. 489–498 (1896)
28. Pearson, K.: On the criterion that a given system of deviations from the probable in the case of correlated system of variables is such that it can be reasonably supposed to have arisen from random sampling. Phil. Mag. **50**, 157–175 (1900)
29. Rajbhandari, S., Rana, O.F., Wootten, I.: A fuzzy model for calculating workflow trust using provenance data. In: MG'08, pp. 1–8. ACM (2008)
30. Rajbhandari, S., Wootten, I., Ali, A.S., Rana, O.F.: Evaluating provenance-based trust for scientific workflows. In: CCGRID 06, pp. 365–372. IEEE (2006)
31. Sabater, J., Sierra, C.: Review on computational trust and reputation models. Artif. Intell. Rev. **24**, 33–60 (2005)
32. U.S. Institute of Museum and Library Service. Steve Social Tagging Project, June 2012. http://www.steve.museum/
33. Venanzi, M., Guiver, J., Kazai, G., Kohli, P., Shokouhi, M.: Community-based bayesian aggregation models for crowdsourcing. In: WWW, pp. 155–164 (2014)
34. W3C. PROV-O, June 2012. http://www.w3.org/TR/prov-o//
35. W3C. Resource description framework (rdf): concepts and abstract data model, June 2012. www.w3.org/TR/2002/WD-rdf-concepts-20020829/
36. Wien, T.: e1071: Misc functions of the department of statistics (e1071), June 2012. http://cran.r-project.org/web/packages/e1071/

37. Wilcoxon, F.: Individual comparisons by ranking methods. Biom. Bull. **1**, 80–83 (1945)
38. Zaihrayeu, I., da Silva, P.P., McGuinness, D.L.: IWTrust: improving user trust in answers from the web. In: Herrmann, P., Issarny, V., Shiu, S.C.K. (eds.) iTrust 2005. LNCS, vol. 3477, pp. 384–392. Springer, Heidelberg (2005)

Bridging Gaps Between Subjective Logic and Semantic Web

Davide Ceolin$^{(\boxtimes)}$, Archana Nottamkandath, and Wan Fokkink

VU University, Amsterdam, The Netherlands
{d.ceolin,a.nottamkandath,w.j.fokkink}@vu.nl

Abstract. Subjective logic is a powerful probabilistic logic which is useful to handle data in case of uncertainty. Subjective logic and the Semantic Web can mutually benefit from each other, since subjective logic is useful to handle the inner noisiness of the Semantic Web data, while the Semantic Web offers a means to obtain evidence useful for performing evidential reasoning based on subjective logic. In this chapter we describe three extensions and applications of subjective logic in the Semantic Web, namely: the use of deterministic and probabilistic semantic similarity measures for weighing subjective opinions, a way for accounting for partial observations, and "open world opinion", i.e. subjective opinions based on Dirichlet processes, which extend multinomial opinions. For each of these extensions, we provide examples and applications to prove their validity.

Keywords: Subjective logic · Semantic similarity · Dirichlet process · Partial observations

1 Introduction

Subjective logic [12] is a probabilistic logic widely adopted in the trust management domain, based on evidential reasoning and statistical principles. This logic focuses on the representation and the reasoning on assertions of which the truth value is not fully determined, but estimated on the basis of the observed evidence. The logic comes with a variety of operators that allow to combine such assertions and to derive the truth values of the consequences.

Subjective logic is well-suited for the management of uncertainty within the Semantic Web. For instance, the incremental access to these data (as a consequence of crawling) can give rise to uncertainty issues which can be dealt with using this logic. Furthermore, the fact that the fulcrum of this logic is the concept of "subjective opinion" (which represents a logic proposition, its corresponding belief and the source of this evidence), allows correctly representing how the estimated truth value of an assertion is bound to the source of the corresponding evidence and allows to easily keep lightweight provenance information. Finally, evidential reasoning allows to limit the typical noisiness of Semantic Web data. On the other hand, we also believe that the Semantic Web can be

© Springer International Publishing Switzerland 2014
F. Bobillo et al. (Eds.): URSW 2011-2013, LNAI 8816, pp. 242–264, 2014.
DOI: 10.1007/978-3-319-13413-0_13

beneficial to this logic, as an immeasurably important source of information: since the truth value of assertions is based on availability of observations, the more data is available (hopefully of high enough quality), the closer we can get to the correct truth value for our assertions. We believe that this mutual relationship can be improved. This chapter proposes extensions and applications of subjective logic that aim at the Semantic Web, namely: the use of deterministic and probabilistic semantic similarity measures for weighing subjective opinions, a method for accounting for partial observations, and "open world opinions", that are subjective opinions based on Dirichlet processes. Open world opinions allow modeling categorical data for which categories are partially known. Only the latter is a proper extension of the logic, while the first two items are representations within the logic of external elements, with proper mappings and, when necessary, specific representations.

This chapter revises and extends the paper "Subjective Logic Extensions for the Semantic Web" [5], presented at the 8th International Workshop on Uncertainty Reasoning for the Semantic Web, at the 11th International Semantic Web Conference 2012. Here, we add to that paper two methods to map probabilistic semantic similarity measures and subjective opinions.

The rest of the chapter is organized as follows. Section 2 gives an overview of subjective logic. Sections 3 and 4 show how to combine subjective logic with deterministic and probabilistic semantic similarity measures respectively. Section 5 introduces a method for dealing with partial observations of evidence, Sect. 6 describes the concept of "open world opinion". Section 7 describes related work, and Sect. 8 provides a final conclusion about the work presented.

2   Subjective Logic

In subjective logic, so-called "subjective opinions" express the belief that source x owns with respect to the value of proposition y. The values of y are chosen among the elements of the set Θ ("frame of discernment"). For instance, if y is a binomial proposition, then $\Theta = \{true, false\}$. A subjective opinion describes the belief in the elements of the power set of Θ (2^{Θ}). In symbols, an opinion is represented as

$$\omega_y^x(b, d, u, a)$$

when $|\Theta| = 2$ (binomial opinion) or as

$$\omega_y^x(\overrightarrow{B}, u, \overrightarrow{A})$$

when $|\Theta| > 2$ (multinomial opinion). Throughout the paper we refer to vectors with the following notation: $\overrightarrow{B}$. Its elements are represented as b_x. In the binomial opinion, b represents the belief in y being $true$ and d the belief in y being $false$, i.e., the disbelief. The uncertainty u represents a part of probability mass that we are unable to assign to either $true$ or $false$ and it therefore corresponds to the belief in Θ. In the case of the multinomial opinion there is no disbelief because there is no specific *false* value, since y can assume multiple ones. a represent the prior probability that y has to be true, while A represents the vector of prior probabilities for each of the possible truth values of y. The values b, d, u are determined by observing pieces of evidence. a is given a priori. The positive and negative evidence is represented as p and n respectively. The belief (b), disbelief (d), uncertainty (u), and a priori values (a) for binomial opinions are computed as:

$$b = \frac{p}{p+n+2} \quad d = \frac{n}{p+n+2} \quad u = \frac{2}{p+n+2} \quad a = \frac{1}{2}. \qquad (1)$$

The value 2 indicates the cardinality of Θ, i.e., the number of values that y can take. A subjective opinion is equivalent to a Beta probability distribution (binomial opinion) or to a Dirichlet distribution (multinomial opinion). This probability distribution describes the most likely probability values that y can take. If y has Pr probability to be true, since we determine Pr starting from a limited set of evidence, we estimate the most likely value of Pr by means of a Beta (or Dirichlet) probability distribution.

Opinions can be contextualized. For example, source x provides an observation about assertion y in context c (e.g. about an agent's expertise). The most likely value for y in context c, represented as $t(x, y : c)$, is the expected value of the Beta distribution corresponding to the opinion and computed as:

$$E = t(x, y : c) = b + a \cdot u. \qquad (2)$$

The reason why we rely on this logic is the fact that it makes use of a double probabilistic layer. The probability of each proposition can be represented by means of a Binomial distribution (or by means of a Multinomial distribution if the proposition is multivalued). However, we base our truth estimations on samples of Web data so the parameter p of the Binomial distribution (or the vector of parameters $\overrightarrow{P}$ of the Multinomial) is rather uncertain. In fact, the Web data sample is possibly unreliable, uncertain and partially representative of the entire Web data population. Subjective logic uses a second-order distribution based on the distribution and size of the sample at our disposal to estimate the most likely value that the p (or $\overrightarrow{P}$) value can take. This is the primary reason why we adopt this logic. Also, on the Web, data are exposed by different sources presenting different reliability levels. The ability to keep track of the source that exposes a given piece of data or a subjective opinion is crucial to be able to assess the trust in that piece of data or subjective opinion. Subjective logic allows keeping track of such provenance information and reasoning on subjective opinions weighing them on the reputation of their source. Lastly, subjective logic offers a variety of operators that allow combining subjective opinions in several manners. For instance, operators allow "discounting" an opinion based on the reputation of the source that exposes it, or computing the truth value (expressed as a subjective opinion) of the logical disjunction or conjunction of two opinions held by the same source. This makes the logic a useful tool to reason upon data extracted from the Web. One important remark is that this logic allows reasoning on binomial or multinomial data, that include, for instance URIs. The Beta and the Dirichlet distributions are used because they are "conjugated" [11] with the Binomial and Multinomial distributions respectively, i.e., their computation is particularly manageable. Other kind of data and other probability distributions are outside the scope of this logic.

2.1 Base Rate Discounting Operator in Subjective Logic

An important class of operators of subjective logic is the so-called "discounting" operator. In fact, a subjective opinion allows keeping track of the source of the opinion itself. This permits the reuse of the opinion by third parties, because these third parties, knowing where the opinion comes from, can decide to use it. However, before using it, these third parties may require to "smoothen" the opinion to take into account the limited reliability of the source or its possible maliciousness. Therefore, in subjective logic there exists a variety of discounting operators: for instance, one favors disbelief (to be used if the source is known to be malicious), and one that favors uncertainty (to be used when no specific intention of the source is known). We can also make use of the base-rate sensitive discounting operator in the case we just have a probability (i.e., the expected value of an opinion), instead of having at our disposal a complete subjective opinion for a source. The base-rate sensitive discounting of opinion of source B on y by opinion of source A on B ω_B^A,

$$\omega_y^B = (b_y^B, d_y^B, u_y^B, a_y^B)$$

by opinion

$$\omega_B^A = (b_B^A, d_B^A, u_B^A, a_B^A)$$

of source A produces transitive belief

$$\omega_y^{A:B} = (b_y^{A:B}, d_y^{A:B}, u_y^{A:B}, a_y^{A:B})$$

where

$$
\begin{aligned}
b_y^{A:B} &= E(\omega_B^A) b_y^B \\
d_y^{A:B} &= E(\omega_B^A) d_y^B \\
u_y^{A:B} &= 1 - E(\omega_B^A)(b_y^B + d_y^B) \\
a_y^{A:B} &= a_y^B.
\end{aligned}
\tag{3}
$$

3 Combining Subjective Logic with Deterministic Semantic Similarity Measures

We saw in the previous section that opinions can be contextualized. Setting the context is important, because it allows delimiting the validity of an opinion and increasing the precision of the corresponding evaluation. For instance, if we gather evidence about the expertise of a user in a given topic, let us say, flowers, then it is important to delimit the validity of the corresponding opinion to the topic "flowers". However, contexts may also impede the use of evidence about a given subject, if the context differs from the context where the evidence was collected. Therefore, we propose to "bridge" contexts by using semantic similarity measures to import evidence from a context to another, after having weighed them on the similarity of the two contexts. Many semantic similarity measures have been developed (see the work of Budanitsky and Hirst [2]). These basically split into two main classes: deterministic and probabilistic semantic similarity measures. The deterministic ones are based on deterministic computations made, for example, on word graphs (e.g. WordNet [19]). The probabilistic ones apply probabilistic reasoning to derive semantic relatedness between words based, for instance, on the occurrence and co-occurrence of these two words in large document corpora. We extend the logic to incorporate these measures by representing semantic similarity measures by means of subjective opinions (discounting), or by using the similarity measures to weigh items of evidence before using them to build subjective opinions (weighing). The first extension of subjective logic that we propose regards the use of deterministic semantic similarity measures and is described as follows.

3.1 Wu and Palmer Semantic Similarity Measure

Among all deterministic semantic similarity measures, our attention focuses on those computed from *WordNet*. *WordNet* groups words into sets of synonyms called synsets that describe semantic relationships between them. It is a directed and acyclic graph in which each vertex v, is an integer that represents a synset, and each directed edge from v to w represents that w is a hypernym of v. In particular, we use the Wu and Palmer similarity measure [28], which calculates semantic relatedness in a deterministic way by considering the depths between two synsets in the WordNet taxonomies, along with the depth of the Least Common Subsumer (lcs) as follows:

$$score(s1, s2) = \frac{2 \cdot depth(lcs(s1, s2))}{depth(s1) + depth(s2)}. \tag{4}$$

This means that $score \in [0, 1]$. For deriving the opinions about a concept where no evidence is available, we incorporate $score$, which represents the semantic similarity $(sim(c, c'))$ in our trust assessment, where c and c' are concepts belonging to synset $s1$ and $s2$ respectively which represent two contexts.

3.2 Using Semantic Similarity Measures Within Subjective Logic

We propose two means to import deterministic semantic similarity measures into subjective logic, by mapping them with subjective opinions.

Deriving an Opinion About a New or Unknown Context. Since we compute opinions based on contexts, it is possible that evidence required to compute the opinion for a particular context is unavailable. For example, suppose that source x owns observations about a proposition in a certain context (e.g. the expertise of an agent about tulips), but needs to evaluate them in a new context (e.g. the agent's expertise about sunflowers), of which it owns no observations. The semantic similarity measure between two contexts, $sim(c, c')$, can be used for obtaining the opinion about an agent y on an unknown or new context through two different methods. We can weigh the evidence at our disposal, and for every piece of evidence, use only the part that corresponds to the semantic similarity between the two contexts. If I have one observation in the known context c' and the similarity between the two contexts is 0.5, then I can use that observation in the new context c as 0.5 piece of evidence. Otherwise, I could compute a subjective opinion in the known context c' and then use it in the unknown context c after having "discounted" it (using the subjective logic discounting operator). The discounting factor would be a subjective opinion that represents the semantic similarity between the two contexts. The reason why we have these two different approaches is that weighing operates directly on the evidence, while discounting applies on the subjective opinion. In the latter case uncertainty has already been quantified (and therefore some probability mass has been assigned to it), while in the first case not. Hence, the choice between the two alternative depends on the strategy chosen (it could be that operating on the opinion is more computationally efficient, and hence discounting is preferable), or on case study constraints (e.g., if the evidence from a given context are already expressed as a subjective opinion, then it is simpler to use it than to revert it to the pieces of evidence on which it is computed). Below we provide more details about the two methods.

Evidence weighing. We weigh the positive and negative evidence belonging to a certain context (e.g. *Tulips*) on the corresponding semantic similarity to the new context (e.g. *Sunflowers*), *sim(Tulips, Sunflowers)*. We then perform this for all the contexts for which source x has already provided an opinion, $\forall c' \in C$, by weighing all the positive (p) and negative (n) evidence of c' with the similarity measure $sim(c, c')$ to obtain an opinion about y in c (see the work of Ceolin et al. [4]).

Opinion discounting. In the second approach, every opinion source x has about other related contexts c', where $c' \in C$, is discounted with the corresponding semantic similarity measure $sim(c, c')$ using the Discounting operator in subjective logic. The discounted opinions are then aggregated to form the final opinion of x about y in the new context c.

Discounting Operators and Semantic Similarity. Subjective logic offers a variety of operators for "discounting", i.e. for smoothing opinions given by third parties, provided that we have at our disposal an opinion about the source itself. "Smoothing" is meant as reducing the belief provided by the third party, depending on the opinion on the source (the worse the opinion, the higher the reduction). Moreover, since the components of the opinion always sum to one, reducing the belief implies an increase of (one of) the other components: hence there exists a discounting operator favoring uncertainty and one favoring disbelief. Finally, there exists a discounting operator that makes use of the expected value E of the opinion. Following this line of thought, we can use the semantic similarity as a discount factor for opinions imported from contexts related to the one of interest, in case of a lack of opinions in it, to handle possible variations in the validity of the statements due to the change of context.

So, we need to choose the appropriate discounting operator that allows us to use the semantic similarity value as a discounting factor for opinions. The disbelief favoring discounting is an operator that is employed whenever one believes that the source considered might be malicious. This is not our case, since the discounting is used to import opinions own by ourselves but computed in different contexts than the one of interest. Hence we do not make use of the disbelief favoring operator.

In principle, we would have no specific reason to choose one between the uncertainty favoring discounting and the base rate discounting. Basically, having that only rarely the belief (and hence the expected value) is equal to 1, the two discounting operators decrease the belief of the provided opinion, one by multiplying it by the belief in the source, the other one by the expected value of the opinion about the source. In practice, we will see that, thanks to Theorem 1 these two operators are almost equivalent in this context.

Theorem 1 (Semantic relatedness measure is a dogmatic opinion). *Let $sim(c, c')$ be the semantic similarity between two contexts c and c' obtained by computing the semantic relatedness between the contexts in a graph through deterministic measurements (e.g. [28]). Then, $\forall\ sim(c, c') \in [0,1]$,*

$$\omega_{c=c'}^{measure} = (b_{c=c'}^{measure}, d_{c=c'}^{measure}, u_{c=c'}^{measure}, a_{c=c'}^{measure})$$

is equivalent to a dogmatic opinion in subjective logic, i.e., a subjective opinion with uncertainty equal to zero.

Proof. A binomial opinion is a dogmatic opinion if the value of *uncertainty* is *0*. The semantic similarity measure can be represented as an opinion about the similarity of two contexts c and c'. However, since we restrict our focus on *WordNet*-based measures, the similarity is inferred by graph measurements, and not by probabilistic means. This means that, according to the source, this is a "dogmatic" opinion, since it does not provide any indication of uncertainty: $u_{c=c'}^{measure} = 0$. The opinion is not based on evidence observation, rather on actual deterministic measurements.

$$E(\omega_{c=c'}^{measure}) = b_{c=c'}^{measure} + u_{c=c'}^{measure} \cdot a = sim(c, c'), \tag{5}$$

where *measure* indicates the procedure used to obtain the semantic relatedness, e.g. Wu and Palmer Measure. The values of belief and disbelief are obtained as:

$$b_{c=c'}^{measure} = sim(c, c') \qquad d_{c=c'}^{measure} = 1 - b_{c=c'}^{measure}. \quad \square \tag{6}$$

Corollary 1 (Discounting an opinion with a dogmatic opinion). *Let A be a source who has an opinion about y in context c' expressed as*

$$\omega^A_{y:c'} = (b^A_{y:c'}, d^A_{y:c'}, u^A_{y:c'}, a^A_{y:c'})$$

and let the semantic similarity between the contexts c and c' be represented as a dogmatic opinion

$$\omega^{measure}_{c=c'} = (b^{measure}_{c=c'}, d^{measure}_{c=c'}, 0, a^{c'}_{c=c'}).$$

Since the source A does not have any prior opinion about the context c, we derive the opinion of A about c represented as

$$\omega^{A:c'}_c = (b^{A:c'}_c, d^{A:c'}_c, u^{A:c'}_c, a^{A:c'}_c)$$

using the base rate discounting operator on the dogmatic opinion.

$$a^{A:B}_y = a^B_y \qquad b^{A:B}_y = sim(c, c') \cdot b^B_y$$
$$u^{A:B}_y = 1 - sim(c, c') \cdot (b^B_y + d^B_y) \qquad d^{A:B}_y = sim(c, c') \cdot d^B_y. \tag{7}$$

Definition 1 (Weighing operator). *Let C be the set of contexts c' of which a source A has an opinion derived from the positive and negative evidence in the past. Let c be a new context for which A has no opinion yet. We can derive the opinion of A about facts in c, by weighing the relevant evidences in set C with the semantic similarity measure $sim(c, c') \, \forall c' \in C$. The belief, disbelief, uncertainty and a priori obtained through the weighing operation are expressed below.*

$$b^A_c = \frac{sim(c,c') \cdot p^A_{c'}}{sim(c,c')(p^A_{c'}+n^A_{c'})+2} \qquad d^A_c = \frac{sim(c,c') \cdot n^A_{c'}}{sim(c,c')(p^A_{c'}+n^A_{c'})+2}$$
$$u^A_c = 1 - \frac{sim(c,c') \cdot (p^A_{c'}+n^A_{c'})}{sim(c,c')(p^A_{c'}+n^A_{c'})+2} \qquad a^A_c = a^A_{c'}. \tag{8}$$

Theorem 2 (Approximation of the weighing and discounting operators). *Let*

$$\omega^{A:c'}_{y:c} = (b^{A:c'}_{y:c}, d^{A:c'}_{y:c}, u^{A:c'}_{y:c}, a^{A:c'}_{y:c})$$

be a discounted opinion which source A has about y in a new or unknown context c, derived by discounting A's opinion on known contexts $c' \in C$ represented as $\omega^A_{c'} = (b^A_{c'}, d^A_{c'}, u^A_{c'}, a^A_{c'})$ with the corresponding dogmatic opinions (e.g. $sim(c, c')$). Let source A also obtain an opinion about the unknown context c based on the evidence available from the earlier contexts c', by weighing the evidence (positive and negative) with semantic similarity between c and c', $sim(c, c') \, \forall c' \in C$. Then the difference between the results from the weighing and from the discount operator in subjective logic are statistically insignificant.

Proof. We substitute the values of belief, disbelief, uncertainty values in Eq. (9) for Base Rate Discounting with the values from Eq. (1) and expectation value from Eq. (5). We obtain the new value of the discounted base rate opinion as follows:

$$b_c^{A:c'} = \frac{sim(c,c') \cdot p_{c'}^A}{(p_{c'}^A + n_{c'}^A + 2)} \qquad d_c^{A:c'} = \frac{sim(c,c') \cdot n_{c'}^A}{(p_{c'}^A + n_{c'}^A + 2)}$$

$$u_c^{A:c'} = 1 - \frac{sim(c,c') \cdot (p_{c'}^A + n_{c'}^A)}{(p_{c'}^A + n_{c'}^A + 2)} \qquad a_c^{A:c'} = a_{c'}^A. \tag{9}$$

Equations (9) and (8) are pretty similar, except for the $sim(c,c').(p_{c'}^A + n_{c'}^A)$ factor in the weighing operator. In the following section we use a 95 % t-student and Wilcoxon signed-rank statistical test to prove that the difference due to that factor is not statistically significant for large values of $sim(c,c')$ (at least 0.5).

3.3   Evaluations

We show empirically the similarity between the weighing and the discounting.[1]

First Experiment: Discounting and Weighing in a Real-Life Case. We propose here a first validation of the similarity between weighing and discounting by using both of them in the process of estimation of the trustworthiness of a series of tags derived from a cultural heritage crowdsourcing project.

Steve social tagging project dataset. For the purpose of our evaluations, we use the "Steve Social Tagging Project" [25] data (in particular, the "Researching social tagging and folksonomy in the ArtMuseum"), which is a collaboration of museum professionals and others aimed at enhancing social tagging. In our experiments, we used a sample of tags which the users of the system provided for the 1784 images of the museum available online. Most of the tags were evaluated by the museum professionals to assess their trustworthiness. The tags can be single words or a string of words provided by the user regarding any objective aspect of the image displayed to them for the tagging. We used only the evaluated tags for our experiments.

Gathering evidence for evaluation. We select a very small set of semantically related tags, by using a Web-based *WordNet* interface [23]. We then gather the list of users who provided the tags regarding the chosen words and count the number of positive and the negative evidence. The chosen tags are only three (Asian, Chinese and Buddhist), and they correspond to 206 entries in total (i.e., they are associated 206 times to one or more pictures by one or more users). This represents a small sample compared to the total number of tag entries (0.5 %). However, this experiment is meant only to exemplify the use of the semantic similarity measure when one needs to compute an opinion about a new context (e.g., "Chinese"), given two existing ones (e.g., "Asian" and "Buddhist"). Therefore, we consider the *Chinese-Asian* pair (semantic similarity 0.933) and the *Chinese-Buddhist* pair (semantic similarity 0.6667). We refer to the second experiment for a

---

[1] Complete results are available at http://trustingwebdata.org/books/URSW_III/ slsw.zip.

more indicative evaluation. The opinions are calculated using two different methods. First by weighing the evidence with the semantic relatedness using Eq. (8) and the second method is by discounting the evidence with the semantic relatedness using Eq. (9).

Results. We employ the Student's t-test and the Wilcoxon signed-rank test to assess the statistical significance of the difference between two sample means. At 95 % confidence level, both tests show a statistically significant difference between the two means. This difference, for the *Chinese-Asian* pair is 0.025, while for the *Chinese-Buddhist* pair is 0.11, thanks also to the high similarity (higher than 0.5) between the considered topics. Having removed the average difference from the results obtained from discounting (which, on average, are higher than those from weighing), both the tests assure that the results of the two methods distribute equally.

Second Experiment: Discounting and Weighing on a Large Simulated Dataset. In the Steve.Museum dataset, the average number of annotations provided by a given user is limited (about 20). To check if the two methods for building subjective opinions using semantic similarity measures are significantly different, we built a large dataset consisting of 1000 sample tags and we treated the tags as if they were contributed by the same user. In this manner, we could check if the two methods present relevant differences both when the evidence amount is small or large. We perform the Student's t-test and the Wilcoxon signed-rank test to evaluate the hypothesis that the two methods are not statistically significantly different. For semantic relatedness values $sim(c, c') > 0.7$, the mean difference between the belief values obtained by weighing and discounting is 0.092. Thus with 95 % confidence interval, both tests assure that both the weighing operator and the discounting operator produce similar results. The semantic similarity threshold $sim(c, c') > 0.7$ is relevant and reasonable, because it becomes more meaningful to compute opinions for a new context based on the opinions provided earlier for the most semantically related contexts, while also in case of lack of evidence for a given context, evidence about a very diverse context can not be very significant.

4 Combining Probabilistic Semantic Similarity Measures Within Subjective Logic

The second extension that we propose regards the use of probabilistic semantic similarity measures within subjective logic.

Wikipedia Relatedness Measure. The Wu and Palmer measure introduced above is a deterministic semantic similarity measure, because it is deterministically computed based on the position of the two examined words in WordNet. We want to exemplify the adoption in subjective logic of semantic similarity measures belonging to another class, that is the probabilistic class of measures.

These measures determine the semantic similarity between two words in a statistical manner, by checking the occurrence and co-occurrence of the two words within a large corpora of documents. A famous example of this kind of similarity measures is the Normalized Google Distance [9], which uses Google as a corpus of documents.

We use the Wikipedia [27] relatedness measure, as defined by Milne et al. [20,21] because of its easiness of use. This distance adapts the Normalized Google Distance to use Wikipedia as a corpus of reference for computation. The Wikipedia similarity distance is defined as follows:

$$sim(c, c') = \frac{log(max(|A|, |B|)) - log(|AB|)}{log(|W|) - log(min(|A|, |B|))} \qquad (10)$$

where $|A|$ and $|B|$ are the cardinalities of the set of documents containing s_1 and s_2 respectively, and $|W|$ is the size of Wikipedia.

Moreover, Milne et al. provide a disambiguation confidence score for the measure, that ranges between zero and one.

4.1   Wikipedia Relatedness Measure as a Subjective Opinion

As in the previous section, given two synsets (s_1 and s_2), we name c and c' the respective context identified by them. To differentiate from the previous section, we use *measure'* as a placeholder for probabilistic similarity measures.

The elements at our disposal from the Wikipedia distance are:

- $sim(c, c') \in [0, 1]$ is the semantic relatedness between two synsets a and b;
- $conf(c, c') \in [0, 1]$ is the confidence in the semantic relatedness between a and b.

To represent the Wikipedia distance in subjective logic, we need to map all its elements to specific elements (or combinations of elements) of subjective logic, while taking into account the logic's constraints and mechanisms (e.g., the fact that $b + d + u = 1$). We provide a mapping for each of the elements above, and we provide a motivation for them as follows.

1. $conf(s1, s2) = 1 - u_{c=c'}^{measure'}$ because the confidence value determines exactly the portion of probability mass that is certain. Therefore, the remaining part of the probability mass is assigned to the uncertainty element of subjective opinions.
2. $E_{c=c'}^{measure'} = sim(c, c')$. That is, the expected value of the subjective opinion should coincide with the similarity between the two synsets considered.
3. $b_{c=c'}^{measure'} = conf(c, c') \cdot sim(c, c')$ because the certain part of an opinion $(1 - u)$ is assigned $b + d$. Thus, we assign this mass proportionally to the value of the similarity measure, to represent our belief in the two synsets being semantically related.

However, given the constraints of subjective logic, by virtue of Eq. (2) that we report as follows,

$$E_{c=c'}^{measure'} = b_{c=c'}^{measure'} + a_{c=c'}^{measure'} \cdot u_{c=c'}^{measure'}$$

we obtain

$$sim(c, c') = a_{c=c'}^{measure'}$$

which is, of course, wrong. The similarity value might depend on the subjective opinion's prior, but if the equation above holds, then we do not even need to compute the opinion, since the a priori value would already give the similarity value.

We propose, then, two mappings between subjective opinions and probabilistic semantic similarity measures, each of them satisfying two of the three requirements above. Of the three requirements, only the first one is considered as unavoidable, because of the definition of the uncertainty of subjective opinions.

Definition 2 (Wikipedia relatedness measure of two synsets as a subjective opinion (expected value as semantic similarity)). *We define a subjective opinion capturing the similarity between synset$_a$ and synset$_b$ using the Wikipedia distance as follows:*

$$sim(c, c') \equiv \omega_{c=c'}^{mesaure'}(b_{c=c'}^{measure'}, d_{c=c'}^{measure'}, u_{c=c'}^{measure'}) \tag{11}$$

where

$$b_{c=c'}^{measure'} = sim(c, c') - a_{c=c'}^{measure'} + a_{c=c'}^{measure'} \cdot conf(c, c')$$

$$d_{c=c'}^{measure'} = sim(c, c') + a_{c=c'}^{measure'} - a_{c=c'}^{measure'} \cdot conf(c, c') + conf(c, c') \tag{12}$$

$$u_{c=c'}^{measure'} = 1 - conf(c, c'),$$

hence

$$E_{c=c'}^{mesaure'} \equiv sim(c, c'). \tag{13}$$

We provide here motivation for the mapping that we propose. We treat the confidence value $conf(c, c')$ as the inverse of the uncertainty of a subjective opinion. In fact, we interpret the confidence as the percentage of probability mass confidently assigned by the semantic relatedness: the semantic relatedness ranges between zero and one, but we are confident on only $conf(c, c')\%$ of that mass. The rest of the probability mass $(1 - conf(c, c'))$ is, indeed, uncertain.

We also set the expected value of the opinion to coincide with the similarity value, that is:

$$E_{c=c'}^{measure'} = sim(c, c')$$

From this, given Eq. (2), and having set $u_{c=c'}^{measure'} = 1 - conf(c, c')$, follows that:

$$b_{c=c'}^{measure'} = E_{c=c'}^{measure'} - a_{c=c'}^{measure'} \cdot (1 - conf(c, c')) =$$
$$= sim(c, c') - a_{c=c'}^{measure'} \cdot (1 - conf(c, c')) =$$
$$= sim(c, c') - a_{c=c'}^{measure'} + a_{c=c'}^{measure'} \cdot conf(c, c')$$

and

$$d_{c=c'}^{measure'} = 1 - b_{c=c'}^{measure'} - u_{c=c'}^{measure'} =$$
$$= 1 - (sim(c, c') - a_{c=c'}^{measure'} + a_{c=c'}^{measure'} \cdot conf(c, c')) - (1 - conf(c, c')),$$

so

$$d_{c=c'}^{measure'} = sim(c, c') + a_{c=c'}^{measure'} - a_{c=c'}^{measure'} \cdot conf(c, c') + conf(c, c').$$

In this manner we define an opinion that reflects our constraints, that is: (1) uncertainty as inverse of the confidence of the semantic similarity value and (2) semantic similarity value as expected value of the subjective opinion. However, this mapping has the undesirable consequence that the belief $b_{c=c'}^{measure'}$ and the disbelief $d_{c=c'}^{measure'}$ depend on the a priori value $a_{c=c'}^{measure'}$. So, we propose an alternative mapping.

Definition 3 (Wikipedia relatedness measure of two synsets as a subjective opinion (belief as semantic similarity times confidence)). *We propose here an alternative mapping that allows a subjective opinion to capture the similarity between synset a and b using the Wikipedia distance. The mapping is defined as follows:*

$$sim(c, c') \equiv \omega_{c=c'}^{measure'}(b_{c=c'}^{measure'}, d_{c=c'}^{measure'}, u_{c=c'}^{measure'}), \qquad (14)$$

where

$$b_{c=c'}^{measure'} = conf(c, c') \cdot sim(c, c')$$
$$d_{c=c'}^{measure'} = conf(c, c') \cdot (1 - sim(c, c')) \qquad (15)$$
$$u_{c=c'}^{measure'} = 1 - conf(c, c').$$

Again, we set the constraint $u_{c=c'}^{measure'} = 1 - conf(c, c')$, however we do not bind the expected value of the opinion to be equal to $sim(c, c')$.

4.2   Using Wikipedia Relatedness Measure as a Subjective Opinion

We have shown in the previous subsection that we can represent Wikipedia relatedness measures between two synsets or two words by means of subjective opinion. As with many other subjective logic operators [17], we propose two possible mappings for the probabilistic semantic similarity measure. In particular, the second mapping that we propose does not present the undesirable characteristic shown by the first one, that is a dependency between a priori value and belief in the mapped opinion. Of course, these two mappings are different, so we do not check their equivalence, like we did in the previous section for the mapping between subjective logic and probabilistic semantic similarity measures.

Our goal is to show how to represent semantic similarity measures in subjective logic, to import externally defined elements in the logic and increase its capabilities. The choice of the mapping is dependent on the specific constraints given by a domain or an application where the logic is used in combination with the similarity measure, although our preference goes for the second mapping, because the first one presents an already mentioned undesirable dependency between belief and a priori value. The same reasoning applies to the choice of the semantic similarity measure to adopt. Each semantic similarity measure has specific limitations, like the requirement that words are present in a given graph or corpus of documents. The choice of a specific semantic similarity measure is beyond our focus, because it is a very domain- and application-dependent choice.

5 Partial Evidence Observation

The Web and the Semantic Web are pervaded of data that can be used as evidence for a given purpose, but that constitute partially positive/negative evidence for others. Think about the *Waisda?* tagging game [22]. Here, users challenge each other about video tagging. The more users insert the same tag about the same video within the same time frame, the more the tag is believed to be correct. Matching tags can be seen as positive observations for a specific tag to be correct. However, consider the orthogonal issue of the user reputation. User reputation is based on past behavior, hence on the trustworthiness of the tags previously inserted by him/her. Now, the trustworthiness of each tag is not deterministically computed, since it is roughly estimated from the number of matching tags for each tag inserted by the user. The expected value of each tag, which is less than one, can be considered as a partial observation of the trustworthiness of the tag itself. Vice-versa, the remainder can be seen as a negative partial observation. After having considered tag trustworthiness, one can use each evaluation as partial evidence with respect to the user reliability: no tag (or other kind of observation) is used as a fully positive or fully negative evidence, unless its correctness has been proven by an authority or by another source of validation. However, since only rarely the belief (and therefore, the expected value) is equal to one, these observations almost never count as a fully positive or fully negative evidence. We propose an operator for building opinions based on indirect observations, i.e., on observations used to build these opinions, each of which counts as an evidence.

Theorem 3 (Partial evidence-based opinions). *Let $\overrightarrow{P}$ be a vector of positive observations (e.g. a list of "hits" or "match" counts) about distinct facts related to a given subject s. Let l be the length of $\overrightarrow{P}$. Let each opinion based on each entry of $\overrightarrow{P}$ have an a priori value of $\frac{1}{2}$. Then we can derive an opinion about the reliability of the subject in one of the following two manners.*

- *By cumulating the expected values (counted as partial positive evidence) of each opinion based on each element of p:*

$$b_s = \frac{1}{l+2}\Sigma_{i=1}^{l}\frac{p_i+1}{p_i+2} \quad d_s = \frac{1}{l+2}\Sigma_{i=1}^{l}\frac{1}{p_i+2} \quad u_s = \frac{2}{l+2}. \quad (16)$$

– *By averaging the expected values of the opinions computed on each of the elements of p:*

$$b_s = \frac{1}{3l} \Sigma_{i=1}^{l} \frac{p_i + 1}{p_i + 2} \quad d_s = \frac{1}{3} - \frac{1}{3l} \Sigma_{i=1}^{l} \frac{1}{p_i + 2} \quad u_s = \frac{2}{3}. \tag{17}$$

Proof. For each "fact" about s we have at our disposal a count of positive pieces of evidence. We treat each fact as an observation about the trustworthiness of s. Examples of these observations are tags inserted by s in a crowdsourcing platform, and the items of evidence are the approvals or matches that these tags obtain. We do not set an upper limit to the amount of positive evidence. Rather, we convert it into a subjective opinion and we compute its expected value as follows (remember that no negative evidence is registered):

$$E_i = b_i + a_i \cdot u = \frac{p_i}{p_i + 2} + \frac{1}{2} \cdot \frac{2}{p_i + 2} = \frac{p_i + 1}{p_i + 2}. \tag{18}$$

E is considered as partial positive evidence. If p is an extremely high number, then E is approximated to 1. Otherwise, $1 - E$ is considered as partial negative evidence. Given that we have l pieces of partial evidence (because we have l distinct elements in $\overrightarrow{P}$), we compute the opinion about s following Eq. (1). Here we have two possibilities. If l contains evidence about distinct and independent facts, then we can cumulate all the pieces of evidence (represented as $E_i, 1 - E_i$) and by setting:

$$p_s = \Sigma_{i=1}^{l} \frac{p_i + 1}{p_i + 2} \quad n_s = \Sigma_{i=1}^{l} \frac{1}{p_i + 2},$$

we obtain Eq. (16). In fact, we consider each item of $\overrightarrow{P}$ as providing an observation about s.

If, instead, $\overrightarrow{P}$ contains dependent observations, then it makes sense to average them in order to uniformly represent the evidence about s. In this case, we set:

$$p_s = \frac{1}{l} \Sigma_{i=1}^{l} \frac{p_i + 1}{p_i + 2} \quad n_s = \frac{1}{l} \Sigma_{i=1}^{l} \frac{1}{p_i + 2}.$$

Following again Eq. (1), we obtain Eq. (17). Note that, in this case, we use only the average of the observation as item of evidence. Therefore, we have only one item of evidence. This justifies the fact that in Eq. 17 we always have 3 as denominator: following Eq. (1), $p + n + 2 = 1 + 2 = 3$.   □

More often, we use Eq. 16, because we consider the cases where evidence from different and independent facts about the same individual are provided.

For instance, consider the *Waisda?* tagging game [22]. Here, users challenge each other in tagging videos. Whenever two tag entries for the same video from two different users match, the users get a score. Indeed a matched tag has a higher probability to be correct, and the goal of the game is to collect reliable tags by incentivizing the users. How can we estimate a user reliability? Suppose that a user *user* added two different tags about two different videos. One of them got five matches, the other got two. We can compute a subjective opinion about *user* that represents his reputation using Eq. (16) and we obtain:

$$\omega_{user}\left(\frac{1}{l+2}\Sigma_{i=1}^{l}\frac{p_i+1}{p_i+2},\frac{1}{l+2}\Sigma_{i=1}^{l}\frac{1}{p_i+2},u_s=\frac{2}{l+2}\right)=$$

$$\omega_{user}\left(\frac{1}{4}\left(\frac{6}{7}+\frac{3}{4}\right),\frac{1}{4}\left(\frac{1}{7}+\frac{1}{4}\right),\frac{1}{4}\left(\frac{2}{4}\right)\right)=$$

$$\omega_{user}\left(\frac{55}{112},\frac{11}{112},\frac{1}{2}\right).$$

If, instead, the two tags got the same scores as before, but they were inserted for the same video in different matches, we can average their contribution, since they provide indications about the user reliability in the same situation. What we obtain using Eq. (17) is:

$$\omega_{user}\left(\frac{1}{3l}\Sigma_{i=1}^{l}\frac{p_i+1}{p_i+2},\frac{1}{3}-\frac{1}{3l}\Sigma_{i=1}^{l}\frac{1}{p_i+2},\frac{2}{3}\right)=$$

$$\omega_{user}\left(\frac{1}{6}\left(\frac{6}{7}+\frac{3}{4}\right),\frac{1}{3}-\frac{1}{6}\left(\frac{6}{7}+\frac{3}{4}\right),\frac{2}{3}\right)=$$

$$\omega_{user}\left(\frac{55}{168},\frac{1}{168},\frac{2}{3}\right).$$

This method, and in particular Eq. (16), has been adopted and implemented in a work of ours [3].

6 Dirichlet Process-Based Opinions: Open World Opinions

We present here an extension of subjective logic that allows using Dirichlet processes called "open world opinions". We start from introducing Dirichlet processes, and then describe the extension.

6.1 Preliminaries: Dirichlet Process

The Dirichlet process [10] is a stochastic process representing a probability distribution whose domain is a random probability distribution. As we previously saw, the binomial and multinomial opinions are equivalent to Beta and Dirichlet probability distributions. The Dirichlet distribution represents an extension of the Beta distribution from a two-category situation to a situation where one among n possible categories has to be chosen. A Dirichlet process over a set S is a stochastic process whose sample path (i.e. an infinite-dimensional set of random variables drawn from the process) is a probability distribution on S. The finite dimensional distributions are from the Dirichlet distribution: if H is a finite measure on S, α is a positive real number and X is a sample path drawn from a Dirichlet process, written as

$$X \sim DP(\alpha, H), \tag{19}$$

then for any partition of S of cardinality m, say $\{B_i\}_{i=1}^m$

$$(X(B_1), \ldots, X(B_m)) \sim Dirichlet(\alpha H(B_1), \ldots, \alpha H(B_m)). \tag{20}$$

Moreover, given n draws from X, we can predict the next observation as:

$$obs_{n+1} = \begin{cases} x_i^*(i \in [1 \ldots k]) & \text{with probability } \frac{n(x_i^*)}{n+\alpha} \\ H & \text{with probability } \frac{\alpha}{n+\alpha} \end{cases} \tag{21}$$

where x_i^* is one of the k unique values among the observations gathered.

6.2   Open World Opinions

Having to deal with real data coming from the Web, which are accessed incrementally, the possibility to update the relative probabilities of possible outcomes might not be enough to deal with them. We may need to handle unknown categories of data which should be accounted and manageable anyway. Ceolin et al. [8] show how it is important to account for unseen categories, when dealing with Web data. Here, we propose a particular subjective opinion called "open world opinion" which accounts for partial knowledge about the possible outcomes. A subjective opinion resembles personal opinion provided by sources with respect to facts. Open world opinions represent the case when something about a given fact has been observed, but the evidence allows also for some other (not yet observed) outcome to be considered as plausible. With this extension we allow the frame of discernment to have infinite cardinality. In practice, open world opinions allow to represent situations when the unknown outcome of an event can be equal to one among a list of already observed values (proportionally to the amount of observations for each of them), but it is also possible that (and so some probability mass is reserved for cases where) the outcome is different from what has been observed so far, and is drawn from an infinitely large domain.

Definition 4 (Open world opinion). *Let: X be a frame of infinite cardinality, $\alpha \in \mathbb{R}^+$, k be the number of categories observed, $\overrightarrow{P}$ be the array of evidence per category, $\overrightarrow{B}$ be a belief function over X, and H be a continuous function representing the prior probabilities for all the categories considered. New observations will belong to the previously observed categories with probability determined by the previous observations and to a new category with a probability determined by the parameter α (that determines the uncertainty u). New categories are drawn from H. We define the open world opinion ω_x as:*

$$\omega_x(\overrightarrow{B}, u, H)$$

$$b_{x_i} = \frac{p_{x_i}}{\alpha + \Sigma_{i=1}^k p_{x_i}} \qquad u = \frac{\alpha}{\alpha + \Sigma_{i=1}^k p_{x_i}} \qquad 1 = u + \Sigma_{x_i} b_{x_i}. \tag{22}$$

Definition 5 (Expected value of open world opinion). *The expected value of a category x_i given an open world opinion is computed as follows:*

$$E(x_i, \omega_x(\overrightarrow{B}, u, H))) = b_{x_i} + H(x_i) \cdot u = \frac{p_{x_i} + \alpha \cdot H(x_i)}{\alpha + \Sigma p_{x_t}} = \frac{p_{x_i}}{\alpha + \Sigma p_{x_t}}, \quad (23)$$

where $\overrightarrow{P}$ is the array of evidence observed, α is the concentration parameter, that determines how frequently observations belonging to new categories are likely to appear (we often set this parameter to 1 as default value; higher values imply that observations belonging to new categories are drawn with high probability), and H is the base distribution of the Dirichlet process, that is the probability distribution from which new categories are drawn with probability $\frac{\alpha}{\alpha + \Sigma p_{x_t}}$. Given that H is a continuous probability distribution (and hence with an infinite number of values), the probability of x_i determined by H is zero.

Theorem 4 (Equivalence between the subjective and Dirichlet process notation). *Let $\omega_X^{bn} = (\overrightarrow{B}, U, H)$ be an open world opinion expressed in belief notation, and $\omega_X^{pn} = (\overrightarrow{P}, \alpha, H)$ be an opinion expressed in probabilistic notation (i.e., using the notation from a Dirichlet process; $\overrightarrow{P}$ is an array of evidence), both over the same frame X. ω_X^{bn} and ω_X^{pn} are equivalent when the following mappings holds:*

$$\begin{cases} b_{x_i} = \dfrac{p_{x_i}}{\alpha + \Sigma_{i=1}^k p_{x_i}} \\[2mm] u = \dfrac{\alpha}{\alpha + \Sigma_{i=1}^k p_{x_i}} \end{cases} \Leftrightarrow \begin{cases} p_{x_i} = \dfrac{\alpha b_{x_i}}{U} \\[2mm] 1 = u + \Sigma b_{x_i}. \end{cases} \quad (24)$$

Proof. Each step of the Dirichlet process can be seen as a Dirichlet distribution. Hence the mapping between Dirichlet distributions and multinomial opinions [14] holds also here. □

Theorem 5 (Mapping between open world opinion and multinomial opinion). *Let $\omega 1_y^x(\overrightarrow{B}, u, H)$ be an open world opinion and let $\omega 2_y^x(\overrightarrow{B}, u, \overrightarrow{a})$ be a multinomial opinion. Let X_2 and Θ_2 be the frame and the frame of discernment of $\omega 2_y^x$. Let $\{b_i\}_{i=1}^k$ be the result of the partition of $dom(H)$ such that:*

1. $|\Theta_2| = |\{b_i\}|$;
2. $\bigcup \{b_i\}_{i=1}^k = dom(H)$;
3. $\forall \{x_i\}[(\{x_i\} \in X_2 \wedge |\{x_i\}| = 1 \wedge x_i \in b_j) \Rightarrow \nexists x_{k \neq j} \in b_i]$.
4. $W = k$, *where W is the non-informative constant of multinomial opinions*

Then there exists a function $D : Dom(H) \to \{b_i\}$ such that $D(\omega 1_y^x) = \omega 2_y^x$.

Proof. The equivalence between the discretized open world opinion and the multinomial opinion is proven by showing that:

– given Eq. (20), since the partition $\{b_i\}_{i=1}^k$ covers the entire $dom(H)$, then the partition distributes like the corresponding Dirichlet distribution;
– to each category of $\omega 2_y^x$ corresponds one and only one partition of $\{b_i\}$ as per item 2 of Theorem 5.   □

In other words, open world opinions extend multinomial opinions by allowing the frame of discernment Θ to be infinite. However, by properly discretizing an open world opinion, what we obtain is an equivalent multinomial opinion.

6.3   Example: Using Open World Opinions

Here we illustrate an example of the use of open world opinions. Piracy at sea is a well-known problem. Every year, several ships are attacked, hijacked, etc. by pirates. The International Chamber of Commerce has created a repository of reports about ship attacks.[2] van Hage et al. [26] have created an enriched Semantic Web version of such a repository, the Linked Open Piracy (LOP).[3] On the basis of LOP, one might think to be able to predict the frequency of attacks from one year based on the previously available data. However, a problem arises in this situation, since new attack types appear every year and this makes that frequencies vary. Ceolin et al. [8] have shown how the Dirichlet process can be employed to model such situations. Having the possibility to represent this information by means of an open world opinion adds the power of subjective logic to the Dirichlet process based representation. We can merge contributions from different sources, taking into account their reliability. Moreover, we can combine these facts with others in a logical way and then estimate the opinion (and the corresponding probability to be true) of the consequent facts. By using open world opinions, we can easily apply usual subjective logic operators to these data and easily represent them in a way that takes into account basic provenance information (e.g. data source) when applying fusing or discounting operators. For instance, if according to LOP, in Asia in 2010 we had ten hijacking events and ten attempted boardings, then we would represent this as:

$$\omega^{LOP}_{Attacks\ in\ Asia\ in\ 2010}([0.48, 0.48], 0.04, U(0, 1)).$$

If our opinion about LOP is that is a reliable but not fully accountable source (e.g. $\omega^{us}_{LOP}(0.8, 0.1, 0.1)$), then we can take this information into account by weighing the opinion given by LOP as follows:

$$\omega^{us}_{LOP}(0.8, 0.1, 0.1) \otimes \omega^{LOP}_{Attacks\ in\ Asia\ in\ 2010}([0.48, 0.48], 0.04, U(0, 1)) =$$

$$= \omega^{us:LOP}_{Attacks\ in\ Asia\ in\ 2010}([0.384, 0.384], 0.232, U(0, 1)).$$

The resulting weighted opinion is more uncertain than the initial one, because, even though the two observed types are more likely to happen, the small uncertainty about the source reliability makes the other probabilities to rise.

---

[2] http://www.icc-ccs.org
[3] http://semanticweb.cs.vu.nl/lop

Having represented this information in subjective logic allows us to reason on the data that we gathered. For instance, we could estimate the cost of the insurance premium for a ship that goes along Asian routes, given that insurance companies relate their premiums to the attack predicted to happen. By means of Dirichlet processes these attacks are quite precisely predictable. The determination of the cost of the premium given the happening of some attacks is representable by means of logical statements. In the future, we will develop subjective logic operators that allow to combine logically open world opinions. However, representing these opinions is the first step towards the possibility to permit this kind of reasoning.

Another useful consequence of this representation is the fact that, as we saw in the example above, we take into account the reputation of the source when modeling the opinion. We could also merge contributions from different sources, once we have developed extensions of the fusion operators tailored for open world opinions.

So, in the future we plan to develop subjective operators that extend those currently existing in order to handle open world opinions. The logical operators will allow combining propositions in logical manners (conjunction, disjunction, etc.) and will probably allow ontological reasoning, although this needs to be investigated further, for instance, with respect to the feasibility of subsumption computation et simila. The fusion and discounting operators will allow handling opinions from different sources and accounting for their reliability.

7 Related Work

The core element of subjective logic is the concept of "opinion" that is, the representation that a given source holds with respect to the truth value of a given proposition. Subjective logic's operators allow combining opinions in different manners, and their development has been widely investigated. Remarkably, the averaging and cumulative fusion [13,14] (i.e., operators that allow averaging or cumulating opinions about the same proposition from different sources) and the discounting [16] (i.e., the operator that allows weighing a source's opinion based on the source's reputation) operators are among the most generic and useful operators for this logic. These operators provide the foundations for the work proposed in this chapter. The connections between subjective logic and the (Semantic) Web are increasing. Ceolin et al. [7] adopt this logic for computing trust values of annotations provided by experts, using DBpedia and other Web sources as evidence. Unlike this work, they do not use semantic similarity measures. Ceolin et al. [4,6] and Bellenger et al. [1] provide applications of the combination of evidential reasoning with semantic similarity measures and Semantic Web technologies. In this chapter we provide the theoretical foundations for these approaches, and we generalize them. Sensoy et al. [24] use semantic similarity in combination with subjective logic to import knowledge from one context to another. They use the semantic similarity measure to compute a prior value for the imported data, while we use it to weigh all the available evidence.

Kaplan et al. [18] focus on the exploration of uncertain partial observations used for building subjective opinions. Unlike their work, we restrict our focus on partial observations of Web-like data and evaluations, which comprise the number of "likes", links and other similar indicators related to a given Web item. The weighing and discounting based on semantic similarity measures can resemble the work of Jøsang et al. [13], although the additional information that we include in our reasoning (that is, semantic similarity) is related only to the frame of discernment in subjective logic, and not to the belief assignment function.

8   Conclusion

We show the potential for employing subjective logic as a basis for reasoning on Web and Semantic Web data. We show that it can be really powerful for handling uncertainty and how little extensions can help in improving the mutual benefit that Semantic Web and subjective logic obtain from cooperating together. We propose the use of semantic similarity measures, both deterministic (in particular, the Wu and Palmer similarity measure) and probabilistic ones (in particular, the Wikipedia semantic relatedness), within subjective logic. Part of this work is based on previously mentioned practical applications that show the usefulness of it, and here we provide theoretical foundations for it.

Second, we propose a means to represent subjective opinions on the basis of partial evidence, which is a common phenomenon on the Web (e.g. number of hits or number of tweets). This operator has been employed in a few empirical works already, but here we provide a formal definition for it.

Lastly, we extend subjective opinions to model Dirichlet processes. These have shown to be particularly useful to represent at least some Web datasets. We introduce open world opinions to incorporate Dirichlet processes in subjective logic.

We plan to investigate further the integration of semantic similarity measures in subjective logic, to make it more uniform, and possibly provide best practices that help choosing the right measure and mapping for a given set of requirements. Also, we will provide additional operators for managing open world opinions. We foresee that other extensions will be possible as well like, for instance, the usage of hyperopinions [15] to handle subsumption reasoning about uncertain data.

Acknowledgements. This research was supported by the Data2Semantics and SEALINC Media projects in the Dutch national program COMMIT.

References

1. Bellenger, A., Gatepaille, S., Abdulrab, H., Kotowicz, J.-P.: An evidential approach for modeling and reasoning on uncertainty in semantic applications. In: URSW, pp. 27–38. CEUR-WS.org (2011)
2. Budanitsky, A., Hirst, G.: Evaluating Wordnet-based measures of lexical semantic relatedness. Comput. Linguist. **32**(1), 13–47 (2006)

3. Ceolin, D., Groth, P., van Hage, W., Nottamkandath, A., Fokkink, W.: Trust evaluation through user reputation and provenance analysis. In: 8th International Workshop on Uncertainty Reasoning for the Semantic Web, pp. 15–26 (2012)
4. Ceolin, D., Nottamkandath, A., Fokkink, W.: Automated evaluation of annotators for museum collections using subjective logic. In: Dimitrakos, T., Moona, R., Patel, D., McKnight, D.H. (eds.) IFIPTM 2012. IFIP AICT, vol. 374, pp. 232–239. Springer, Heidelberg (2012)
5. Ceolin, D., Nottamkandath, A., Fokkink, W.: Subjective logic extensions for the semantic web. In: URSW, vol. 900, pp. 27–38. CEUR-WS.org (2012)
6. Ceolin, D., Nottamkandath, A., Fokkink, W.: Semi-automated assessment of annotation trustworthiness. In: PST, pp. 325–332. IEEE Computer Society (2013)
7. Ceolin, D., Van Hage, W., Fokkink, W.: A trust model to estimate the quality of annotations using the Web. In: WebSci10. Web Science Trust (2010)
8. Ceolin, D., van Hage, W., Fokkink, W.: Estimating the uncertainty of categorical web data. In: URSW, pp. 15–26. CEUR-WS.org (2011)
9. Cilibrasi, R.L., Vitanyi, P.M.B.: The Google similarity distance. IEEE Trans. Knowl. Data Eng. **19**(3), 370–383 (2007)
10. Ferguson, T.S.: A Bayesian analysis of some nonparametric problems. Ann. Stat. **2**, 209–230 (1973)
11. Fink, D.: A compendium of conjugate priors. Technical report, Cornell University (1995)
12. Jøsang, A.: A logic for uncertain probabilities. Int. J. Uncertainty Fuzziness Knowl. Based Syst. **9**(3), 212–279 (2001)
13. Jøsang, A., Daniel, M., Vannoorenberghe, P.: Strategies for combining conflicting dogmatic beliefs. In: FUSION, pp. 1133–1140. IEEE (2003)
14. Jøsang, A., Diaz, J., Rifqi, M.: Cumulative and averaging fusion of beliefs. Inf. Fusion **11**(2), 192–200 (2010)
15. Jøsang, A., Hankin, R.: Interpretation and fusion of hyper opinions in subjective logic. In: FUSION, pp. 1225–1232. IEEE (2012)
16. Jøsang, A., Marsh, S., Pope, S.: Exploring different types of trust propagation. In: Stølen, K., Winsborough, W.H., Martinelli, F., Massacci, F. (eds.) iTrust 2006. LNCS, vol. 3986, pp. 179–192. Springer, Heidelberg (2006)
17. Jøsang, A., McAnally, D.: Multiplication and comultiplication of beliefs. Int. J. Approximate Reasoning **38**(1), 19–51 (2005)
18. Kaplan, L., Chakraborty, S., Bisdikian, C.: Subjective logic with uncertain partial observations. In: FUSION, pp. 565–572. IEEE (2012)
19. Miller, G.A.: WordNet: a lexical database for English. Commun. ACM **38**(11), 39–41 (1995)
20. Milne, D., Witten, I.H.: An effective, low-cost measure of semantic relatedness obtained from Wikipedia Links. In: Wikipedia and Artificial Intelligence: An Evolving Synergy, pp. 25–30. AAAI Press (2008)
21. Milne, D., Witten, I.H.: Learning to link with Wikipedia. In: CIKM, pp. 509–518. ACM (2008)
22. Netherlands Inst. for Sound and Vision. Waisda? (2012). http://wasida.nl
23. Princeton University. Wordnet::Similarity (2012). http://marimba.d.umn.edu/cgi-bin/similarity/similarity.cgi
24. Sensoy, M., Pan, J., Fokoue, A., Srivatsa, M., Meneguzzi, F.: Using subjective logic to handle uncertainty and conflicts. In: TrustCom, pp. 1323–1326. IEEE (2012)
25. U.S Institute of Museum and Library Service. Steve Social Tagging Project (2012)
26. van Hage, W., Malaisé, V., van Erp, M.: Linked Open Piracy: a story about e-Science, linked data, and statistics. J. Data Semant. **1**(3), 187–201 (2012)

27. Wikimedia Foundation. Wikipedia: The Free Encyclopedia (2014). http://www.
    wikipedia.org
28. Wu, Z., Palmer, M.: Verbs semantics and lexical selection. In: ACL '94, pp. 133–
    138. ACL (1994)

Uncertainty Estimation and Analysis
of Categorical Web Data

Davide Ceolin[1]($\boxtimes$), Willem Robert van Hage[2], Wan Fokkink[1],
and Guus Schreiber[1]

[1] VU University, Amsterdam, The Netherlands
{d.ceolin,w.j.fokkink,guus.schreiber}@vu.nl
[2] Synerscope B.V., Eindhoven, The Netherlands
willem.van.hage@synerscope.com

Abstract. Web data often manifest high levels of uncertainty. We focus on categorical Web data and we represent these uncertainty levels as first- or second-order uncertainty. By means of concrete examples, we show how to quantify and handle these uncertainties using the Beta-Binomial and the Dirichlet-Multinomial models, as well as how take into account possibly unseen categories in our samples by using the Dirichlet process. We conclude by exemplifying how these higher-order models can be used as a basis for analyzing datasets, once at least part of their uncertainty has been taken into account. We demonstrate how to use the Battacharyya stastistical distance to quantify the similarity between Dirichlet distributions, and use such results to analyze a Web dataset of piracy attacks both visually and automatically.

Keywords: Uncertainty · Bayesian statistics · Non-parametric statistics · Beta-Binomial · Dirichlet-Multinomial · Dirichlet process · Bhattacharyya distance

1 Introduction

The World Wide Web and the Semantic Web offer access to an enormous amount of data and this is one of their major strengths. However, the uncertainty about these data is quite high, due to the multi-authoring nature of the Web itself and to its time variability: some data are accurate, some others are incomplete or inaccurate, and generally, such a reliability level is not explicitly provided.

We focus on the real distribution of these Web data, in particular of categorical Web data, regardless of whether they are provided by documents, RDF [33] statements or other means. Categorical data are among the most important types of Web data, because they include also URIs. We assume that any kind of reasoning that might produce new statements (e.g. subsumption) has already taken place. Hence, unlike for instance Fukuoe et al. [16], that apply probabilistic reasoning in parallel to OWL [32] reasoning, we propose some models to address uncertainty issues on top of that kind of reasoning layers. These models, namely the parametric Beta-Binomial and Dirichlet-Multinomial, and the

© Springer International Publishing Switzerland 2014
F. Bobillo et al. (Eds.): URSW 2011-2013, LNAI 8816, pp. 265–288, 2014.
DOI: 10.1007/978-3-319-13413-0_14

non-parametric Dirichlet process, use first- and second-order probabilities and the generation of new classes of observations, to derive safe conclusions on the overall populations of our data, given that we are deriving those from possibly biased samples. These models are chosen exactly because they allow modeling categorical datasets, while taking into account the uncertainty related to the fact that we observe these datasets through samples that are possibly misleading or only partially representative.

Our goal is twofold. On the one hand, we want to show that higher-order probability distributions are useful to model categorical Web datasets while coping with their uncertainty. Hence we compare them with first-order probability distributions and show that taking uncertainty into account is preferable, for instance, when such distributions are used as a basis for prediction. On the other hand, we also show that it is possible to use higher-order probability distributions as basis for data analyses, rather than necessarily focusing on the raw data.

This chapter revises and extends the paper "Estimating Uncertainty of Categorical Web Data" [6], presented at the 7th International Workshop on Uncertainty Reasoning for the Semantic Web at the 10th International Semantic Web Conference 2011. The extension regards mainly the demonstration of use of higher-order probability distributions as a basis for categorical Web data analysis. In particular, we show how to use statistical distances (specifically, the Bhattacharyya statistical distance) to identify patterns and relevant changes in our data.

The chapter continues as follows. First we describe the scope of these models (Sect. 2), second we introduce the concept of conjugate prior (Sect. 3), and then two classes of models: parametric and non-parametric (Sect. 4). We show how it is possible to utilize such models to analyze dataset from the Web (Sect. 5) and, finally, we discuss the results and conclude (Sect. 6).

2   Scope of This Work

We define here the scope of the work presented in this chapter.

2.1   Empirical Evidence from the Web

Uncertainty is often an issue in case of empirical data. This is especially the case with empirical Web data, because the nature of the Web increases the relevance of this problem but also offers means to address it, as we see in this section. The relevance of the problem is related to the utilization of the mass of data that any user can find over the Web: can one safely make use of these data? Lots of data are provided on the Web by entities the reputation of which is not surely known. In addition to that, the fact that we access the Web by crawling, means that we should reduce our uncertainty progressively, as long as we increment our knowledge. Moreover, when handling our sample it is often hard to determine how representative such a sample is of the entire population, since often we do not own enough sure information about it.

On the other hand, the huge amount of Web data gives also a solution for managing this reliability issue, since it can provide the evidence necessary to limit the risk when using a certain data set.

Of course, even within the Web it can be hard to find multiple sources asserting about a given fact of interest. However, the growing dimension of the Web makes it reasonable to believe in the possibility to find more than one data set about the given focus, at least by means of implicit and indirect evidence.

This work aims to show how it is possible to address the described issues by handling such empirical data, categorical empirical data in particular, by means of the Beta-Binomial, Dirichlet-Multinomial and Dirichlet process models.

2.2 Requirements

Our approach needs to be quite elastic in order to cover several issues, as described below. The non-triviality of the problem comes in a large part from the impossibility to directly handle the sampling process from which we derive our conclusions. The requirements that we need to meet are:

Ability to handle incremental data acquisition. The model should be incremental, in order to reflect the process of data acquisition: as long as we collect more data (even by crawling), our knowledge should reflect that increase.

Prudence. It should derive prudent conclusions given all the available information. In case not enough information is available, the wide range of possible conclusions derivable should clearly make it harder to set up a decision strategy.

Cope with biased sampling. The model should deal with the fact that we are not managing a supervised experiment, that is, we are not randomly sampling from the population. We are using an available data set to derive safe consequences, but these data could, in principle, be incomplete, inaccurate or biased, and we must take this into account.

Ability to handle samples from mixtures of probability distributions. The data we have at our disposal may have been drawn from diverse distributions, so we cannot use the central limit theorem, because it relies on the fact that the sequence of variables is identically distributed. This implies the impossibility to make use of estimators that approximate by means of the Normal distribution.

Ability to handle temporal variability of parameters. Data distributions can change over time, and this variability has to be properly accounted.

Complementarity with higher-order layers. The aim of the approach is to quantify the intrinsic uncertainty in the data provided by the reasoning layer, and, in turn, to provide to higher-order layers (time series analysis, decision strategy, trust, etc.), reliable data and/or metadata.

2.3 Related Work

The models adopted here are applied in a variety of fields. For the parametric models, examples of applications are: topic identification and document clustering [12,24], quantum physics [20], and combat modeling in the naval domain [23]. What these heterogeneous fields have in common is the presence of multiple levels of uncertainty (for more details about this, see Sect. 4.1).

Also non-parametric models are applied in a wide variety of fields. Examples of these applications include document classification [9] and haplotype inference [36]. These heterogeneous fields have in common with the applications mentioned above the presence of several layers of uncertainty, but they also show a lack of prior information about the number of parameters. These concepts are treated in Sect. 4.2 where the Wilcoxon sign-ranked test [35], used for validation purposes, falls into the non-parametric models class.

Our focus is on the statistical modeling of categorical Web data. The analysis of categorical data is a widespread and well consolidated topic (see, for instance, the work of Davis and Koch [7] or Agresti [1]). About the statistical analysis of Web datasets, Auer et al. [3] present a statement-stream-based approach for gathering comprehensive statistics about RDF datasets that differs from our approach as we do not focus on streams. To our best knowledge, the chosen models have not been applied to categorical Web data yet. We propose to adopt them, because, as the following sections show, they fit the requirements previously listed. Moreover, we see models such as SCOVO [18], RDF Data Cube [8] and VoID [2] as complementary to our work, since these would allow modeling and publishing the results of our analyses.

3 Prelude: Conjugate Priors

To tackle the requirements described in the previous section, we adopt some Bayesian parametric and non-parametric models in order to be able to answer questions about Web data.

Conjugate priors [17] are the "leit motiv", common to all the models adopted here. The basic idea starts from the Bayes theorem (1): given a prior knowledge and our data, we update the knowledge into a posterior probability.

$$P(A|B) = \frac{P(B|A) * P(A)}{P(B)} \tag{1}$$

This theorem describes how it is possible to compute the posterior probability, $P(A|B)$, given the prior probability of our data, $P(A)$, the likelihood of the model, given the data, $P(B|A)$, and the probability of the model itself, $P(B)$.

When dealing with continuous probability distributions, the computation of the posterior distribution by means of Bayes theorem can be problematic, due to the need to possibly compute complicated integrals. Conjugate priors allow us to overcome this issue: when prior and posterior probability distributions belong to the same exponential family, the posterior probability can be obtained by updating the prior parameters with values depending on the observed sample [15]. Exponential families are classes of probability distributions with a density function of the form $f(x) = e^{a(q)b(x)+c(q)+d(x)}$, with q a known parameter and a, b, c, d known functions. Exponential families include many important probability distributions, like the Normal, Binomial, Beta, etc., [11]. So, if X is a random variable that distributes as defined by the function $P(p)$ (for some parameter or vector of parameters p) and, in turn, p distributes as $Q(\alpha)$ for some parameter (or vector of parameters α called "hyperparameter"), and P belongs to the same exponential family as Q,

$$p \sim Q(\alpha), \; X \sim P(p)$$

then, after having observed obs,

$$p \sim Q(\alpha')$$

where $\alpha' = f(\alpha, obs)$, for some function f. For example, the Beta distribution is the conjugate of the Binomial distribution. This means that the Beta, shaped by the prior information and by the observations, defines the range within which the parameter p of the Binomial is probably situated, instead of directly assigning to it the most likely value. Other examples of conjugate priors are: Dirichlet, which is conjugate to the Multinomial, and Gaussian, which is conjugate to itself.

Conjugacy guarantees ease of computation, which is a desirable characteristic when dealing with very big data sets as Web data sets often are. Moreover, the model is incremental, and this makes it fit the crawling process with which Web data are obtained, because crawling, in turn, is an incremental process. Both the heterogeneity of the Web and the crawling process itself increase the uncertainty of Web data. The probabilistic determination of the parameters of the distributions adds a smoothing factor that helps to handle this uncertainty.

4   Higher-Order Probability Distributions for Modeling Categorical Web Data

This section presents three higher-order probability distributions that are useful to model uncertain categorical Web data. We present them in order of growing complexity, and then we outline a procedure for employing these models.

4.1   Parametric Bayesian Models for Categorical Web Data

Here we handle situations where the number of categories is known a priori, by using the Dirichlet-Multinomial model and its special case with two categories, i.e. the Beta-Binomial model [15]. Since we handle categorical data, the Binomial and the Multinomial distributions could be the natural choice to model them, depending on whether these data are divided into two or more categories. The Binomial and the Multinomial distributions allow modeling n draws from these datasets. These presume that the frequency of the categories is known, but this is not possible in our case, because we have at our disposal only a data sample which representativity is unknown. So we still model the data distributions by means of Binomial or Multinomial distributions (depending on the number of categories that we have), but we also model the parameters of these distributions by means of Beta or Dirichlet distributions respectively, since these are conjugated with the Binomial and with the Multinomial distributions. The shape of the Beta and of the Dirichlet distribution is determined by both the size and the distribution of the sample observed. The resulting models (Beta-binomial and Dirichlet-Multinomial) allow us to model the data distribution even though we base our modeling on samples that are uncertain and limited in size.

These models are parametric, since the number and type of parameters is given a priori, and they can also be classified as "empirical Bayesian models". This further classification means that they can be seen as an approximation of a full hierarchical Bayesian model, where the prior hyperparameters are set to their maximum likelihood values according to the analyzed sample.

Case Study 1 - Ratio Estimation. Suppose that a museum has to annotate a particular item I of its collection. Suppose further, that the museum does not have expertise in the house about that particular subject and, for this reason, in order to correctly classify the item, it seeks judgments from outside people, in particular from Web users that provide evidence of owning the desired expertise.

After having collected judgments, the museum faces two possible classifications for the item, $C1$ and $C2$. $C1$ is supported by four experts, while $C2$ by only one expert. We can use these numbers to estimate a probability distribution that resembles the correct distribution of $C1$ and $C2$ among all possible annotations.

A basic decision strategy that could make use of this probability distribution, could accept a certain classification only if its probability is greater or equal to a given threshold (e.g. 0.75). If so, the Binomial distribution representing the sample would be treated as representative of the population, and the sample proportions would be used as parameters of a Bernoulli distribution about the possible classifications for the analyzed item: $P(class(I) = C1) = 4/5 = 0.8$, $P(class(I) = C2) = 1/5 = 0.2$. (A Bernoulli distribution describes the possibility that one of two alternative events happens. One of these events happens with probability p, the other one with probability $1 - p$. A Binomial distribution with parameters n, p represents the outcome of a sequence of n Bernoulli trials having all the same parameter p.)

However, this solution shows a manifest leak. It provides to the decision strategy layer the probabilities for each of the possible outcomes, but these probabilities are based on the current available sample, with the assumption that it correctly represents the complete population of all existing annotations. This assumption is too ambitious. (Flipping a coin twice, obtaining a heads and a tails, does not guarantee that the coin is fair, yet.)

In order to overcome this limitation, we should try to quantify how much we can rely on the computed probability. In other words, if the previously computed probability can be referred to as a "first-order" probability, what we need to compute now is a "second-order" probability [20]. Given that the conjugate prior for the Binomial distribution representing our data is the Beta distribution, the model becomes:

$$p \sim Beta(\alpha, \beta), \; X \sim Bin(p, n) \tag{2}$$

where $\alpha = \#evidence_{C1} + 1$ and $\beta = \#evidence_{C2} + 1$.

By analyzing the shape of the conjugate prior Beta(5,2), we can be certain enough about the probability of $C1$ being safely above our acceptance threshold. In principle, our sample could be drawn by a population distributed with a 40 %–60 % proportion. If so, given the threshold of acceptance of 0.75, we would not be able to take a decision based on the evidence. However, the quantification of that proportion would only be possible if we know the population. Given that we do not have such information, we need to estimate it, by computing (3), where we can see how the probability of the parameter p being above the threshold is less than 0.5. This manifests the need for more evidence: our sample suggests to accept the most popular value, but the sample itself does not guarantee to be representative enough of the population.

$$P(p \geq 0.75) = 0.4660645, \; p \sim Beta(5, 2) \tag{3}$$

Table 1 shows how the confidence in the value p being above the threshold grows as long as we increase the size of the sample, when the proportion is kept. By applying the previous strategy (0.75 threshold) also to the second-order probability, we still choose $C1$, but only if supported by a sample of size at least equal to 15. Finally, these considerations could also be based on the Beta-Binomial distribution, which is a probability distribution representing a Binomial which parameter p is randomly drawn from a Beta distribution. The Beta-Binomial summarizes model (2) in one single function (4). We can see from Table 2 that the expected proportion of the probability distribution approaches the ratio of the sample (0.8), as the sample size grows. If so, the sample is regarded as a better representative of the entire population and the Beta-Binomial, as sample size grows, converges to the Binomial representing the sample (see Fig. 1).

$$X \sim BetaBin(n, \alpha, \beta) = p \sim Beta(\alpha, \beta), X \sim Bin(n, p) \tag{4}$$

Table 1. The proportion within the sample is kept, so the most likely value for p is always exactly that ratio. However, given our 0.75 threshold, we are sure enough only if the sample size is 15 or higher.

$\#C1$	$\#C2$	$P(p \geq 0.75) p \sim Beta(\#C1 + 1, \#C2 + 1)$
4	1	0.47
8	2	0.54
12	3	0.88

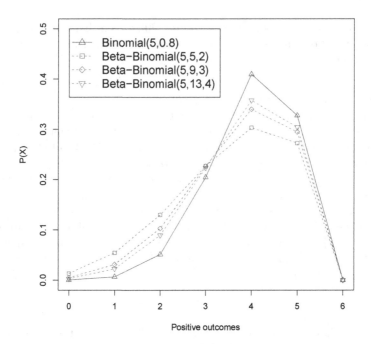

Fig. 1. Comparison between Binomial and Beta-Binomial with increasing sample size. As the sample size grows, Beta-Binomial approaches Binomial.

Case Study 2 - Confidence Intervals Estimation. The Linked Open Piracy (LOP)[1] is a repository of piracy attacks that happened around the world in the period 2005–2011, derived from reports retrieved from the ICC-CCS website.[2] Attack descriptions are provided, in particular covering their type (boarding, hijacking, etc.), place, time, as well as ship type.

Data about attacks is provided in RDF format, and a SPARQL [34] endpoint permits to query the repository. Such a database is very useful, for instance, for insurance companies to properly insure ships. The premium should be related to

Table 2. The sample proportion is kept, but the "expected proportion" p of Beta-Binomial passes the threshold only with a large enough sample. $E(X)$ is the expected value.

X	$E(X)$	$p = E(X)/n$
$BetaBin(5,5,2)$	3.57	0.71
$BetaBin(5,9,3)$	3.75	0.75
$BetaBin(5,13,4)$	3.86	0.77

[1] http://semanticweb.cs.vu.nl/lop
[2] http://www.icc-ccs.org/

both ship conditions and their usual route. The Linked Open Piracy repository allows an insurance company to estimate the probability of a ship to be victim of a particular type of attack, given the programmed route. Different attack types imply different risk levels.

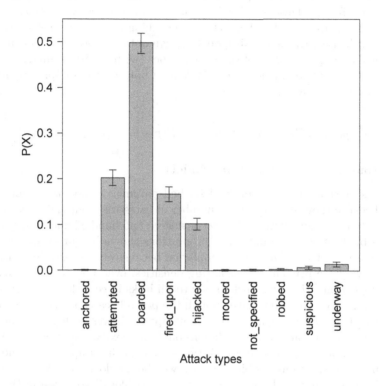

Fig. 2. Attack type proportion and confidence intervals.

However, directly estimating the probability of a new attack given the dataset, would not be correct, because, although derived from data published from an official entity like the Chamber of Commerce, the reports are known to be incomplete. This fact clearly affects the computed proportions, especially because it is likely that this incompleteness is not fully random. There are particular reasons why particular attack types or attacks happening in particular zones are not reported. Therefore, beyond the uncertainty about the type of next attack happening (first-order uncertainty), there is an additional uncertainty order due to the uncertainty in the proportions themselves. This can be handled by a parametric model that allows estimating the parameters of a Multinomial distribution. The model that we adopt is the multivariate version of the model described in Subsect. 4.1, i.e., the Dirichlet-Multinomial model [12,23,24]:

$$Attacks \sim \text{Multinom}(params), \ \ params \sim \text{Dirichlet}(\overrightarrow{\alpha}) \qquad (5)$$

where $\vec{\alpha}$ is the vector of observations per attack type (incremented by one unit each, as the α and β parameters of Beta probability distribution). By adopting this model, we are able to properly handle the uncertainty carried by our sample, due to either time variability (over the years, attack type proportions could have changed) or biased samples. Drawing the parameters of our Multinomial distribution from a Dirichlet distribution instead of directly estimating them, allows us to compensate for this fact, by smoothing our attacks distribution. As a result of the application of this model, we can obtain an estimate of confidence intervals for the proportions of the attack types (with 95 % of significance level, see Eq. (6)). These confidence intervals depend both on the sample distribution and on its dimension (Fig. 2).

$$\forall p \in param, CI_p = (p - \theta_1, p + \theta_2), P(p - \theta_1 \leq p \leq p + \theta_2) = 0.95 \qquad (6)$$

4.2   Non-parametric Bayesian Models

In some situations, the previously described parametric models do not fit our needs, because they set a priori the number of categories, but this is not always possible. In the previous example, we considered and handled uncertainty due to the possible bias of our sample. The proportions shown by our sample could be barely representative of the entire population because of a non-random bias, and therefore we were prudent in estimating densities, even not discarding entirely those proportions. However, such an approach lacks in considering another type of uncertainty: we could not have seen all the possible categories and we are not allowed to know all of them a priori. Our approach was to look for the prior probability to our data in the n-dimensional simplex, where n is the number of categories, that is, possible attack types. Now such an approach is no more sufficient to address our problem. What we should do is to add yet another hierarchical level and look for the right prior Dirichlet distribution in the space of the probability distributions over probability distributions (or space of simplexes). Non-parametric models differ from parametric models in that the model structure is not specified a priori but is instead determined from data. The term non-parametric is not meant to imply that such models completely lack parameters, but that the number and nature of the parameters are flexible and not set in advance. Hence, these models are also called "distribution free".

Dirichlet Process. Dirichlet processes [14] are a generalization of Dirichlet distributions, since they correspond to probability distributions of Dirichlet probability distributions. They are stochastic processes, that is, sequences of random variables (distributed as Dirichlet distributions) which value depends on the previously seen ones. Using the so-called "Chinese Restaurant Process" representation [26], it can be described as follows:

$$X_n = \begin{cases} X_k^* & \text{with probability } \frac{num_{n-1}(X_k^*)}{n-1+\alpha} \\ \text{new draw from } H & \text{with probability } \frac{\alpha}{n-1+\alpha} \end{cases} \qquad (7)$$

where H is the continuous probability measure ("base distribution") from which new values are drawn, representing our prior best guess. Each draw from H returns a different value with probability 1. α is an aggregation parameter, inverse to the variance: the higher α, the smaller the variance, which can be interpreted as the confidence value in the base distribution H: the higher the α value is, the more the Dirichlet process resembles H. The lower the α is, the more the value of the Dirichlet process tends to the value of the empirical distribution observed. Each realization of the process is discrete and is equivalent to a draw from a Dirichlet distribution, because, if

$$G \sim DP(\alpha, H) \tag{8}$$

is a Dirichlet process, and $\{B\}_{i=1}^n$ are partitions of the domain of H, S, we have that

$$(G(B_1)...G(B_n)) \sim Dirichlet(\alpha H(B_1)...\alpha H(B_n)) \tag{9}$$

If our prior Dirichlet process is (8), given (9) and the conjugacy between Dirichlet and Multinomial distribution, our posterior Dirichlet process (after having observed n values θ_i) can assume one of the following representations:

$$(G(B_1)...G(B_n))|\theta_1...\theta_n \sim Dirichlet(\alpha H(B_1) + n_{\theta_1}...\alpha H(B_n) + n_{\theta_n}). \tag{10}$$

$$G \mid \theta_1...\theta_n \sim DP\left(\alpha + n, \frac{\alpha}{\alpha + n}H + \frac{n}{\alpha + n}\frac{\sum_{i=1}^n \delta_{\theta_i}}{n}\right) \tag{11}$$

where δ_{θ_i} is the Dirac delta function [10], i.e., the function having density only in θ_i. The new base function is therefore a merge of the prior H and the empirical distribution, represented by means of a sum of Dirac delta's. The initial status of a Dirichlet process posterior to n observations, is equivalent to the nth status of the initial Dirichlet process that produced those observations (using the De Finetti theorem [19]).

The Dirichlet process, starting from a (possibly non-informative) "best guess", as long as we collect more data, approximates the real probability distribution. Hence, it correctly represents the population in a prudent (smoothed) way, exploiting conjugacy like the Dirichlet-Multinomial model, that approximates well the real Multinomial distribution only with a large enough data set (see Subsect. 4.1). The improvement of the posterior base distribution is testified by the increase of the α parameter, proportional to the number of observations.

Case Study 3: Unseen Categories Generation. We aim at predicting the type distributions of incoming attack events. In order to build an "infinite category" model, we need to allow for event types to be randomly drawn from an infinite domain. Hence, we map already observed attack types with random numbers in $[0..1]$ and, since all events are a priori equally likely, then new events are drawn from the Uniform distribution, $U(0, 1)$, that is our base distribution (and is a measure over $[0..1]$). The model is:

– $type_1 \sim DP(\alpha, U(0,1))$: the prior over the first attack type in region R;
– $attack_1 \sim Categorical(type_1)$: type of the first attack in R during $year_y$.

After having observed $attack_{1...n}$ during $year_y$, our posterior process becomes:

$$type_{n+1} \mid attack_{1...n} \sim DP\left(\alpha + n, \frac{\alpha}{\alpha + n} U(0,1) + \frac{n}{\alpha + n} \frac{\Sigma_{i=1}^{n} \delta_{attack_i}}{n}\right).$$

where α is a low value, given the low confidence in $U(0,1)$, and $type_{n+1}$ is the prior of $attack_{n+1}$, that happens during $year_{y+1}$. A Categorical distribution is a Bernoulli distribution with more than two possible outcomes (see Subsect. 4.1).

Results. Focusing on each region at time, we simulate all the attacks that happened there in $year_{y+1}$. Names of new types generated by simulation are matched to the actual $year_{y+1}$ names, that do not occur in $year_y$, in order of decreasing probability. The simulation is compared with a projection of the proportions of $year_n$ over the actual categories of $year_{n+1}$. The comparison is made by measuring the distance of our simulation and of the projection from the real attack types proportions of $year_{y+1}$ using the Manhattan distance [22]. This metric simply sums, for each attack type, the difference between the real $year_{y+1}$ probability and the one we forecast. Hence, it can be regarded as an error measure. Table 3 summarizes the results over the entire dataset.[3] Our simulation reduces the distance (i.e. the error) with respect to the projection, as confirmed by a Wilcoxon signed-rank test [35] at 95 % significance level. (This non-parametric statistical hypothesis test is used to determine whether one of the means of the population of two samples is smaller/greater than the other.) The simulation improves when a large amount of data is available and the category cardinality varies, as in case of Region India, which results are reported in Figs. 3 and 4a.

Table 3. Averages and variances of the prediction errors. The simulation gets a better performance.

	Simulation	Projection
Average distance	**0.29**	0.35
Variance	**0.09**	0.21

4.3 Model Selection and Utilization

Here we provide generic indication about the choice and use of the models described before.

---

[3] The code is available at http://trustingwebdata.org/books/URSW_III/DP.zip.

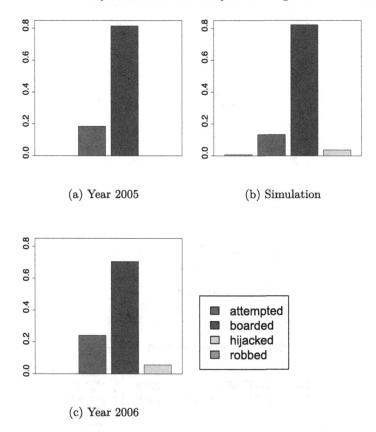

(a) Year 2005   (b) Simulation

(c) Year 2006

Fig. 3. Comparison between the projection forecast and the simulation forecast with the real-life year 2006 data of region India.

Model Selection. The models presented above are closely related each other, since each of them represents a generalization of the preceding model. Algorithm 1 is the algorithm that we propose for choosing the right model to apply when handling categorical Web data. It is rather simple and, under the assumption that we are handling categorical data, determines the choice of the model to use based on the number of categories that are known to be present in the data.

Model Building. Once the model has been selected, we build it based on the observations at our disposal as follows:

Beta-Binomial. The Beta-binomial model has three parameters: n, i.e. the number of draws performed and α and β, that are the frequencies of the two categories. In case no prior knowledge is available, we add the uninformative prior 1 to each frequency parameter. Otherwise, we add the prior frequency to each parameter.

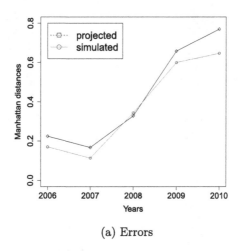

(a) Errors

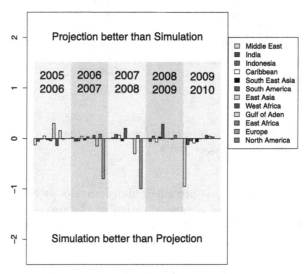

Overall dataset (each bar is one year of a region)

(b) Distances differences

Fig. 4. Error distance from real distribution of the region India (Fig. 4a) and differences of the error of forecast based on simulation and on projection (Fig. 4b). Positive difference means that the projection predicts better than our simulation.

```
if the number of categories is known then
    if the number of categories is two then
    |   return Beta-Binomial
    else
    |   return Dirichlet-Multinomial
    end
else
|   return Dirichlet Process
end
```
Algorithm 1. Model Selection Algorithm.

Dirichlet-Multinomial. This model has a vector $\vec{\alpha}$ of frequency parameters, plus the same n parameter indicating the number of draws to perform, as above. The frequency parameters need to be populated with the absolute frequencies observed. In case no prior knowledge is available, we add the uninformative prior 1 to each frequency parameter. Otherwise, we add the prior frequency to each parameter.

Dirichlet Process. The Dirichlet Process is determined by two parameters: the concentration parameter α and the base distribution H. If no prior information is available, we set $\alpha = 1$ and $H = U(0, 1)$, where U stands for the Uniform probability distribution. Then, after n categorical observations, we obtain the process described in Eq. 11.

Model Utilization. In the examples above, we used the process for prediction. In the following analyses we use them for comparison. To compare models, we compute similarity measures between probability distributions and analyze them. We present a detailed description of this utilization of the models in the following section. To perform predictions, i.e., to draw from the probability distributions, we proceed as follows:

Beta-binomial (and Dirichlet-multinomial). Randomly draw a parameter p (or a vector of parameters $\vec{p}$ in the case of the Dirichlet-multinomial) from a Beta (Dirichlet) distribution shaped by the frequency parameters α and β ($\vec{\alpha}$) described above. Then, randomly draw from a Binomial (Dirichlet) distribution shaped by the parameter p ($\vec{p}$).

Dirichlet Process. Draw from the Dirichlet distribution described above and in Eq. 11. If the drawn value has not been observed yet, then draw again from the base distribution H. Then update the process in order to obtain a new Dirichlet distribution representing its updated state.

5 Analyzing Datasets Using Higher-Order Probability Distributions

In the previous sections we have shown that higher-order probability distributions are useful to model Web data and account for their uncertainty. Here we want to show that higher-order probability distributions, despite the fact that they introduce a computational layer in the data management process, are easily utilizable as a basis for data analyses. The analyses presented here aim at showcasing how a data analyst could use the models presented before to derive insights from the data, for summarizing them and for extracting potentially useful information from large datasets. Besides the uncertainty management advantage, these models provide a means to abstract the data we analyze, thus allowing us to identify interesting patterns and regularities that would be hidden otherwise.

We apply our analyses on the LOP dataset introduced before. In the previous section, and in particular in Case Study 3 (Subsect. 4.2), we represent piracy attacks spread over the world by means of Dirichlet processes that "generate" the attacks over time. Each step of the Dirichlet process is represented by a Dirichlet distribution. We analyze the type distribution of the attacks with respect to time and regions. So, we use the data at our disposal to build one Dirichlet distribution per region per year, to represent the attack types distributions while taking into account the uncertainty in the data. Then, we use a statistical similarity to measure the likeness of distributions over time and regions. Clearly, we could have used different methods (e.g., mixture models), but we prefer this approach for its flexibility and simplicity (Fig. 5).

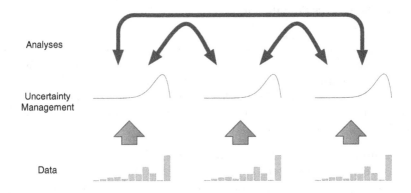

Fig. 5. Data abstraction and analysis overview.

5.1   Bhattacharyya Distance

We adopt the Bhattacharyya distance [4] to quantify the similarity between attack types distributions. The Bhattacharyya distance is a measure of divergence between probability distributions, that allows measuring the dissimilarity between two continuous or discrete probability distributions. As such, it goes from zero (when the compared distributions are identical) to infinite (when there is no overlap between the compared distributions). For continuous probability distributions, it is defined as follows:

$$D_B(dist_a, dist_b) = -\ln\left(\int \sqrt{dist_a(x)dist_b(x)}dx\right)$$

When applied to the Dirichlet Distributions (Rauber et al. [28]), the Bhattacharyya distance becomes:

$$D_B(Dir_a(x_1,\ldots,x_n), Dir_b(y_1,\ldots,y_n)) =$$

$$\Gamma\left(\frac{1}{2}\sum_{i\in\{1,\ldots,n\}} x_i + \frac{1}{2}\sum_{i\in\{1,\ldots n\}} y_i\right) + \frac{1}{2}\sum_{i\in\{1,\ldots,n\}}(\Gamma(x_i)) + \frac{1}{2}\sum_{i\in\{1,\ldots,n\}}(\Gamma(y_i)) -$$

$$\sum_{i\in\{1\ldots n\}}\left(\Gamma\left(\frac{1}{2}(x_i + y_i)\right)\right) - \frac{1}{2}\Gamma\left(\sum_{i\in\{1,\ldots,n\}}(x_i)\right) + \frac{1}{2}\Gamma\left(\sum_{i\in\{1,\ldots,n\}}(y_i)\right)$$

An advantage of the adopted approach is that the computation of the Bhattacharyya distance is particularly convenient. The only change we apply to the distance is to apply the logarithm base 2 to the result of the measure when the value is different from zero. This allows us to handle large numbers without any problem. Thus, the formula becomes:

$$sim(Dir_a, Dir_b) = \begin{cases} 0, & \text{if } D_B(Dir_a, Dir_b) = 0 \\ log_2(D_B(Dir_a, Dir_b)), & \text{otherwise} \end{cases}$$

5.2   Analysis of the Distribution of Piracy Attacks

We measure the Bhattacharyya distance for all the possible combinations of regions and years in the LOP dataset at our disposal. Since the set obtained by computing such similarities is rather big, we split it in two manners: first we look at the similarity of the attack type distributions of different regions, year by year, and second, we analyze the temporal evolution of such similarities, region by region. In this manner, we aim at identifying: (1) similarities in the type distribution across different regions and, (2) patterns related to the temporal distribution of attack types across different regions.

Attack Type Distribution Analysis per Year. We start by grouping our distances per year, and by analyzing their distribution across different regions of the world. In this manner, we aim at identifying similarities between regional attack distributions, while taking into account the temporal evolution of the attacks. Figure 6 shows six heatmaps representing the similarity between all possible combinations of regions for the years considered. Here, we can identify a few peculiar facts. For instance, Indonesia happens to be a region particularly different from the others (due to the presence of a high number of "boarded" and "attempted" attacks in that region), although this difference reduces in the last years of the period considered. With respect to previous analyses based on the actual piracy attacks counts [31], this difference is higher. From a manual investigation, we note that, besides a difference in the data distribution, Indonesia presents a difference in the total number of attacks registered. The higher-order models that we propose allow taking into account both aspects at the same time. Also, we note that the region that comprises India and Bengal differentiates from the rest in the first three years considered, while an important change in the similarity trend happens in Gulf of Aden in 2008 and continues afterwards.

Attack Type Distribution Analysis per Region. Figures 7 and 8 show a series of heatmaps representing the yearly distribution of piracy attacks, grouped by region. Here we can see, for instance, that North America (and, in part, also Europe) is characterized by being quite uniform in its distributions (thanks also to the fact that piracy attacks are quite rare in this region). Also, 2009 and 2010 are two years representing a changing point in several regions (e.g., Gulf of Aden, South America). Given the extension of such changes, we suppose this might be due to one or more events causing the global distribution of piracy attacks to change, although we are not aware of any.

5.3 Automating Piracy Attacks Analysis

The previous section proposes a combination of automatic and visual analysis of the data. Here we finally propose a procedure for automating the process of identifying potential interesting pieces of data in our datasets.

compute_distance_matrix. This procedure computes a similarity matrix that contains a distance between higher-order probabilistic models (e.g., the Bhattacharyya distance defined above in our case).

aggregate_data. This procedure aggregates the data with respect to a feature of interest. For instance, we can aggregate the data by time or region of the world, to see variation of attack types over either time or space.

changepoint. This procedure relies on the R package "ChangePoint" [21] and identifies points in the aggregated series of piracy attacks distributions that significantly differ from the rest. In particular, we make use of the *cpt.meanvar* function of the package, that determines the change point on the basis of a change in the mean and variance with respect to the rest of the series.

Results. We run the above procedure on the LOP dataset, and we obtained:

Regional Aggregation. India and Bengal (in 2005), Indonesia (in 2006), West Africa (in 2007, 2008 and 2010) are the regions identified as change points;

Yearly Aggregation. 2005 (in East Africa, North America and West Africa), 2006 (in Europe and Gulf of Aden), 2007 (in East Asia), 2008 (in Caribbean, India and Bengala, Indonesia, Middle East, South-East Asia and South America) are the years indicated as change point.

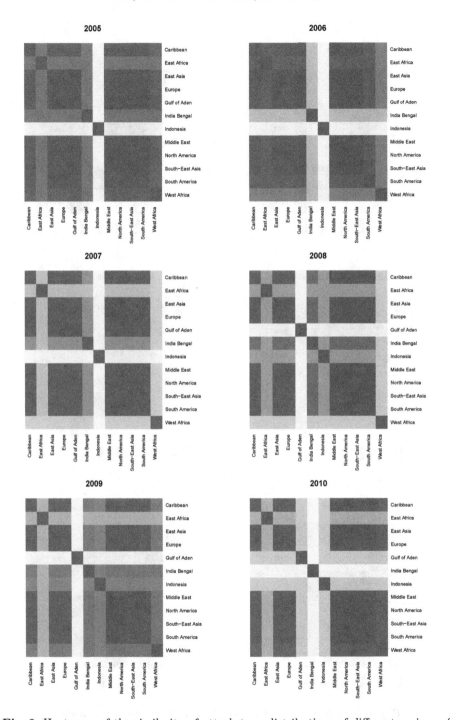

Fig. 6. Heatmaps of the similarity of attack type distributions of different regions of the world, divided by years.

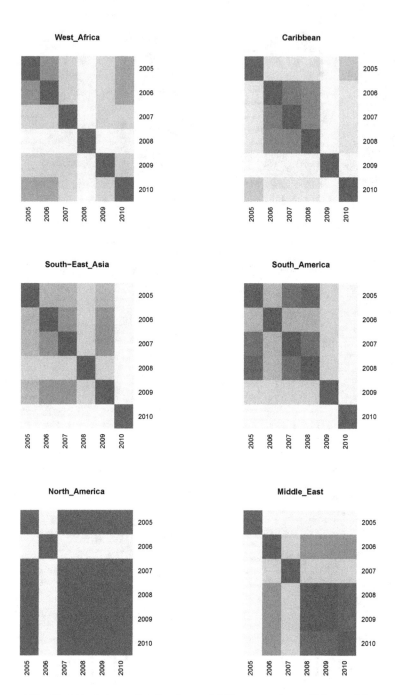

Fig. 7. Attack type distributions of different regions of the world.

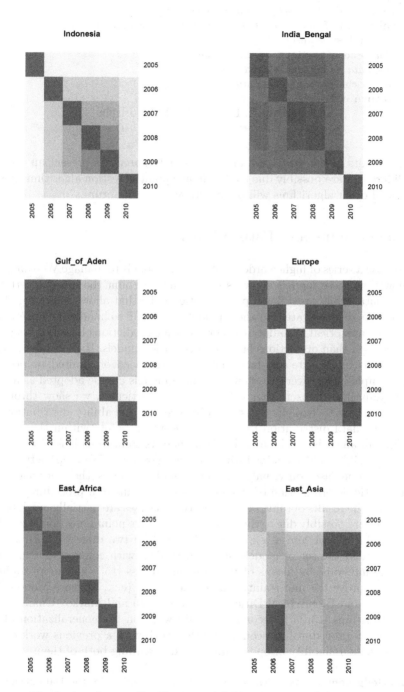

Fig. 8. Attack type distributions of different regions of the world.

Data: A dataset (*dataset*) of piracy attacks
Result: A set of change points in the piracy attacks dataset
Data_Analysis *dataset*

> $dm \leftarrow$ `compute_distance_matrix` (*dataset*);
> $agg\_data \leftarrow$ `aggregate_data` (*dm*);
> $res \leftarrow$ `changepoint` (agg_data);
> **return** *res*;

Algorithm 2. Data Analysis Algorithm.

The results of the visual and of the automated analyses present an overlap. The differences are possibly due to the change point detection algorithm chosen. The use of other algorithms will be investigated in the future.

6   Conclusions and Future Work

We propose a series of higher-order probabilistic models to manage Web data and we show that these models allow us to take into account the inner uncertainty of these data, while providing a probabilistic model that allows reasoning about the data. We demonstrate that these models are useful to handle the uncertainty present in categorical Web data by showing that predictions based on them are more accurate than predictions based on first-order models. Higher-order models allow us to compensate the fact that they are based on limited or possibly biased samples. Moreover, we show how these models can be adopted as a basis for analyzing the datasets that they model. In particular, we show through a case study, how to exploit statistical distances of probability distributions to analyze the data distribution to identify interesting points within the dataset. This kind of analysis can be used by data analysts to have an insight about the dataset, possibly to be combined with domain knowledge. We propose two kinds of analyses, one based on visual interpretation of heatmaps, the other one based on automatic determination of change points by means of a procedure that we introduce. The results obtained with these two analyses are partially overlapping. Differences are possibly due to the choice of the change point detection algorithm.

In the future, we aim at expanding this work in two directions. Firstly, we plan to extend the set of models adopted, to deal with concrete domain data (e.g. time intervals, by means of the Poisson process [15]), and more sophisticated to improve the uncertainty management part (e.g., Mixture Models [27], Nested [29] and Hierarchical Dirichlet processes [30] and Markov Chain Monte Carlo algorithms [13,25]). Secondly, we will work on the generalization of the data analysis procedures, by combining this work with a previous work on the analysis of the reliability of open data [5] and extending both of them.

Acknowledgments. This research was partially supported by the Data2Semantics Media project in the Dutch national program COMMIT.

References

1. Agresti, A.: Categorical Data Analysis, 3rd edn. Wiley, Hoboken (2013)
2. Alexander, K., Cyganiak, R., Hausenblas, M., Zhao, J.: Describing linked datasets with the void vocabulary. Technical report, W3C (2011)
3. Auer, S., Demter, J., Martin, M., Lehmann, J.: LODStats – an extensible framework for high-performance dataset analytics. In: ten Teije, A., Völker, J., Handschuh, S., Stuckenschmidt, H., d'Acquin, M., Nikolov, A., Aussenac-Gilles, N., Hernandez, N. (eds.) EKAW 2012. LNCS, vol. 7603, pp. 353–362. Springer, Heidelberg (2012)
4. Bhattacharyya, A.: On a measure of divergence between two statistical populations defined by their probability distributions. Bull. Calcutta Math. Soc. **35**, 99–109 (1943)
5. Ceolin, D., Moreau, L., O'Hara, K., van Hage, W.R., Fokkink, W.J., Maccatrozzo, V., Schreiber, G., Shadbolt, N.: Two procedures for estimating the reliability of open government data. In: Laurent, A., Strauss, O., Bouchon-Meunier, B., Yager, R.R. (eds.) IPMU 2014. CCIS, vol. 442, pp. 15–24. Springer, Heidelberg (2014)
6. Ceolin, D., van Hage, W.R., Fokkink, W.J., Schreiber, G.: Estimating Uncertainty of Categorical Web Data. In: URSW, pp. 15–26, November 2011. CEUR-WS.org
7. Koch, G., Davis, C.: Categorical Data Analysis Using SAS, 3rd edn. SAS Institute, Norwood (2012)
8. Cyganiak, R., Reynolds, D., Tennison, J.: The RDF data cube vocabulary. Technical report, W3C (2014)
9. Davy, M., Tourneret, J.: Generative supervised classification using dirichlet process priors. IEEE Trans. Pattern Anal. Mach. Intell. **32**, 1781–1794 (2010)
10. Dirac, P.: Principles of Quantum Mechanics. Oxford at the Clarendon Press, Oxford (1958)
11. Andersen, E.: Sufficiency and exponential families for discrete sample spaces. J. Am. Stat. Assoc. **65**, 1248–1255 (1970)
12. Elkan, C.: Clustering documents with an exponential-family approximation of the Dirichlet compound multinomial distribution. In: ICML, pp. 289–296. ACM (2006)
13. Escobar, M.D., West, M.: Bayesian density estimation and inference using mixtures. J. Am. Stat. Assoc. **90**, 577–588 (1994)
14. Ferguson, T.S.: A Bayesian analysis of some nonparametric problems. Ann. Stat. **1**(2), 209–230 (1973)
15. Fink, D.: A compendium of conjugate priors. Technical report, Cornell University (1995)
16. Fokoue, A., Srivatsa, M., Young, R.: Assessing trust in uncertain information. In: Patel-Schneider, P.F., Pan, Y., Hitzler, P., Mika, P., Zhang, L., Pan, J.Z., Horrocks, I., Glimm, B. (eds.) ISWC 2010, Part I. LNCS, vol. 6496, pp. 209–224. Springer, Heidelberg (2010)
17. Schlaifer, R., Raiffa, H.: Applied Statistical Decision Theory. M.I.T Press, Cambridge (1968)
18. Hausenblas, M., Halb, W., Raimond, Y., Feigenbaum, L., Ayers, D.: SCOVO: using statistics on the Web of data. In: Aroyo, L., Traverso, P., Ciravegna, F., Cimiano, P., Heath, T., Hyvönen, E., Mizoguchi, R., Oren, E., Sabou, M., Simperl, E. (eds.) ESWC 2009. LNCS, vol. 5554, pp. 708–722. Springer, Heidelberg (2009)
19. Hazewinkel, M.: Encyclopaedia of Mathematics. In: Chapter De Finetti theorem. Springer, New York (2001)

20. Hilgevoord, J., Uffink, J.: Uncertainty in prediction and in inference. Found. Phys. **21**, 323–341 (1991)
21. Killick, R., Eckley, I.A.: Changepoint: An R Package for Changepoint Analysis (2013). http://cran.r-project.org/package=changepoint
22. Krause, E.F.: Taxicab Geometry. Dover, New York (1987)
23. Kvam, P., Day, D.: The multivariate polya distribution in combat modeling. Naval Res. Logistics (NRL) **48**(1), 1–17 (2001)
24. Madsen, R.E., Kauchak, D., Elkan, C.: Modeling word burstiness using the Dirichlet distribution. In: ICML, pp. 545–552. ACM (2005)
25. Neal, R.M.: Markov chain sampling methods for Dirichlet process mixture models. J. Comput. Graph. Stat. **9**(2), 249–265 (2000)
26. Pitman, J.: Exchangeable and partially exchangeable random partitions. Probab. Theor. Relat. Fields **102**(2), 145–158 (1995)
27. Rasmussen, C.E.: The Infinite Gaussian Mixture Model. Advances in Neural Information Processing Systems, vol. 12, pp. 554–560. MIT Press, Cambridge (2000)
28. Rauber, T.W., Conci, A., Braun, T., Berns, K.: Bhattacharyya probabilistic distance of the dirichlet density and its application to split-and-merge image segmentation. In: WSSIP08, pp. 145–148 (2008)
29. Rodriguez, A., Dunson, D.B., Gelfand, A.E.: The nested Dirichlet process. J. Am. Stat. Assoc. **103**(483), 1131–1144 (2008)
30. Teh, Y.W., Jordan, M.I., Beal, M.J., Blei, D.M.: Hierarchical Dirichlet processes. J. Am. Stat. Assoc. **101**(476), 1566–1581 (2006)
31. van Hage, W.R., van Erp, M., Malaisé, V.: Linked open piracy: a story about e-science, linked data, and statistics. J. Data Seman. **1**(3), 187–201 (2012)
32. W3C. OWL Reference, August 2011. http://www.w3.org/TR/owl-ref/
33. W3C. Resource Definition Framework, August 2011. http://www.w3.org/RDF/
34. W3C. SPARQL, August 2011. http://www.w3.org/TR/rdf-sparql-query/
35. Wilcoxon, F.: Individual comparisons by ranking methods. Biometrics Bull. **1**(6), 80–83 (1945)
36. Xing, E.: Bayesian haplotype inference via the Dirichlet process. In: ICML, pp. 879–886. ACM Press (2004)

Ontological CP-Nets

Tommaso Di Noia[1], Thomas Lukasiewicz[2]([✉]), Maria Vanina Martinez[2],
Gerardo I. Simari[2], and Oana Tifrea-Marciuska[2]

[1] Department of Electrical and Information Engineering,
Polytechnic University of Bari, Bari, Italy
tommaso.dinoia@poliba.it
[2] Department of Computer Science, University of Oxford, Oxford, UK
{thomas.lukasiewicz,vanina.martinez,
gerardo.simari,oana.tifrea}@cs.ox.ac.uk

Abstract. Representing and reasoning about preferences is a key issue in many real-world scenarios in which personalized access to information is required. Many approaches have been proposed and studied in the literature that allow a system to work with qualitative or quantitative preferences; among the qualitative models, one of the most prominent are CP-nets. Their clear graphical structure unifies an easy representation of user preferences with good computational properties when computing the best outcome. In this paper, we show how to reason with CP-nets when the attributes modeling the knowledge domain are structured via an underlying domain ontology. We show how the computation of all undominated feasible outcomes of an ontological CP-net can be reduced to the solution of a constraint satisfaction problem, and study the computational complexity of the basic reasoning problems in ontological CP-nets.

1 Introduction

During the recent years, several revolutionary changes have been taking place in the classical Web. First, the so-called Web of Data is increasingly being realized as a special case of the Semantic Web. Second, as part of the Social Web, users are acting more and more as first-class citizens in the creation and delivery of contents on the Web. The combination of these two technological waves is called the *Social Semantic Web* (or also *Web 3.0*), where the classical Web of interlinked documents is gradually turning into (i) semantic data and tags constrained by ontologies, and (ii) social data, such as connections, interactions, reviews, and tags. The Web is thus shifting away from data on linked Web pages and towards more semantic and social data. This requires new technologies for search and query answering, where the ranking of search results is no longer solely based on the link structure between Web pages, but rather on the information available in the Social Semantic Web—in particular, the underlying ontological knowledge and the preferences of the users. Given a query, the latter play a fundamental role when a crisp yes/no answer is not enough to satisfy a user's needs, as there might be a certain degree of uncertainty in the possible answers [14].

© Springer International Publishing Switzerland 2014
F. Bobillo et al. (Eds.): URSW 2011-2013, LNAI 8816, pp. 289–308, 2014.
DOI: 10.1007/978-3-319-13413-0_15

There exist two main ways of modeling preferences: (a) *quantitative preferences* are associated with a number (or a quantity) representing their worth (e.g., "my preference for wi-fi connection is 0.8" and "my preference for cable connection is 0.4"), while (b) *qualitative preferences* are related to each other via pairwise comparisons (e.g., "I prefer wi-fi over cable connection"). The two approaches can also be combined (see, e.g., [24]). In many applications in practice, it is more natural to use a qualitative approach, as humans are not always comfortable or capable of expressing their preferences via a meaningful numerical value. To have a quantitative representation of her preferences, the user needs to explicitly determine a value for a large number of alternatives, usually described by more than one attribute. It is generally much easier to provide information about preferences as pairwise qualitative comparisons [14]. One of the most powerful qualitative preference formalisms are perhaps CP-nets [5], which are described in a graphical way and unifies an easy representation of user desires with nice computational properties when computing the best outcome.

Previous work on CP-nets, and more generally on preference representation approaches, mainly deals with a propositional representation of preferences. In this paper, we propose an enhancement of CP-nets by adding ontological information associated with preferences. We especially aim at using the resulting ontology-based partial strict orders in personalized semantic search on the Social Web. This is a first step towards a new type of ranking technologies, which are based on ontological and personalized information, and which go beyond PageRank and similar rankings. They will exploit ontological background knowledge as well as social information (e.g., from social networks and other platforms) and model them as semantic-enabled user preferences.

The main contributions of this paper can be briefly summarized as follows.

- We introduce ontological CP-nets, which combine CP-nets with description logics (DLs) so that variable values (which can be non-Boolean) correspond to DL concepts relative to an underlying domain TBox. We define the notions of feasible outcomes, dominance between such outcomes, and consistency in this context.
- We define the semantics of ontological CP-nets by a reduction to the notion of constrained CP-nets, which allows the problem of finding optimal outcomes in ontological CP-nets to be reduced to a constraint satisfaction problem.
- We study the complexity of the main reasoning problems for ontological CP-nets, namely consistency checking, whether a given outcome is undominated, and dominance testing, in relation to both the complexity of checking satisfiability of the underlying ontological language and the structure of the CP-net:
  - For tractable ontology languages, we show that the complexity is determined by that of CP-nets; i.e., the problems are complete for PSPACE.
  - For EXP- (resp., NEXP-) complete ontology languages, the complexity of these problems is dominated by the complexity of the ontology language.
  - Finally, if the CP-net is a polytree, and the ontology language is tractable, we show that dominance can be decided in polynomial time.

The rest of this paper is organized as follows. In Sect. 2, we briefly recall the background on description logics (DLs) and on CP-nets. Section 3 introduces ontological CP-nets, i.e., CP-nets enriched with ontological descriptions, while Sect. 4 describes how to compute optimal outcomes of ontological CP-nets. In Sects. 5 and 6, we provide complexity results and discuss related work, respectively. Finally, we give a summary of the results presented in this paper and an outlook on future work.

2   Preliminaries

In this section, we briefly recall from the literature the basics of description logics (DLs) and of CP-nets (a graphical representation for conditional preferences), which both form the main components of the formalism that we present in this paper.

2.1   Description Logics

Intuitively, description logics (DLs) [1] model a domain of interest in terms of concepts and roles, which represent classes of individuals and binary relations on classes of individuals, respectively. A DL knowledge base (or ontology) encodes in particular (i) subsumption relationships between concepts, (ii) subsumption relationships between roles, (iii) instance relationships between individuals and concepts, and (iv) instance relationships between pairs of individuals and roles, which represent (i) subset relationships between classes of individuals, (ii) subset relationships between binary relations on classes of individuals, (iii) the membership of individuals to classes, and (iv) the membership of pairs of individuals to binary relations on classes, respectively. There are many different DLs of different expressiveness [1]. In this section, we recall the DLs $\mathcal{SHIF}(\mathbf{D})$ and $\mathcal{SHOIN}(\mathbf{D})$, which stand behind the web ontology languages OWL Lite and OWL DL [18], respectively. Note, however, that the approach in this paper does not depend on a specific DL and may, e.g., also be applied to the very expressive $\mathcal{SROIQ}(D)$, which is the logic language behind OWL 2 [30].

Syntax. We now recall the syntax of $\mathcal{SHIF}(\mathbf{D})$ and $\mathcal{SHOIN}(\mathbf{D})$. We first describe the syntax of the latter, which has the following datatypes and elementary components. We assume a set of *elementary datatypes* and a set of *data values*. A *datatype* is an elementary datatype or a set of data values (called *datatype oneOf*). A *datatype theory* $\mathbf{D} = (\Delta^{\mathbf{D}}, \cdot^{\mathbf{D}})$ consists of a *datatype domain* $\Delta^{\mathbf{D}}$ and a mapping $\cdot^{\mathbf{D}}$ that assigns to each elementary datatype a subset of $\Delta^{\mathbf{D}}$ and to each data value an element of $\Delta^{\mathbf{D}}$. We extend $\cdot^{\mathbf{D}}$ to all datatypes by $\{v_1, \ldots\}^{\mathbf{D}} = \{v_1^{\mathbf{D}}, \ldots\}$. Let $\mathbf{A}$, $\mathbf{R}_A$, $\mathbf{R}_D$, and $\mathbf{I}$ be pairwise disjoint sets of *atomic concepts*, *abstract roles*, *datatype roles*, and *individuals*, respectively. We denote by $\mathbf{R}_A^-$ the set of inverses R^- of all $R \in \mathbf{R}_A$.

Roles and concepts are defined as follows. A *role* is any element of $\mathbf{R}_A \cup \mathbf{R}_A^-$ $\cup\,\mathbf{R}_D$. *Concepts* are inductively defined as follows. Each $\phi \in \mathbf{A}$ is a concept, and if $o_1, \ldots, o_n \in \mathbf{I}$, then $\{o_1, \ldots, o_n\}$ is a concept (called *oneOf*). If ϕ, ϕ_1, and ϕ_2 are concepts and if $R \in \mathbf{R}_A \cup \mathbf{R}_A^-$, then also $\neg\phi$, $(\phi_1 \sqcap \phi_2)$, and $(\phi_1 \sqcup \phi_2)$ are concepts (called *negation*, *conjunction*, and *disjunction*, respectively), as well as $\exists R.\phi$, $\forall R.\phi$, $\geqslant nR$, and $\leqslant nR$ (called *existential*, *value*, *atleast*, and *atmost restriction*, respectively) for an integer $n \geqslant 0$. Note that for decidability reasons, number restrictions will be restricted to simple abstract roles (see below). If D is a datatype and $U \in \mathbf{R}_D$, then $\exists U.D$, $\forall U.D$, $\geqslant nU$, and $\leqslant nU$ are concepts (called *datatype existential*, *value*, *atleast*, and *atmost restriction*, respectively) for an integer $n \geqslant 0$. We use $\top$ (resp., $\bot$) to abbreviate the *top* (resp., *bottom*) *concept* $\phi \sqcup \neg\phi$ (resp., $\phi \sqcap \neg\phi$). Furthermore, we write $\exists R$ to abbreviate $\exists R.\top$, and we eliminate parentheses as usual.

We next define axioms and knowledge bases. An *axiom* is an expression of one of the following forms: (1) $\phi \sqsubseteq \psi$ (called *concept inclusion axiom*), where ϕ and ψ are concepts; (2) $R \sqsubseteq S$ (called *role inclusion axiom*), where either $R, S \in \mathbf{R}_A \cup \mathbf{R}_A^-$ or $R, S \in \mathbf{R}_D$; (3) $\mathrm{Trans}(R)$ (called *transitivity axiom*), where $R \in \mathbf{R}_A$; (4) $\phi(a)$ (called *concept membership axiom*), where ϕ is a concept and $a \in \mathbf{I}$; (5) $R(a, b)$ (resp., $U(a, v)$) (called *role membership axiom*), where $R \in \mathbf{R}_A$ (resp., $U \in \mathbf{R}_D$) and $a, b \in \mathbf{I}$ (resp., $a \in \mathbf{I}$ and v is a data value); and (6) $a = b$ (resp., $a \neq b$) (*equality* (resp., *inequality*) *axiom*), where $a, b \in \mathbf{I}$. Two axioms $\phi \sqsubseteq \psi$ and $\psi \sqsubseteq \phi$ of the form (1) are also abbreviated as $\phi \equiv \psi$ (called *definition axiom*). Note that such axioms can be used to define a new atomic concept ϕ used as synonym of a concept ψ. A *TBox* $\mathcal{T}$ is a finite set of axioms of the form (1), (2), (3), and (6), while an *ABox* $\mathcal{A}$ is a finite set of axioms of the form (4) and (5). A *knowledge base* (or *ontology*) *KB* is a finite set of axioms (1)–(6).

We next define simple abstract roles. For abstract roles $R \in \mathbf{R}_A$, we define $\mathrm{Inv}(R) = R^-$ and $\mathrm{Inv}(R^-) = R$. Let $\sqsubseteq_{KB}^\star$ denote the reflexive and transitive closure of $\sqsubseteq$ on $\bigcup\{\{R \sqsubseteq S, \mathrm{Inv}(R) \sqsubseteq \mathrm{Inv}(S)\} \mid R \sqsubseteq S \in KB, R, S \in \mathbf{R}_A \cup \mathbf{R}_A^-\}$. An abstract role S is *simple* relative to *KB* iff for each abstract role R such that $R \sqsubseteq_{KB}^\star S$, it holds that (i) $\mathrm{Trans}(R) \notin KB$ and (ii) $\mathrm{Trans}(\mathrm{Inv}(R)) \notin KB$. Informally, an abstract role S is simple iff it is neither transitive nor has transitive subroles. For decidability, number restrictions in *KB* are restricted to simple abstract roles [19].

In $\mathcal{SHOIN}(\mathbf{D})$, concept and role membership axioms can also be expressed in terms of concept inclusion axioms, since $\phi(a)$ can be expressed by $\{a\} \sqsubseteq \phi$, while $R(a, b)$ (resp., $U(a, v)$) can be expressed by $\{a\} \sqsubseteq \exists R.\{b\}$ (resp., $\{a\} \sqsubseteq \exists U.\{v\}$).

The syntax of $\mathcal{SHIF}(\mathbf{D})$ is as the above syntax of $\mathcal{SHOIN}(\mathbf{D})$, but without the oneOf constructor and with the atleast and atmost constructors limited to 0 and 1.

Example 1 (Conference Organization). A simple TBox $\mathcal{T}_{conf}$ may describe a conference organization website and consist of the following axioms:

$$Hotel \sqsubseteq \exists hasRoom;$$
$$Hotel \sqsubseteq \exists hasBuilding;$$
$$Room \sqsubseteq \exists hasFeature;$$
$$Room \sqsubseteq \exists hasRoomPack;$$
$$Old \sqsubseteq Building;$$
$$New \sqsubseteq Building;$$
$$Old \sqsubseteq \neg New;$$
$$Wifi \sqsubseteq Feature;$$
$$VideoConference \sqsubseteq Feature;$$

$$\exists hasRoom^- \sqsubseteq Room;$$
$$\exists hasBuilding^- \sqsubseteq Building;$$
$$\exists hasFeature^- \sqsubseteq Feature;$$
$$\exists hasRoomPack^- \sqsubseteq Package;$$
$$Small \sqsubseteq \neg Medium;$$
$$Small \sqsubseteq \neg Large;$$
$$Medium \sqsubseteq \neg Large;$$
$$LuxuryPackage \sqsubseteq Package;$$
$$StandardPackage \sqsubseteq Package;$$
$$StandardPackage \sqsubseteq \neg LuxuryPackage;$$

$$\exists hasFeature.Wifi \sqsubseteq \exists hasFeature.VideoConference;$$
$$\exists hasRoomPack.StandardPackage \sqsubseteq \neg\exists hasRoomPack.LuxuryPackage.$$

Roughly speaking, the above axioms describe that hotels have either new or old buildings, and they have either small, medium, or large rooms. In turn, rooms have associated features such as wi-fi, video conference, etc.; if a room has wi-fi, then one can use the wi-fi connection for video conferences. Finally, rooms have associated stay packages; there are two types of packages; luxury packages in general (though not always) provide features that standard packages do not, for instance wi-fi connection. ∎

Semantics. We now define the semantics of $\mathcal{SHIF}(\mathbf{D})$ and $\mathcal{SHOIN}(\mathbf{D})$ in terms of general first-order interpretations, as usual. An *interpretation* $\mathcal{I} = (\Delta^{\mathcal{I}}, \cdot^{\mathcal{I}})$ relative to a datatype theory $\mathbf{D} = (\Delta^{\mathbf{D}}, \cdot^{\mathbf{D}})$ consists of a nonempty (*abstract*) *domain* $\Delta^{\mathcal{I}}$ disjoint from $\Delta^{\mathbf{D}}$, and a mapping $\cdot^{\mathcal{I}}$ that assigns to each atomic concept $\phi \in \mathbf{A}$ a subset of $\Delta^{\mathcal{I}}$, to each individual $o \in \mathbf{I}$ an element of $\Delta^{\mathcal{I}}$, to each abstract role $R \in \mathbf{R}_A$ a subset of $\Delta^{\mathcal{I}} \times \Delta^{\mathcal{I}}$, and to each datatype role $U \in \mathbf{R}_D$ a subset of $\Delta^{\mathcal{I}} \times \Delta^{\mathbf{D}}$. We extend $\cdot^{\mathcal{I}}$ to all roles and concepts as usual (where $\#S$ denotes the cardinality of a set S):

- $(R^-)^{\mathcal{I}} = \{(y,x) \mid (x,y) \in R^{\mathcal{I}}\}$;
- $\{o_1, \ldots, o_n\}^{\mathcal{I}} = \{o_1^{\mathcal{I}}, \ldots, o_n^{\mathcal{I}}\}$; $(\neg\phi)^{\mathcal{I}} = \Delta^{\mathcal{I}} \backslash \phi^{\mathcal{I}}$;
- $(\phi_1 \sqcap \phi_2)^{\mathcal{I}} = \phi_1^{\mathcal{I}} \cap \phi_2^{\mathcal{I}}$; $(\phi_1 \sqcup \phi_2)^{\mathcal{I}} = \phi_1^{\mathcal{I}} \cup \phi_2^{\mathcal{I}}$;
- $(\exists R.\phi)^{\mathcal{I}} = \{x \in \Delta^{\mathcal{I}} \mid \exists y : (x,y) \in R^{\mathcal{I}} \wedge y \in \phi^{\mathcal{I}}\}$;
- $(\forall R.\phi)^{\mathcal{I}} = \{x \in \Delta^{\mathcal{I}} \mid \forall y : (x,y) \in R^{\mathcal{I}} \to y \in \phi^{\mathcal{I}}\}$;
- $(\geqslant nR)^{\mathcal{I}} = \{x \in \Delta^{\mathcal{I}} \mid \#(\{y \mid (x,y) \in R^{\mathcal{I}}\}) \geqslant n\}$;
- $(\leqslant nR)^{\mathcal{I}} = \{x \in \Delta^{\mathcal{I}} \mid \#(\{y \mid (x,y) \in R^{\mathcal{I}}\}) \leqslant n\}$;
- $(\exists U.D)^{\mathcal{I}} = \{x \in \Delta^{\mathcal{I}} \mid \exists y : (x,y) \in U^{\mathcal{I}} \wedge y \in D^{\mathbf{D}}\}$;
- $(\forall U.D)^{\mathcal{I}} = \{x \in \Delta^{\mathcal{I}} \mid \forall y : (x,y) \in U^{\mathcal{I}} \to y \in D^{\mathbf{D}}\}$;
- $(\geqslant nU)^{\mathcal{I}} = \{x \in \Delta^{\mathcal{I}} \mid \#(\{y \mid (x,y) \in U^{\mathcal{I}}\}) \geqslant n\}$;
- $(\leqslant nU)^{\mathcal{I}} = \{x \in \Delta^{\mathcal{I}} \mid \#(\{y \mid (x,y) \in U^{\mathcal{I}}\}) \leqslant n\}$.

The *satisfaction* of an axiom F in an interpretation $\mathcal{I} = (\Delta^{\mathcal{I}}, \cdot^{\mathcal{I}})$ relative to a datatype theory $\mathbf{D} = (\Delta^{\mathbf{D}}, \cdot^{\mathbf{D}})$, denoted $\mathcal{I} \models F$, is defined as follows: (1) $\mathcal{I} \models \phi \sqsubseteq \psi$ iff $\phi^{\mathcal{I}} \subseteq \psi^{\mathcal{I}}$; (2) $\mathcal{I} \models R \sqsubseteq S$ iff $R^{\mathcal{I}} \subseteq S^{\mathcal{I}}$; (3) $\mathcal{I} \models \mathrm{Trans}(R)$ iff $R^{\mathcal{I}}$ is transitive; (4) $\mathcal{I} \models \phi(a)$ iff $a^{\mathcal{I}} \in \phi^{\mathcal{I}}$; (5) $\mathcal{I} \models R(a,b)$ iff $(a^{\mathcal{I}}, b^{\mathcal{I}}) \in R^{\mathcal{I}}$; (6) $\mathcal{I} \models U(a,v)$ iff $(a^{\mathcal{I}}, v^{\mathbf{D}}) \in U^{\mathcal{I}}$; (7) $\mathcal{I} \models a = b$ iff $a^{\mathcal{I}} = b^{\mathcal{I}}$; and (8) $\mathcal{I} \models a \neq b$ iff $a^{\mathcal{I}} \neq b^{\mathcal{I}}$.

The interpretation $\mathcal{I}$ *satisfies* the axiom F, or $\mathcal{I}$ is a *model* of F, iff $\mathcal{I} \models F$. We say that $\mathcal{I}$ *satisfies* a knowledge base KB, or $\mathcal{I}$ is a *model* of KB, denoted $\mathcal{I} \models KB$, iff $\mathcal{I} \models F$ for all $F \in KB$. We say that KB is *satisfiable* (resp., *unsatisfiable*) iff KB has a (resp., no) model. An axiom F is a *logical consequence* of KB, denoted $KB \models F$, iff each model of KB satisfies F. We say that ϕ is *subsumed* (resp., *not subsumed*) by ψ relative to KB, denoted $\phi \sqsubseteq_{KB} \psi$ (resp., $\phi \not\sqsubseteq_{KB} \psi$), iff $KB \models \phi \sqsubseteq \psi$ (resp., $KB \not\models \phi \sqsubseteq \psi$). We say that ϕ is *unsatisfiable* (resp., *satisfiable*) relative to KB, denoted $\phi \sqsubseteq_{KB} \bot$ (resp., $\phi \not\sqsubseteq_{KB} \bot$), iff $KB \models \phi \sqsubseteq \bot$ (resp., $KB \not\models \phi \sqsubseteq \bot$).

2.2 CP-Nets

CP-nets [5] are a widespread formalism to represent and reason with qualitative preferences. More specifically, they are a graphical representation for conditional *ceteris paribus* (all else being equal) preference statements, which allows the specification of preferences based on the notion of conditional preferential independence (CPI) [20].

We assume a finite set of variables $\mathbf{V}$, where each variable $X^i \in \mathbf{V}$ has a finite *domain of values*, denoted $Dom(X^i)$. A *value* for a set of variables $\mathbf{X} = \{X^1, \ldots, X^n\} \subseteq \mathbf{V}$ is a mapping $\mathbf{x} \colon \mathbf{X} \to \bigcup_{i=1}^{n} Dom(X^i)$ such that $\mathbf{x}(X^i) \in Dom(X^i)$ for all $i \in \{1, \ldots, n\}$; the *domain* of $\mathbf{X}$, denoted $Dom(\mathbf{X})$, is the set of all values for $\mathbf{X}$. If $\mathbf{x}$ and $\mathbf{y}$ are values for disjoints sets of variable $\mathbf{X}, \mathbf{Y} \subseteq \mathbf{V}$, then $\mathbf{xy}$ denotes the combination of $\mathbf{x}$ and $\mathbf{y}$. A *preference relation* $\succ$ is a strict partial order (an irreflexive and transitive binary relation). We write $o_1 \succeq o_2$ iff either $o_1 \succ o_2$ or $o_1 = o_2$. We say that o_1 is *strictly preferred* (resp., strictly or equally preferred) to o_2 iff $o_1 \succ o_2$ (resp., $o_1 \succeq o_2$). We say that o_2 is *dominated* by o_1 iff $o_1 \succ o_2$, and that o_2 is *directly dominated* by o_1, denoted $o_1 \succ_d o_2$ iff (i) o_2 is dominated by o_1, and (ii) no o exists such that $o_1 \succ o$ and $o \succ o_2$. We say that o_1 is *undominated* iff no o exists with $o \succ o_1$.

A CP-net $\mathcal{N}$ over $\mathbf{V}$ consists of a directed graph $G = (\mathbf{V}, E)$, where the nodes are variables in $\mathbf{V}$ and if there is a directed edge $(X^j, X^i) \in E$, this shows that the preferences over values of X^i is influenced by the values of X^j. Each node $X^i \in \mathbf{V}$ has an annotated conditional preference table (CPT), denoted $CPT(X^i)$, which associates a total (or partial) order $\succ_{\mathbf{u}}^i$ over the values of X^i with each value $\mathbf{u}$ of X^i's parents in G, denoted $Pa(X^i)$. Intuitively, given a particular value assignment to $Pa(X^i)$, one is able to determine a preference order for the values of X^i, all other things being equal.

An *outcome* of $\mathcal{N}$ is a value $\mathbf{o} \in Dom(\mathbf{V})$. A preference relation $\succ$ on the set of all outcomes of $\mathcal{N}$ is defined via the notion of worsening flip, which informally combines the preference relations on the values of the variables of $\mathcal{N}$ (as encoded by their CPTs) to a preference relation on the outcomes of $\mathcal{N}$. More specifically, given two outcomes $\mathbf{o}$ and $\mathbf{o}'$ of $\mathcal{N}$, it holds $\mathbf{o} \succ_{wf} \mathbf{o}'$ in $\mathcal{N}$ iff (i) there exist a variable $X^i \in \mathbf{V}$, values $x, x' \in Dom(X^i)$, and a value $\mathbf{u} \in Dom(Pa(X^i))$ such that (i) $\mathbf{o}(X^i) = x$, (ii) $\mathbf{o}'(X^i) = x'$, (iii) $\mathbf{o}(X^j) = \mathbf{o}'(X^j)$ for all $X^j \in \mathbf{V} \backslash \{X^i\}$, (iv) $\mathbf{o}(Pa(X^i)) = \mathbf{u}$, and (v) $x \succ_{\mathbf{u}}^i x'$; here, the single change from x to x', moving from $\mathbf{o}$ to $\mathbf{o}'$, is called a *worsening flip*. The preference relation $\succ$ is then defined as the transitive closure of $\succ_{wf}$. Intuitively, $\mathbf{o} \succ \mathbf{o}'$ iff there exists a sequence of worsening flips from $\mathbf{o}$ to $\mathbf{o}'$.

CP-net actually assumes that its directed graph encodes conditional preferential independences (CPIs). In detail, let $\mathbf{X}$, $\mathbf{Y}$, and $\mathbf{Z}$ partition $\mathbf{V}$, and let $\succ$ be a preference relation over $Dom(\mathbf{V})$. Then, $\mathbf{X}$ is *conditionally preferentially independent* of $\mathbf{Y}$ given $\mathbf{Z}$ iff for all $\mathbf{x}, \mathbf{x}' \in Dom(\mathbf{X}), \mathbf{y}, \mathbf{y}' \in Dom(\mathbf{Y}), \mathbf{z} \in Dom(\mathbf{Z})$, we have that $\mathbf{xyz} \succ \mathbf{x'yz}$ iff $\mathbf{xy'z} \succ \mathbf{x'y'z}$. Hence, every variable in a CP-net, given its parents, is conditionally preferentially independent of all the other variables in the CP-net.

Example 2. A CP-net with five variables R, W, B, C, and P that expresses the preferences over different features of hotels is shown in Fig. 1. The variables have the domains $\{r_s, r_m, r_l\}$, $\{w_y, w_n\}$, $\{b_o, b_n\}$, $\{c_y, c_n\}$, and $\{p_s, p_l\}$, respectively. This CP-net tells us that large rooms (r_l) are preferred to medium rooms (r_m), which in turn are preferred to small rooms (r_s). If only small rooms are available then a hotel in an old building is preferred to one in a new building (perhaps for historical reasons). A hotel that has a wi-fi connection (w_y) is preferred to one that does not (w_n). Furthermore, a hotel that facilitates video conferences (c_y) is preferred to one that does not (c_n). If the hotel is a new building, and the rooms facilitate video conferences, then the luxury package (p_l) is preferred to the standard package, otherwise the standard package is preferred (p_s). Here, we have a worsening flip moving from $\mathbf{o_1} = r_l\ w_y\ b_n\ c_y\ p_l$ to $\mathbf{o_2} = r_l\ w_y\ b_o\ c_y\ p_l$, since b_n is preferred to b_o, given r_l. Similarly, we have a worsening flip from $\mathbf{o_2}$ to $\mathbf{o_3} = r_l\ w_n\ b_o\ c_y\ p_l$. Hence, $\mathbf{o_1} \succ \mathbf{o_2}$, $\mathbf{o_2} \succ \mathbf{o_3}$, and $\mathbf{o_1} \succ \mathbf{o_3}$. Given the room size, the building type is conditionally preferentially independent of all the other variables. ∎

The following are the two main computational tasks for CP-nets:

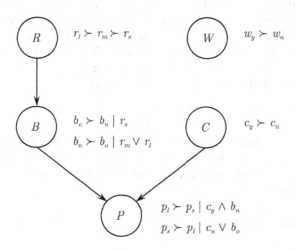

Fig. 1. CP-net for Example 2

- *Dominance query:* given a CP-net $\mathcal{N}$ and two outcomes o_1 and o_2 of $\mathcal{N}$, decide whether $o_1 \succ o_2$ holds in $\mathcal{N}$.
- *Outcome optimization:* given a CP-net $\mathcal{N}$, compute an undominated outcome of $\mathcal{N}$.

Acyclic CP-nets (i.e., the associated directed graph does not have any directed cycles) with total orders in their conditional preference tables have only one undominated outcome, which can be computed in linear time [4]. The algorithm just follows the order among the variables that is represented by the directed graph and assigns values to the variables A_i from top to bottom, satisfying the preference relations in the CPTs corresponding to the variables. For arbitrary CP-nets, deciding dominance queries are PSPACE-complete [16], and computing an optimal outcome is NP-hard [12].

2.3   Constrained CP-Nets

In constrained CP-nets [6, 26], constraints among variables are added to the basic formalism of CP-nets, which may reduce the set of possible outcomes. The approach in [26] to finding all optimal outcomes of a CP-net (if some exist) relies on a reduction of the preferences represented in the CP-net to a set of hard constraints taking into account the variables occurring in the preferences. Given a CP-net $\mathcal{N}$ and a set of constraints $\mathcal{C}$, an outcome o is *feasible* iff it satisfies all the constraints in $\mathcal{C}$. A feasible outcome is *Pareto-optimal* iff it is undominated [6]. In [13], the authors present an approach to finding the Pareto-optimal outcomes by solving a constraint satisfaction problem in the presence of soft and hard constraints. Here, we focus only on the latter. For every variable A and every instantiation γ of its parents in $\mathcal{N}$, the conditional preferences for $Dom(A)$ encoded in $\mathcal{N}$ yield an optimality constraint. The undominated outcomes of $(\mathcal{N}, \mathcal{C})$ are then exactly the solutions of the conjunction of all optimality constraints. For example, consider a variable A in $\mathcal{N}$ with domain $Dom(A) = \{a_1, \dots, a_m\}$ and an instantiation γ of its parents in $\mathcal{N}$. If the conditional preferences encoded in $\mathcal{N}$ are a total order over the values in $Dom(A)$, i.e., $a_1 \succ a_2 \succ \cdots \succ a_m$, which means that a_1 is undominated relative to all other a_i, then the optimality constraint is given by $\gamma \to a_1$. In the most general case, where we have more than one undominated value, the optimality constraint is given by:

$$\gamma \to \bigvee_{a_i \in undom(A|\gamma)} a_i, \tag{1}$$

where $undom(A|\gamma)$ denotes the set of all undominated values of A under γ. In the following, we consider only the case with one undominated value, i.e., $|undom(A|\gamma)| = 1$. The approach can be extended to the general case in a straightforward way.

3   Ontological CP-Nets

We now introduce an approach to ontological CP-net-based preference representation, which combines CP-nets and DLs, harnessing the technologies described in the previous section. Intuitively, the main idea behind this approach is to use certain satisfiable concepts relative to an underlying TBox as values of the variables of a CP-net. More precisely, the values are taken from a finite nonempty set $\mathcal{C}$ of *basic classification concepts* (or *basic c-concepts* for short), which are (not necessarily atomic) concepts C in $\mathcal{SHIF}(\mathbf{D})$ (resp., $\mathcal{SHOIN}(\mathbf{D})$) that are free of individuals from $\mathbf{I}$.

Definition 1 (ontological CP-net). Let $\mathbf{V}$ be a finite set of variables. An *ontological CP-net* $(\mathcal{N}, \mathcal{T})$ over $\mathbf{V}$ consists of a CP-net $\mathcal{N}$ over $\mathbf{V}$ and a TBox $\mathcal{T}$ such that the domain of each variable $A \in \mathbf{V}$ is of the form $Dom(A) = \{\alpha_1, \ldots, \alpha_m\}$, where:

1. every α_i, $i \in \{1, \ldots, m\}$, is a concept from $\mathcal{C}$ that is satisfiable relative to $\mathcal{T}$,
2. $\mathcal{T} \models \alpha_i \sqcap \alpha_j \sqsubseteq \bot$ for all $i, j \in \{1, \ldots, m\}$ with $i < j$, and
3. $\mathcal{T} \models \top \sqsubseteq \alpha_i \sqcup \cdots \sqcup \alpha_m$.

The following example illustrates the above notion of ontological CP-net.

Example 3 (Conference Organization cont'd). An ontological CP-net $(\mathcal{N}, \mathcal{T})$ over $\mathbf{V}$ is given by the TBox $\mathcal{T}$ from Example 1 and the CP-net $\mathcal{N}$ over $\mathbf{V}$ from Example 2, where the values of the variables are now defined as the following DL concepts:

$$r_s = \exists hasRoom.Small;\quad r_m = \exists hasRoom.Medium;$$
$$r_l = \exists hasRoom.Large;$$
$$w_y = \exists hasRoom.(\exists hasFeature.Wifi);\; w_n = \neg w_y;$$
$$b_o = \exists hasBuilding.Old;\, b_n = \exists hasBuilding.New;$$
$$c_y = \exists hasRoom.(\exists hasFeature.VideoConference);\, c_n = \neg c_y;$$
$$p_l = \exists hasRoom.(\exists hasRoomPack.LuxuryPackage);$$
$$p_s = \exists hasRoom.(\exists hasRoomPack.StandardPackage). \qquad \blacksquare$$

Observe here that even if we do not have any explicit hard constraint expressed among the values of the variables of the CP-net, due to their logical structure and the underlying TBox, we have a set of implicit constraints among these values. We will show in Sect. 4.2 below how to explicitly encode these constraints. Hence, due to these constraints among the values of the variables of the CP-net, some outcomes are infeasible, where outcomes are values o of the set $\mathbf{V}$ of all variables of the CP-net. The following definition formally introduces feasible outcomes and undominated feasible outcomes (which are not dominated by any other feasible outcome) as well as the consistency of CP-nets as the existence of at least one undominated feasible outcome.

Definition 2 (feasible outcome, dominance, and consistency). Given an ontological CP-net $(\mathcal{N}, \mathcal{T})$ over $\mathbf{V}$, an outcome $o \in Dom(\mathbf{V})$ is *feasible* iff $\sqcap o(\mathbf{V})$ $(= \sqcap_{A \in \mathbf{V}} o(A))$ is satisfiable relative to $\mathcal{T}$. A feasible outcome o is *undominated* iff no feasible outcome o' exists such that $o' \succ o$. We say that $(\mathcal{N}, \mathcal{T})$ is *consistent* iff it has a feasible outcome.

The following example briefly illustrates the above notions of feasible outcomes, undominated feasible outcomes, and consistent CP-nets.

Example 4 (Conference Organization cont'd). Reconsider the ontological CP-net of the running example. The outcome $r_l\, b_n\, w_y\, c_y\, p_l$ is feasible and undominated, while $r_l\, b_n\, w_y\, c_n\, p_s$ is not feasible (as the availability of wi-fi implies the availability of video conferences). Thus, the ontological CP-net of the running example is consistent.   ∎

4   Computing Optimal Outcomes

The main computational task around ontological CP-nets that we want to solve in this paper is how to determine all undominated feasible outcomes of a consistent ontological CP-net. In this section, we show how to compute them, given an ontological CP-net. The approach mainly relies on the HARD-PARETO algorithm of [26] (see Algorithm 1).

Generalizing the results of Sect. 2.3 to ontologies, we express the undominated feasible outcomes of an ontological CP-net $(\mathcal{N}, \mathcal{T})$, if some exist, in an ontological way as follows. For every variable A and every value γ of its parents in $(\mathcal{N}, \mathcal{T})$, the conditional preferences for $Dom(A)$ encoded in $(\mathcal{N}, \mathcal{T})$ yield the optimality constraint

$$\textstyle\prod \gamma(Pa(A)) \sqsubseteq \bigsqcup undom(A|\gamma).$$

Let $DL\text{-}opt(\mathcal{N})$ denote the set of all these optimality constraints. The undominated feasible outcomes of $(\mathcal{N}, \mathcal{T})$ are then exactly the set of all outcomes $o \in Dom(\mathbf{V})$ such that $\prod o(\mathbf{V})$ is satisfiable relative to $\mathcal{T} \cup DL\text{-}opt(\mathcal{N})$.

4.1   Propositional Compilation of DL Formulas

A TBox can be seen as a set of logical constraints that reduces the set of models for a formula. Given a set of concepts $\mathcal{F}$, we now show how to compute a compact representation of a TBox $\mathcal{T}$ as a set of clauses whose variables have a one-to-one mapping to the concepts in $\mathcal{F}$. Hereafter, we write $\tilde{\phi}$ to denote $\tilde{\phi} \in \{\phi, \neg\phi\}$.

Definition 3 (ontological constraint). Given a TBox $\mathcal{T}$ and a set of satisfiable concepts $\mathcal{F} = \{\phi_i \mid i \in \{1, \ldots, n\}\}$ relative to $\mathcal{T}$, we say that $\mathcal{F}$ is *minimally constrained* relative to $\mathcal{T}$ iff

1. there exists a concept $\tilde{\phi}_1 \sqcup \cdots \sqcup \tilde{\phi}_n$ such that $\mathcal{T} \models \top \sqsubseteq \tilde{\phi}_1 \sqcup \cdots \sqcup \tilde{\phi}_n$, and
2. there is no proper subset $\mathcal{E} \subset \mathcal{F}$ such that the previous condition holds.

The ontological axiom $\top \sqsubseteq \tilde{\phi}_1 \sqcup \cdots \sqcup \tilde{\phi}_n$ is an *ontological constraint*.

An ontological constraint is an explicit representation of the constraints existing among a set of concepts, due to the information encoded in the TBox $\mathcal{T}$.

Definition 4 (ontological closure). Given a TBox $\mathcal{T}$ and a set of satisfiable concepts $\mathcal{F} = \{\phi_1, \ldots, \phi_n\}$ relative to $\mathcal{T}$, the *ontological closure* of $\mathcal{F}$ and $\mathcal{T}$, denoted $\mathcal{OCL}(\mathcal{F}, \mathcal{T})$, is the set of all ontological constraints, if any, for each subset of $\mathcal{F}$.

The ontological closure of $\mathcal{F}$ and $\mathcal{T}$ is an explicit representation of all the logical constraints among a set of concepts $\mathcal{F}$, considering also an underlying TBox $\mathcal{T}$.

Proposition 1. *Given a TBox $\mathcal{T}$ and a set of satisfiable concepts $\mathcal{F} = \{\phi_1, \ldots, \phi_n\}$ relative to $\mathcal{T}$, if $\mathcal{T} \models \bigsqcap_{i=1}^{n} \tilde{\phi}_i \sqsubseteq \bot$, then $\mathcal{OCL}(\mathcal{F}, \mathcal{T}) \models \bigsqcap_{i=1}^{n} \tilde{\phi}_i \sqsubseteq \bot$.*

Proof. Since $\mathcal{T} \models \bigsqcap_{i=1}^{n} \tilde{\phi}_i \sqsubseteq \bot$, this means that we have the corresponding clause $\psi = \bigsqcup_{i=1}^{n} \neg \tilde{\phi}_i$ such that $\mathcal{T} \models \top \sqsubseteq \psi$. If $\mathcal{F} = \{\phi_1, \ldots, \phi_n\}$ is minimally constrained, then $\psi \in \mathcal{OCL}(\mathcal{F}, \mathcal{T})$, otherwise, by definition of $\mathcal{OCL}(\mathcal{F}, \mathcal{T})$, there will be a clause $\psi' \in \mathcal{OCL}(\mathcal{F}, \mathcal{T})$ such that $\psi' \subset \psi$. □

Hence, if we are interested only in the relationships between predefined concepts (due to their logical structure and $\mathcal{T}$), then the corresponding ontological closure is a compact and complete representation. Note that we are only looking for minimal clauses, as we are interested in computing the actual constraints between the formulas representing the domain of a variable in the CP-net. Hence, to compute the final outcomes, it is preferable to deal with a compact representation of all possible constraints.

Example 5 (Conference Organization cont'd). Given the set of concepts $\mathcal{F} = \{w_y, c_y, p_l, p_s\}$, due to the axioms in the TBox, we have the minimally constrained sets $\mathcal{F}' = \{w_y, c_y\}$ and $\mathcal{F}'' = \{p_l, p_s\}$ and the two corresponding ontological constraints $\top \sqsubseteq \neg w_y \sqcup c_y$ (indeed $w_y \sqsubseteq_{\mathcal{T}} c_y$) and $\top \sqsubseteq \neg p_l \sqcup \neg p_s$ (as $p_l \sqcap p_s \sqsubseteq_{\mathcal{T}} \bot$). The corresponding ontological closure is then $\mathcal{OCL}(\mathcal{F}, \mathcal{T}) = \{\top \sqsubseteq \neg w_y \sqcup c_y, \top \sqsubseteq \neg p_l \sqcup \neg p_s\}$. ∎

The set $\tilde{\mathcal{F}} = \{\tilde{\phi}_1, \ldots, \tilde{\phi}_n\}$ is a *feasible* assignment for $\mathcal{F}$ and $\mathcal{T}$ iff

$$\mathcal{OCL}(\mathcal{F}, \mathcal{T}) \not\models \bigsqcap_{i=1}^{n} \tilde{\phi}_i \sqsubseteq \bot.$$

We are interested in feasible assignments, since (as we will show in the following), they represent feasible outcomes for an ontological CP-net. Note that by Proposition 1, if $\tilde{\mathcal{F}} = \{\tilde{\phi}_1, \ldots, \tilde{\phi}_n\}$ is a feasible assignment for $\mathcal{F}$ and $\mathcal{T}$, then $\mathcal{T} \not\models \bigsqcap_{i=1}^{n} \tilde{\phi}_i \sqsubseteq \bot$, i.e., $\bigsqcap_{i=1}^{n} \tilde{\phi}_i$ is satisfiable relative to $\mathcal{T}$. The next proposition shows that there always exist feasible assignments for sets of satisfiable concepts relative to an underlying TBox.

Proposition 2. *Let $\mathcal{T}$ be a TBox, and $\mathcal{F} = \{\phi_1, \ldots, \phi_n\}$ be a set of satisfiable concepts relative to $\mathcal{T}$. Then, there always exists a feasible assignment for $\mathcal{F}$ and $\mathcal{T}$.*

Proof. If $\mathcal{OCL}(\mathcal{F},\mathcal{T}) = \emptyset$, then every $\tilde{\mathcal{F}} = \{\tilde{\phi}_1, \ldots, \tilde{\phi}_n\}$ is a feasible assignment for $\mathcal{F}$ and $\mathcal{T}$. Otherwise, since $\mathcal{T} \models \mathcal{OCL}(\mathcal{F},\mathcal{T})$, every interpretation $\mathcal{I}$ that satisfies $\mathcal{T}$ also satisfies $\mathcal{OCL}(\mathcal{F},\mathcal{T})$. That is, $\mathcal{I}$ satisfies every $\top \sqsubseteq \tilde{\phi}_{i_1} \sqcup \cdots \sqcup \tilde{\phi}_{i_m} \in \mathcal{OCL}(\mathcal{F},\mathcal{T})$, which means that $(\tilde{\phi}_{i_j})^{\mathcal{I}} \neq \emptyset$ for some $j \in \{1, \ldots, m\}$ (as every interpretation $\mathcal{I}$ has a nonempty domain $\Delta^{\mathcal{I}}$). Thus, there exists a feasible assignment for $\mathcal{F}$ and $\mathcal{T}$. $\square$

4.2   Computing Optimal Outcomes

If we have an ontological CP-net $(\mathcal{N}, \mathcal{T})$, the variable values (concepts) in a set $\mathcal{F}$ may constrain each other, and the corresponding constraints are encoded in $\mathcal{OCL}(\mathcal{F}, \mathcal{T})$. The ontological closure of a set of concepts explicitly represents all the logical constraints among them with respect to an underlying ontology. The computation of all undominated feasible outcomes for an ontological CP-net goes through the Boolean encoding of both the ontology $\mathcal{T}$ and of the clauses corresponding to the preferences represented in the CPTs of $\mathcal{N}$ for each variable $A \in \mathbf{V}$. To use HARD-PARETO, we need a few pre-processing steps. Given the ontological CP-net $(\mathcal{N}, \mathcal{T})$:

1. for every $A^i \in \mathbf{V}$ and every $\alpha_j^i \in Dom(A^i) = \{\alpha_1^i, \ldots, \alpha_{m_i}^i\}$, choose a fresh atomic concept V_j^i;
2. define the ontology $\mathcal{T}' = \mathcal{T} \cup \{V_j^i \equiv \alpha_j^i \mid A^i \in \mathbf{V}, j \in \{1, \ldots, m_i\}\}$;
3. define the ontological CP-net $(\mathcal{N}', \mathcal{T}')$, where $\mathcal{N}'$ is obtained from $\mathcal{N}$ by isomorphically replacing every α_j^i by V_j^i, for all $A^i \in \mathbf{V}$ and $j \in \{1, \ldots, m_i\}$;
4. define $\mathcal{F} = \{V_j^i \mid A^i \in \mathbf{V}, j \in \{1, \ldots, m_i\}\}$;
5. compute $\mathcal{OCL}(\mathcal{F}, \mathcal{T}')$;
6. introduce a Boolean variable v_j^i for each $V_j^i \in \mathcal{F}$;
7. transform $\mathcal{OCL}(\mathcal{F}, \mathcal{T}')$ into the corresponding set of Boolean clauses $\mathcal{C}$ by replacing V_j^i by the corresponding binary variable v_j^i (and the ontological constant "$\top$" and the ontological connectives "$\sqcup$" and "$\sqsubseteq$" by the Boolean constant "$true$" and the Boolean connectives "$\vee$" and "$\rightarrow$", respectively);
8. transform $DL\text{-}opt(\mathcal{N}')$ into the set of Boolean clauses $opt(\mathcal{N}')$ by replacing every V_j^i by the corresponding variable v_j^i (and the ontological connectives "$\sqcap$", "$\sqcup$", and "$\sqsubseteq$" by the Boolean connectives "$\wedge$", "$\vee$", and "$\rightarrow$", respectively).

Note that $\mathcal{T}$ is logically equivalent to $\mathcal{T}'$, as we only use equivalence axioms to define new concepts V_j^i as synonyms of complex concepts α_j^i. The same holds for $(\mathcal{N}, \mathcal{T})$ and $(\mathcal{N}', \mathcal{T}')$, as we just replace concepts in $Dom(A^i)$ by equivalent concepts.

Once we have $\mathcal{C}$ and $opt(\mathcal{N}')$, we can compute the optimal outcome of $(\mathcal{N}, \mathcal{T})$ by using the slightly modified version of HARD-PARETO, shown in Algorithm 1. The function $sol(\cdot)$ used in Algorithm 1 computes all the solutions for the Boolean constraint satisfaction problem represented by $\mathcal{C}$, $opt(\mathcal{N}')$ and $\mathcal{C} \cup opt(\mathcal{N}')$. Differently from the original HARD-PARETO, by Proposition 2, we know that $\mathcal{C}$ is always consistent, and so we do not need to check its consistency at the beginning of the algorithm. Moreover, note that the algorithm works with propositional variables although we are computing undominated feasible solutions for an ontological CP-net. That is, the dominance test in line 11 can be computed using well-known techniques for Boolean problems.

Input: $opt(\mathcal{N}')$ and $\mathcal{C}$

1   $S_{opt} \leftarrow sol(\mathcal{C} \cup opt(\mathcal{N}'))$;
2   **if** $S_{opt} = sol(\mathcal{C})$ **then**
3 | **return** S_{opt};
4   **end**
5   **if** $sol(opt(\mathcal{N}')) \neq \emptyset$ *and* $S_{opt} = sol(opt(\mathcal{N}'))$ **then**
6 | **return** S_{opt};
7   **end**
8   $S \leftarrow sol(\mathcal{C}) - S_{opt}$;
9   **repeat**
10   choose $o \in S$;
11   **if** $\forall o' \in sol(\mathcal{C}) - o, o' \not\succ o$ **then**
12 | $S_{opt} \leftarrow S_{opt} \cup \{o\}$;
13   **end**
14      $S \leftarrow S - \{o\}$;
15   **until** $S = \emptyset$;
16   **return** S_{opt}.

Algorithm 1. Algorithm HARD-PARETO adapted to ontological CP-nets

---

The outcomes returned by Algorithm 1 in S_{opt} are *true/false* assignments to the Boolean variables v_j^i. To compute undominated outcomes for the original ontological CP-net $(\mathcal{N}, \mathcal{T})$, we need to revert to a DL setting. Hence, we build the set $DL\text{-}S_{opt}$, where for each outcome $o \in S_{opt}$, we add to $DL\text{-}S_{opt}$ the following value o':

$$o'(A^i) = V_j^i \text{ iff } o(v_j^i) = true, \text{ for all } A_i \in \mathbf{V} \text{ and } j \in \{1, \ldots, m_i\}.$$

The following example shows a trace of Algorithm 1 for our running example.

Example 6 (Conference Organization cont'd). For the CP-net in Fig. 1, we obtain:

- $\mathcal{T}' = \mathcal{T} \cup \{V_1^1 \equiv \exists hasRoom.Small, \ V_2^1 \equiv \exists hasRoom.Medium,$
           $V_3^1 \equiv \exists hasRoom.Large, \ V_1^2 \equiv \exists hasRoom.(\exists hasFeature.Wifi),$
           $V_2^2 \equiv \neg \exists hasRoom.(\exists hasFeature.Wifi),$
           $V_1^3 \equiv \exists hasBuilding.Old, \ V_2^3 \equiv \exists hasBuilding.New,$
           $V_1^4 \equiv \exists hasRoom.(\exists hasFeature.VideoConference),$
           $V_2^4 \equiv \neg \exists hasRoom.(\exists hasFeature.VideoConference),$
           $V_1^5 \equiv \exists hasRoom.(\exists hasRoomPack.LuxuryPackage),$
           $V_2^5 \equiv \exists hasRoom.(\exists hasRoomPack.StandardPackage)\}$;

- $\mathcal{F} = \{V_1^1, V_2^1, V_3^1, V_1^2, V_2^2, V_1^3, V_2^3, V_1^4, V_2^4, V_1^5, V_2^5\}$;

- $\mathcal{OCL}(\mathcal{F}, \mathcal{T}') = \{\top \sqsubseteq \neg V_1^2 \sqcup V_1^4,\ \top \sqsubseteq \neg V_1^1 \sqcup \neg V_2^1,\ \top \sqsubseteq \neg V_2^1 \sqcup \neg V_3^1,$
  $\top \sqsubseteq \neg V_1^1 \sqcup \neg V_3^1,\ \top \sqsubseteq V_1^1 \sqcup V_2^1 \sqcup V_3^1,\ \top \sqsubseteq \neg V_1^2 \sqcup \neg V_2^2,$
  $\top \sqsubseteq V_1^2 \sqcup V_2^2,\ \top \sqsubseteq \neg V_1^3 \sqcup \neg V_2^3,\ \top \sqsubseteq V_1^3 \sqcup V_2^3,$
  $\top \sqsubseteq \neg V_1^4 \sqcup \neg V_2^4,\ \top \sqsubseteq V_1^4 \sqcup V_2^4,\ \top \sqsubseteq \neg V_1^5 \sqcup \neg V_2^5,$
  $\top \sqsubseteq V_1^5 \sqcup V_2^5\}$;

- $\mathcal{C} = \{\neg v_1^2 \vee v_1^4,\ \neg v_1^1 \vee \neg v_2^1,\ \neg v_2^1 \vee \neg v_3^1,\ \neg v_1^1 \vee \neg v_3^1,\ v_1^1 \vee v_2^1 \vee v_3^1,\ \neg v_1^2 \vee \neg v_2^2,$
  $v_1^2 \vee v_2^2,\ \neg v_1^3 \vee \neg v_2^3,\ v_1^3 \vee v_2^3,\ \neg v_1^4 \vee \neg v_2^4,\ v_1^4 \vee v_2^4,\ \neg v_1^5 \vee \neg v_2^5,\ v_1^5 \vee v_2^5\}$;

- $DL\text{-}opt(N') = \{V_3^1,\ V_1^2,\ V_1^1 \sqsubseteq V_1^3,\ V_2^1 \sqsubseteq V_2^3,\ V_1^4,\ V_3^1 \sqsubseteq V_1^3,\ V_2^3 \sqcap V_1^4 \sqsubseteq V_1^5,$
  $V_1^3 \sqcap V_1^4 \sqsubseteq V_2^5,\ V_2^3 \sqcap V_2^4 \sqsubseteq V_2^5,\ V_1^3 \sqcap V_2^4 \sqsubseteq V_2^5\}$;

- $opt(N') = \{v_3^1,\ v_1^2,\ v_1^1 \to v_1^3,\ v_2^1 \to v_2^3,\ v_1^4,\ v_3^1 \to v_2^3,\ v_2^3 \wedge v_1^4 \to v_1^5,$
  $v_1^3 \wedge v_1^4 \to v_2^5,\ v_2^3 \wedge v_2^4 \to v_2^5,\ v_1^3 \wedge v_2^4 \to v_2^5\}$.

Then, $S_{opt} = \{v_3^1\ v_1^2\ v_2^3\ v_1^4\ v_1^5\}$, and $r_l\ w_y\ b_n\ c_y\ p_l$ is the only optimal outcome. ∎

The following theorem shows the correctness of the algorithm.

Theorem 1. *Given an ontological CP-net $(\mathcal{N}, \mathcal{T})$ over* $\mathbf{V}$*, the values $o' \in DL\text{-}S_{opt}$ are all the undominated feasible outcomes for $(\mathcal{N}, \mathcal{T})$.*

Proof. We start by showing that o' is a feasible outcome. If we consider the final assignment $o = \bigwedge_{i=1}^{|\mathbf{V}|} \bigwedge_{j=1}^{m_i} \tilde{v}_j^i$, the corresponding formula $o' = \bigsqcap_{i=1}^{|\mathbf{V}|} \bigsqcap_{j=1}^{m_i} \tilde{V}_j^i$ is a feasible assignment. In fact, if we had $\mathcal{OCL}(\mathcal{F}, \mathcal{T}') \models \bigsqcap_{i=1}^{|\mathbf{V}|} \bigsqcap_{j=1}^{m_i} \tilde{V}_j^i \sqsubseteq \bot$, then we should have the corresponding constraint (or one that implies) $\bigvee_{i=1}^{|\mathbf{V}|} \bigvee_{j=1}^{m_i} \neg \tilde{v}_j^i$ in $\mathcal{C}$, thus not allowing o to be a solution. By Definition 1, for each variable A^i, both $\mathcal{T} \models a_j^i \sqcap a_{j'}^i \sqsubseteq \bot$, for all $j, j' \in \{1, \ldots, m_i\}$ with $j < j'$, and $\mathcal{T} \models \top \sqsubseteq a_1^i \sqcup \cdots \sqcup a_{m_i}^i$. These axioms are encoded in the corresponding binary constraints $\neg v_j^i \vee \neg v_{j'}^i$ and $v_1^i \vee \cdots \vee v_{m_i}^i$ saying that, given A^i, in o, we have all v_j^i negated but one. As a consequence, in o', we have only one V_j^i for each A^i, i.e., o' is an outcome. Overall, o' is a feasible outcome. Finally, as o satisfies all the optimality constraints, o' is an undominated outcome. □

5 Computational Complexity

We now explore the complexity of the main computational problems in ontological CP-nets for underlying ontological languages with typical complexity of deciding knowledge base satisfiability, namely, tractability and completeness for EXP and NEXP. We also provide some special tractable cases of dominance testing in ontological CP-nets.

5.1 General Results

For tractable ontology languages (i.e., those for which deciding knowledge base satisfiability is tractable), the complexity of ontological CP-nets is dominated by the complexity of CP-nets. That is, deciding (a) consistency, (b) whether a given outcome is undominated, and (c) dominance of two given outcomes are all PSPACE-complete. Note also that the same complexity results hold for ontology languages with PSPACE-complete knowledge base satisfiability checks and that even computing the set of all undominated outcomes (generalizing (b)) is PSPACE-complete under the condition that there are only polynomially many of them.

Theorem 2. *Given an ontological CP-net $(\mathcal{N}, \mathcal{T})$ over a tractable ontology language,*

(a) deciding whether $(\mathcal{N}, \mathcal{T})$ is consistent,
(b) deciding whether a given outcome o is undominated,
(c) deciding whether $o \succ o'$ for two given outcomes o and o'

are all PSPACE-complete.

Proof (sketch). The lower bounds follow immediately from the fact that ontological CP-nets generalize CP-nets, for which these problems are all PSPACE-complete [16].

As for the upper bounds, compared to standard CP-nets, these problems additionally involve knowledge base satisfiability checks, which can all be done in polynomial time and thus also in polynomial space. Note that in (a) (resp., (b)), one has to go through all outcomes o' and check that it is not the case that $o \succ o'$ (resp., $o' \succ o$), which can each and thus overall be done in polynomial space.   □

In particular, if the ontological CP-net is defined over a DL of the *DL-Lite* family [9] (which all allow for deciding knowledge base satisfiability in polynomial time, such as *DL-Lite$_{\mathcal{R}}$*, which stands behind the important OWL 2 QL profile [31]), deciding (a) consistency, (b) whether a given outcome is undominated, and (c) dominance of two given outcomes are all PSPACE-complete.

Corollary 1. *Given an ontological CP-net $(\mathcal{N}, \mathcal{T})$ over a DL from the DL-Lite family,*

(a) deciding whether $(\mathcal{N}, \mathcal{T})$ is consistent,
(b) deciding whether a given outcome o is undominated,
(c) deciding whether $o \succ o'$ for two given outcomes o and o'

are all PSPACE-complete.

For EXP (resp., NEXP) complete ontology languages (i.e., those for which knowledge base satisfiability is complete for EXP (resp., NEXP)), the complexity of ontological CP-nets is dominated by the complexity of the ontology languages. That is, deciding (a) inconsistency, (b) whether a given outcome is dominated, and (c) dominance of two given outcomes are all complete for EXP (resp., NEXP). Note that computing the set of all undominated outcomes (generalizing (b)) is also EXP-complete for EXP-complete ontology languages.

Theorem 3. *Given an ontological CP-net $(\mathcal{N}, \mathcal{T})$ over an EXP (resp., NEXP) complete ontology language,*

(a) deciding whether $(\mathcal{N}, \mathcal{T})$ is inconsistent,
(b) deciding whether a given outcome o is dominated,
(c) deciding whether $o \succ o'$ for two given outcomes o and o'

are all complete for EXP (resp., NEXP).

Proof (sketch). The lower bounds follow from the fact that all three problems in ontological CP-nets can be used to decide knowledge base satisfiability in the underlying ontology language.

As for the upper bounds, in (a) and (b), we have to go through all outcomes, which is in EXP (resp., NEXP). Then, we have to perform knowledge base satisfiability checks, which are also in EXP (resp., NEXP), and dominance checks in standard CP-nets, which are in PSPACE, and thus also in EXP (resp., NEXP). Overall, (a) to (c) are thus in EXP (resp., NEXP). □

In particular, if the ontological CP-net is defined over the expressive DL $\mathcal{SHIF}(\mathbf{D})$ (resp., $\mathcal{SHOIN}(\mathbf{D})$) [18] (which stands behind OWL Lite (resp., OWL DL) [30,32], and allows for deciding knowledge base satisfiability in EXP [18,29] (resp., NEXP, for both unary and binary number encoding; see [25,29] and the NEXP-hardness proof for $\mathcal{ALCQIO}$ in [29], which implies the NEXP-hardness of $\mathcal{SHOIN}(\mathbf{D})$)), deciding (a) inconsistency, (b) whether a given outcome is dominated, and (c) dominance of two given outcomes are all complete for EXP (resp., NEXP).

Corollary 2. *Given an ontological CP-net $(\mathcal{N}, \mathcal{T})$ over the DL $\mathcal{SHIF}(\mathbf{D})$ (resp., $\mathcal{SHOIN}(\mathbf{D})$),*

(a) deciding whether $(\mathcal{N}, \mathcal{T})$ is inconsistent,
(b) deciding whether a given outcome o is dominated,
(c) deciding whether $o \succ o'$ for two given outcomes o and o'

are all complete for EXP (resp., NEXP).

5.2   Tractability Results

If the ontological CP-net is a polytree (i.e., the underlying undirected graph has no cycles) and defined over a tractable ontology language, deciding dominance of two outcomes can be done in polynomial time. Note that polytree ontological CP-nets are always consistent.

Theorem 4. *Given an ontological CP-net $(\mathcal{N}, \mathcal{T})$ over a tractable ontology language, where $\mathcal{N}$ is a polytree, deciding whether $o \succ o'$ for two given outcomes o and o' can be done in polynomial time.*

Proof (sketch). We have to decide whether (i) $o \succ o'$ holds in $\mathcal{N}$ and (ii) o and o' are feasible outcomes of $(\mathcal{N}, \mathcal{T})$. The former can be done in polynomial time, as for standard polytree CP-nets, dominance can be decided in polynomial time [5], while the latter can also be done in polynomial time in tractable ontology languages. □

In particular, if the ontological CP-net is a polytree and defined over a DL of the *DL-Lite* family, deciding dominance of two outcomes can be done in polynomial time.

Corollary 3. *Given an ontological CP-net $(\mathcal{N}, \mathcal{T})$ over a DL from the DL-Lite family, where $\mathcal{N}$ is a polytree, deciding whether $o \succ o'$ for two given outcomes o and o' can be done in polynomial time.*

6 Related Work

Modeling and dealing with preferences has traditionally been studied in several areas within computer science. In the databases community, the work of [21] stands out as the seminal work in the area; see [28] for a survey of notable works in this line. Much work has also been carried out in the intersection of databases and knowledge representation and reasoning, such as in preference logic programs [17], incorporation of preferences into formalisms such as answer set programs [8], and answering k-rank queries in ontological languages [22].

On the other hand, in the philosophical tradition, preferences are usually expressed over mutually exclusive "worlds", such as truth assignments to formulas. The work of [2] is framed in this interpretation of preferences, aiming at bridging the gap between several formalisms from the AI community such as CP-nets and those studied traditionally in philosophy. In this regard, CP-nets [5] is one of the most widely known formalisms. More recently, the work of Wang et al. [33] proposes an efficient algorithm and indexing scheme for top-k retrieval in CP-nets.

Constrained CP-nets were originally proposed in [6], along with algorithm SEARCH-CP, which uses branch and bound to compute undominated outcomes. The algorithm has an anytime behavior: it can be stopped at any time, and the set of computed solutions are a subset of the set containing all the undominated outcomes. This means that in case one is interested in any undominated outcome, one can use the first one returned by SEARCH-CP. In [26], HARD-PARETO is presented; the most notable difference is that HARD-PARETO does not rely on topological information like SEARCH-CP, but it exploits only the CP-statements, thus allowing to work also with cyclic CP-nets. Differently from the previous two papers, in our work, we allow the variable domains to contain DL formulas constrained via ontological axioms.

Recently, there has been some interest regarding the combination of Semantic Web technologies with preference representation and reasoning. A combination of conditional preferences (very different from CP-nets) with DL reasoning for ranking objects is introduced in [24]. A ranking function is described that exploits conditional preferences to perform a semantic personalized search and ranking over a set of resources annotated via an ontological description. In [22], Datalog+/− is extended with preference management formalisms closely related to those previously studied for relational databases; the authors further extend this formalism to the case where ontological inferences involve probabilistic uncertainty [23]. Another interesting approach to mixing qualitative preferences with Semantic Web technology is presented in [27], where an extension of SPARQL that can encode user preferences in the query is proposed.

There is, however, very little work on the particular combination of Semantic Web technologies and CP-nets. To our knowledge, the most notable work is that of [3], developed within an information retrieval context; in this work, Wordnet is used to add semantics to CP-net variables.

7 Summary and Outlook

In classical decision theory and analysis, the preferences of decision makers are modeled by utility functions. Unfortunately, the effort needed to obtain a good utility function requires a significant involvement of the user [15]. This is one of the main reasons behind the success obtained by CP-nets since they were originally proposed [4]: they are compact, easily understandable, and well-suited for combinatorial domains, such as multi-attribute ones. In this paper, we have described how to reason with CP-nets that are augmented by assigning description logic axioms to its variable values—such axioms refer to a common underlying ontology and constrain the possible outcomes in the CP-net. Furthermore, we studied the complexity of the problems of consistency checking, whether a given outcome is undominated, and dominance testing for ontological CP-nets, showing how the complexity of checking satisfiability of the underlying ontological language and the structure of the CP-net affects the complexity of solving these problems. The proposed framework is very useful in many semantic retrieval scenarios, among which we distinguish semantic search.

Other formalisms related to the original CP-nets have been subsequently proposed in the literature, such as TCP-nets (Trade-off CP-nets) [7] or CP-theories [34]. TCP-nets extend CP-nets by allowing to express also statements of relative importance between variables. With TCP-nets, the user is allowed to express her preferences over compromises that sometimes may be required. CP-theories generalize (T)CP-nets allowing conditional preference statements on the values of a variable, along with a set of variables that are allowed to vary when interpreting the preference statement. In future work, we plan to enrich these frameworks by introducing ontological descriptions and reasoning, thus allowing the development of more powerful semantic-enabled preference-based retrieval systems.

Acknowledgements. This work was supported by the UK EPSRC grant EP/J0083-46/1 "PrOQAW: Probabilistic Ontological Query Answering on the Web", by a Google

European Doctoral Fellowship, by the ERC (FP7/2007-2013) grant 246858 ("DIADEM"), by a Yahoo! Research Fellowship, and by PON02_00563_3470993 ("VIN-CENTE"). This paper is a significantly extended and revised version of papers that appeared in *Proceedings URSW-2013* [10] and *Proceedings SUM-2013* [11].

References

1. Baader, F., Calvanese, D., McGuinness, D.L., Nardi, D., Patel-Schneider, P.F. (eds.): The Description Logic Handbook. Cambridge University Press, Cambridge (2002)
2. Bienvenu, M., Lang, J., Wilson, N.: From preference logics to preference languages, and back. In: Proceedings of KR, pp. 214–224. AAAI Press (2010)
3. Boubekeur, F., Boughanem, M., Tamine-Lechani, L.: Semantic information retrieval based on CP-nets. In: Proceedings of FUZZ-IEEE, pp. 1–7. IEEE Computer Society (2007)
4. Boutilier, C., Brafman, R.I., Hoos, H.H., Poole, D.: Reasoning with conditional ceteris paribus preference statements. In: Proceedings of UAI, pp. 71–80. Morgan Kaufmann (1999)
5. Boutilier, C., Brafman, R.I., Domshlak, C., Hoos, H.H., Poole, D.: CP-nets: a tool for representing and reasoning with conditional ceteris paribus preference statements. J. Artif. Intell. Res. **21**, 135–191 (2004)
6. Boutilier, C., Brafman, R.I., Domshlak, C., Hoos, H.H., Poole, D.: Preference-based constrained optimization with CP-nets. Comput. Intell. **20**(2), 137–157 (2004)
7. Brafman, R.I., Domshlak, C., Shimony, S.E.: On graphical modeling of preference and importance. J. Artif. Intell. Res. **25**(1), 389–424 (2006)
8. Brewka, G.: Preferences, contexts and answer sets. In: Dahl, V., Niemelä, I. (eds.) ICLP 2007. LNCS, vol. 4670, pp. 22–22. Springer, Heidelberg (2007)
9. Calvanese, D., De Giacomo, G., Lembo, D., Lenzerini, M., Rosati, R.: Tractable reasoning and efficient query answering in description logics: the DL-Lite family. J. Autom. Reasoning **39**(3), 385–429 (2007)
10. Di Noia, T., Lukasiewicz, T.: Introducing ontological cp-nets. In: Proceedings of URSW, CEUR Workshop Proceedings, vol. 900, pp. 90–93. CEUR-WS.org (2012)
11. Di Noia, T., Lukasiewicz, T., Simari, G.I.: Reasoning with semantic-enabled qualitative preferences. In: Liu, W., Subrahmanian, V.S., Wijsen, J. (eds.) SUM 2013. LNCS, vol. 8078, pp. 374–386. Springer, Heidelberg (2013)
12. Domshlak, C., Brafman, R.I.: CP-nets-reasoning and consistency testing. In: Proceedings of KR, pp. 121–132. Morgan Kaufmann (2002)
13. Domshlak, C., Prestwich, S.D., Rossi, F., Venable, K.B., Walsh, T.: Hard and soft constraints for reasoning about qualitative conditional preferences. J. Heuristics **12**(4/5), 263–285 (2006)
14. Domshlak, C., Hüllermeier, E., Kaci, S., Prade, H.: Preferences in AI: an overview. Artif. Intell. **175**(7/8), 1037–1052 (2011)
15. French, S.: Decision Theory: An Introduction to the Mathematics of Rationality. Prentice Hall, Englewood Cliffs (1988)
16. Goldsmith, J., Lang, J., Truszczynski, M., Wilson, N.: The computational complexity of dominance and consistency in CP-nets. J. Artif. Intell. Res. **33**, 403–432 (2008)
17. Govindarajan, K., Jayaraman, B., Mantha, S.: Preference logic programming. In: Proceedings of ICLP, pp. 731–745. MIT Press (1995)

18. Horrocks, I., Patel-Schneider, P.F.: Reducing OWL entailment to description logic satisfiability. In: Fensel, D., Sycara, K., Mylopoulos, J. (eds.) ISWC 2003. LNCS, vol. 2870, pp. 17–29. Springer, Heidelberg (2003)

19. Horrocks, I., Sattler, U., Tobies, S.: Practical reasoning for expressive description logics. In: Ganzinger, H., McAllester, D., Voronkov, A. (eds.) LPAR 1999. LNCS, vol. 1705, pp. 161–180. Springer, Heidelberg (1999)

20. Keeney, R.L., Raiffa, H.: Decisions with Multiple Objectives: Preferences and Value Trade-Offs. Cambridge University Press, Cambridge (1993)

21. Lacroix, M., Lavency, P.: Preferences: putting more knowledge into queries. In: Proceedings of VLDB, pp. 1–4. Morgan Kaufmann (1987)

22. Lukasiewicz, T., Martinez, M.V., Simari, G.I.: Preference-based query answering in Datalog+/− ontologies. In: Proceedings of IJCAI, pp. 1017–1023 (2013)

23. Lukasiewicz, T., Martinez, M.V., Simari, G.I., Tifrea-Marciuska, O.: Preference-based query answering in probabilistic Datalog+/− ontologies. J. Data Semant. 1–21 (2014). http://dx.doi.org/10.1007/s13740-014-0040-x

24. Lukasiewicz, T., Schellhase, J.: Variable-strength conditional preferences for ranking objects in ontologies. J. Web Sem. 5(3), 180–194 (2007)

25. Pratt-Hartmann, I.: Complexity of the two-variable fragment with counting quantifiers. J. Log. Lang. Inf. 14(3), 369–395 (2005)

26. Prestwich, S.D., Rossi, F., Venable, K.B., Walsh, T.: Constraint-based preferential optimization. In: Proceedings of AAAI, pp. 461–466. AAAI Press (2005)

27. Siberski, W., Pan, J.Z., Thaden, U.: Querying the semantic web with preferences. In: Cruz, I., Decker, S., Allemang, D., Preist, C., Schwabe, D., Mika, P., Uschold, M., Aroyo, L.M. (eds.) ISWC 2006. LNCS, vol. 4273, pp. 612–624. Springer, Heidelberg (2006)

28. Stefanidis, K., Koutrika, G., Pitoura, E.: A survey on representation, composition and application of preferences in database systems. ACM Trans. Database Syst. 36(3), 19 (2011)

29. Tobies, S.: Complexity results and practical algorithms for logics in knowledge representation. Ph.D. thesis, RWTH Aachen, Germany (2001)

30. W3C: OWL 2 Web Ontology Language Primer (2nd Edition). W3C Recommendation, 11 December 2012. http://www.w3.org/TR/owl2-primer/

31. W3C: OWL 2 Web Ontology Language Profiles. W3C Recommendation, 27 October 2009. http://www.w3.org/TR/owl2-profiles/

32. W3C: OWL Web Ontology Language Overview. W3C Recommendation, 10 February 2004. http://www.w3.org/TR/2004/REC-owl-features-20040210/

33. Wang, H., Zhou, X., Chen, W., Ma, P.: Top-k retrieval using conditional preference networks. In: Proceedings of CIKM, pp. 2075–2079. ACM Press (2012)

34. Wilson, N.: Extending CP-nets with stronger conditional preference statements. In: Proceedings of AAAI, pp. 735–741. AAAI Press (2004)

Uncertainty Handling in Named Entity Extraction and Disambiguation for Informal Text

Maurice van Keulen and Mena B. Habib[✉]

Faculty of EEMCS, University of Twente,
Enschede, The Netherlands
{m.vankeulen,m.b.habib}@ewi.utwente.nl

Abstract. Social media content represents a large portion of all textual content appearing on the Internet. These streams of user generated content (UGC) provide an opportunity and challenge for media analysts to analyze huge amount of new data and use them to infer and reason with new information. A main challenge of natural language is its ambiguity and vagueness. To automatically resolve ambiguity, the grammatical structure of sentences is used. However, when we move to informal language widely used in social media, the language becomes more ambiguous and thus more challenging for automatic understanding.

Information Extraction (IE) is the research field that enables the use of unstructured text in a structured way. Named Entity Extraction (NEE) is a sub task of IE that aims to locate phrases (mentions) in the text that represent names of entities such as persons, organizations or locations regardless of their type. Named Entity Disambiguation (NED) is the task of determining which correct person, place, event, etc. is referred to by a mention.

The goal of this paper is to provide an overview on some approaches that mimic the human way of recognition and disambiguation of named entities especially for domains that lack formal sentence structure. The proposed methods open the doors for more sophisticated applications based on users' contributions on social media. We propose a robust combined framework for NEE and NED in semi-formal and informal text. The achieved robustness has been proven to be valid across languages and domains and to be independent of the selected extraction and disambiguation techniques. It is also shown to be robust against the informality of the used language. We have discovered a reinforcement effect and exploited it a technique that improves extraction quality by feeding back disambiguation results. We present a method of handling the uncertainty involved in extraction to improve the disambiguation results.

Keywords: Named entity extraction · Named entity disambiguation · Informal text · Uncertainty handling

© Springer International Publishing Switzerland 2014
F. Bobillo et al. (Eds.): URSW 2011-2013, LNAI 8816, pp. 309–328, 2014.
DOI: 10.1007/978-3-319-13413-0_16

1   Introduction

Computers cannot understand natural languages like humans do. Our ability to easily distinguish between multiple word meanings is developed in a lifetime of experience. Using the context in which a word is used, a fundamental understanding of syntax and logic, and a sense of the speaker's intention, we understand what another person is telling us or what we read. It is the aim of the Natural Language Processing (NLP) society to mimic the way humans understand natural languages. Although efforts spent for more than 50 years by linguists and computer scientists to get computers to understand human language, there is still long way to go to achieve this goal.

A main challenge of natural language is its ambiguity and vagueness. The basic definition of ambiguity, as generally used in natural language processing, is *"capable of being understood in more than one way"*. Scientists try to resolve ambiguity, either semantic or syntactic, based on properties of the surrounding context. Examples include, Part Of Speech (POS) tagging, morphology analysis, Named Entity Recognition (NER), and relations (facts) extraction. To automatically resolve ambiguity, typically the grammatical structure of sentences is used, for instance, which groups of words go together (phrases) and which words are the subject or object of a verb. However, when we move to informal language widely used in social media, the language becomes more ambiguous and thus more challenging for automatic understanding.

The rapid growth in the IT in the last two decades leads to the growth in the amount of information available on the World Wide Web (WWW). Social media content represents a big part of all textual content appearing on the Internet. According to an eMarketer report [1], nearly one in four people worldwide will use social networks in 2013. The number of social network users around the world rose to 1.73 billion in 2013. By 2017, the global social network audience will total 2.55 billion. Twitter as an example of highly active social media network, has 140 million active users publishing over 400 million tweet every day[1].

These streams of user generated content (UGC) provide an opportunity and challenge for media analysts to analyze huge amount of new data and use them to infer and reason with new information. Making use of social media content requires measuring, analyzing and interpreting interactions and associations between people, topics and ideas. An example of a main sector for social media analysis is the area of customer feedback through social media. With so many feedback channels, organizations can mix and match them to best suit corporate needs and customer preferences.

Another beneficial sector is social security. Communications over social networks have helped to put entire nations to action. Social media played a key role in The Arab Spring that started in 2010 in Tunisia. The riots that broke out across England during the summer of 2011 also showed the power of social media. The growing criminality associated with social media has been an alarm to government security agencies. There is a growing demand to automatically

---

[1] https://blog.twitter.com/2012/twitter-turns-six

monitor the discussions on social media as a source of intelligence. Nowadays, increasing numbers of people within investigative agencies are being deployed to monitor social media. Unfortunately, the existing tools and technologies used are limited because they are based on simple keyword selection and classification instead of reasoning with meaningful information. Furthermore, the processes followed are time and resources consuming. There is also a need for new tools and technologies that can deal with the informal language widely used in social media.

Information Extraction (IE) is the research field that enables the use of such a vast amount of unstructured distributed data in a structured way. IE systems analyze human language in order to extract information about different types of events, entities, or relationships. Named Entity Extraction (NEE) is a sub task of IE that aims to locate phrases (mentions) in the text that represent names of persons, organizations or locations regardless of their type. It differs from the term Named Entity Recognition (NER) which involves both extraction and classification to one of the predefined set of classes. Named Entity Disambiguation (NED) is the task of exploring which correct person, place, event, etc. is referred to by a mention. NEE and NED have become a basic steps of many technologies like Information Retrieval (IR), Question Answering (QA).

Although state-of-the-art NER systems for English produce near-human performance [2], their performance drops when applied to informal text of UGC where the ambiguity increases. It this the aim of this paper to study the interdependency of NEE and NED on the domain of informal text, and to show how one could be used to improve the other and vice versa. We call this potential for mutual improvement, the *reinforcement effect*. It mimics the way humans understand natural language. Natural language processing (NLP) tasks are commonly split into a set of pipelined sub tasks. The residual error produced in any sub task propagates, adversely affecting the end objectives. This is why we believe that back propagation would help improving the overall system quality. We show the benefit of using this *reinforcement effect* on two domains: NEE and NED for toponyms in semi-formal text that represents advertisements for holiday properties; and for arbitrary entity types in informal short text in tweets. We proved that this mutual improvement makes NEE and NED robust across languages and domains. This improvement is also independent on what extractions and disambiguation techniques are used. Furthermore, we developed extraction methods that consider alternatives and uncertainties in text with less dependency on formal sentence structure. This leads to more reliability in cases of informal and noisy UGC text.

2   Examples of Application Domains

Information extraction has applications in a wide range of domains. There are many stakeholders that could benefit from UGC on social media. Here, we give some examples for applications of information extraction:

- Security agencies typically analyze large amounts of text manually to search for information about people involved in criminal or terrorism activities.

Social media is a continuously instantly updated source of information. Football hooligans sometimes start their fight electronically on social media networks even before the sport event. Another real life example is the Project X Haren[2]. Project X Haren was an event that started out as a public invitation to a birthday party by a girl on Facebook, but ended up as a gathering of thousands of youths causing riots in the town of Haren, Groningen. Automatic monitoring and gathering of such information could be helpful to take actions to prevent such violent, and destructive behaviors. As an example for real application, we contribute to the TEC4SE project[3]. The aim of the project is to improve the operational decision-making within the security domain by gathering as much information available from different sources (like cameras, police officers on field, or social media posts). Then these information is linked and relationships between different information streams are found. The result is a good overview of what is happening in the field of security in the region. Our contribution to this project is to the enrich Twitter stream messages by extracting named entities at run time. The amount and the nature of the flowing data is beyond the possibility of manually tracking. This is why we need new technologies that is capable of dealing with such huge noisy amounts of data.

– As users become more involved in creating contents in a virtual world, more and more data is generated in various aspects of life for studying user attitudes and behaviors. Social sciences study human behavior by studying their physical space and belongings. Now, it is possible to investigate users by studying their online activities, postings, and behavior in a virtual space. This method can be a replacement for traditional surveys and experiments [3]. Prediction and understanding of the attitudes and behaviors of individuals and groups based on the sentiment expressed within online virtual communities is a natural area of research in the Internet era. To reach this goal, social scientists are in dire need of stronger tools to provide them with the required data for their studies.

– Financial experts always look for specific information to help their decision making. Social media can be a very important source of information about the attitudes and behaviors of stakeholders. In general, if extracted and analyzed properly, the data on social media can lead to useful predictions of certain human related events. Such prediction has great benefits in many realms, such as finance, product marketing and politics [4]. For example, a finance company may want to know the stakeholders' reaction towards some political action. Automatically finding such information from user posts on social media requires special information extraction technologies to analyze the noisy social media streams and capture such information.

– With the fast growth of the Web, search engines have become an integral part of people's daily lives, and users search behaviors are much better understood now. Search based on bag-of-words representation of documents can no longer

---

[2] http://en.wikipedia.org/wiki/Project_X_Haren
[3] http://www.tec4se.nl/

provide satisfactory results. More advanced information needs such as entity search, and question answering can provide users with better search experience. To facilitate these search capabilities, information extraction is often needed as a pre-processing step to enrich the document with information in structured form.

3   Challenges

NEE and NED in informal text are challenging. Here we summarize the challenges of NEE and NED for tweets as an example of informal text:

- The informal nature of tweets makes the extraction process more difficult. For example, in Table 1 case 1, it is hard to extract the mentions (phrases that represent NEs) using traditional NEE methods because of the ill-formed sentence structure. Traditional NEE methods might extract '*Grampa*' as a mention because of it capitalization. Furthermore, it is hard to extract the mention '*Speechless*', which is a name of a song, as it requires further knowledge about '*Lady Gaga*' songs.
- The limited length (140 characters) of tweets forces the senders to provide dense information. Users resort to acronyms to reserve space. Informal language is another way to express more information in less space. All of these problems make both the extraction and the disambiguation processes more complex. For example, in Table 1 case 2 shows two abbreviations ('*Qld*' and '*Vic*'). It is hard to infer their entities without extra information.

Table 1. Some challenging cases for NEE and NED in tweets (NE mentions are written in bold).

Case #	Tweet Content
1	– **Lady Gaga** - **Speechless** live @ **Helsinki** 10/13/2010 http://www.youtube.com/watch?v=yREociHyijk. . . @ladygaga also talks about her Grampa who died recently
2	**Qld** flood victims donate to **Vic** bushfire appeal
3	**Laelith Demonia** has just defeated **liwanu Hird**. Career wins is 575, career losses is 966.
4	Adding **Win7Beta**, **Win2008**, and **Vista** x64 and x86 images to munin. #wds
5	history should show that **bush jr** should be in jail or at least never should have been president
6	RT @BBCClick: Joy! **MS Office** now syncs with **Google Docs** (well, in beta anyway). We are soon to be one big happy (cont) http://tl.gd/73t94u
7	"Even Writers Can Help..An Appeal For **Australian** Bushfire Victims" http://cli.gs/Zs8zL2

- The limited coverage of a Knowledge Base (KB) is another challenge facing NED for tweets. According to [5], 5 million out of 15 million mentions on the web cannot be linked to Wikipedia. This means that relying only on a KB for NED leads to around 33 % loss in disambiguated entities. This percentage is higher on Twitter because of its social nature where users discuss information about infamous entities. For example, Table 1 case 3 contains two mentions for two users on the '*My Second Life*' social network. It is very unlikely that one could find their entities in a KB. However, their profile pages ('https://my.secondlife.com/laelith.demonia' and 'https://my.secondlife.com/liwanu.hird') can be found easily by a search engine.
- Named entity (NE) representation in KB implies another NED challenge. YAGO KB [6] uses Wikipedia anchor text as possible mention representation for named entities. However, there might be more representations that do not appear in Wikipedia anchor text. Either because of misspelling or because of a new abbreviation of the entity. For example, in Table 1 case 4, the mentions '*Win7Beta*' and '*Win2008*' do not appear in YAGO KB mention-entity look-up table, although they refer to the entities 'http://en.wikipedia.org/wiki/Windows_7' and 'http://en.wikipedia.org/wiki/Windows_Server_2008' respectively.
- The processes of NEE and NED involve degrees of uncertainty. For example, in Table 1 case 5, it is uncertain whether the word *jr* should be part of the mention *bush* or not. Same for '*Office*' and '*Docs*' in case 6 which some extractors may miss. Another example, in case 7, it is hard to assess whether '*Australian*' should refer to 'http://en.wikipedia.org/wiki/Australia' or 'http://en.wikipedia.org/wiki/Australian_people'[4]. Both might be correct. This is why we believe that it is better to consider possible alternatives in the processes of NEE and NED.
- Another challenge is the freshness of the KBs. For example, the page of '*Barack Obama*' on Wikipedia was created on 18 March 2004. Before that date '*Barack Obama*' was a member of the Illinois Senate and you could find his profile page on 'http://www.ilga.gov/senate/Senator.asp?MemberID=747'. It is very common on social networks that users talk about some infamous entity who might become later a public figure.
- Informal nature of language used in social media implies many different random representations of the same fact. This adds new challenges to machine learning approaches which need regular patterns for generalization. We need new methods that require less training data and generalize well at the same time.

Semi-formal text is text lacking the formal structure of the language but follows some pattern or format like product descriptions and advertisements. Although semi-formal text involves some regularity in representing information, this regularity implies some challenges.

In Table 2, cases 1 and 2 show two examples for true toponyms included in a holiday description. Any machine learning approach uses cases 1 and 2 as training samples will annotate '*Airport*' as a toponym following the same

---

[4] Some NER datasets consider nationalities as NEs [7].

2-room apartment 55 m2: living/dining room with 1 sofa bed and satellite-TV, exit to the balcony. 1 room with 2 beds (90 cm, length 190 cm). Open kitchen (4 hotplates, freezer). Bath/bidet/WC. Electric heating. Balcony 8 m2. Facilities: telephone, safe (extra). Terrace Club: Holiday complex, 3 storeys, built in 1995 2.5 km from the centre of **Armacao de Pera**, in a quiet position. For shared use: garden, swimming pool (25 x 12 m, 01.04.-30.09.), paddling pool, children's playground. In the house: reception, restaurant. Laundry (extra). Linen change weekly. Room cleaning 4 times per week. Public parking on the road. Railway station "**Alcantarilha**" 10 km. Please note: There are more similar properties for rent in this same residence. Reception is open 16 hours (0800-2400 hrs). Lounge and reading room, games room. Daily entertainment for adults and children. Bar-swimming pool open in summer. Restaurant with Take Away service. Breakfast buffet, lunch and dinner(to be paid for separately, on site). Trips arranged, entrance to water parks. Car hire. Electric cafetiere to be requested in adavance. Beach football pitch. IMPORTANT: access to the internet in the computer room (extra). The closest beach (350 m) is the "**Sehora da Rocha**", **Playa de Armacao de Pera** 2.5 km. Please note: the urbanisation comprises of eight 4 storey buildings, no lift, with a total of 185 apartments. Bus station in **Armacao de Pera** 4 km.

Fig. 1. Example of EuroCottage holiday home descriptions (toponyms in bold).

Table 2. Some challenging cases for toponyms extraction in semi-formal text (toponyms are written in bold).

Case #	Semi-formal Text Samples
1	**Bargecchia** 9 km from **Massarosa**
2	**Olšova Vrata** 5 km from **Karlovy Vary**
3	Bus station in **Armacao de Pera** 4 km
4	Airport 1.5 km (2 planes/day)

pattern of having a capitalized word followed by a number and the word '*km*'. Furthermore, the state-of-the-art approaches performs poorly on this type of text. Figure 2 shows the results of the application of three of the leading Stanford NER models[5] on a holiday property description text (see Fig. 1). Regardless of NE classification, even the extraction (determining if a phrase represents a NE or not) is performing poorly. Problems vary between (*a*) extracting false positives (like '*Electric*' and '*Trips*' in Fig. 2a); or (*b*) missing some true positives (like '*Sehora da Rocha*' in Fig. 2b, c); or (*c*) partially extracting the NE (like '*Sehora da Rocha*' in Figs. 2a and '*Armacao de Pera*' in Fig. 2b).

4 General Approach

Natural language processing (NLP) tasks are commonly composed of a set of chained sub tasks that form the processing pipeline. The residual error produced in these sub tasks propagates, affecting the final process results. In this paper we are concerned with NEE and NED which are two common processes in many NLP applications.

---

[5] http://nlp.stanford.edu:8080/ner/process

2-room apartment 55 m2: living/dining room with 1 sofa bed and satellite- TV, exit to the balcony. 1 room with 2 beds (90 cm, length 190 cm). Open kitchen (4 hotplates, freezer). Bath/bidet/WC. Electric heating. Balcony 8 m2. Facilities: telephone, safe (extra). Terrace Club: Holiday complex, 3 storeys, built in 1995 2.5 km from the centre of Armacao de Pera, in a quiet position. For shared use: garden, swimming pool (25 x 12 m, 01.04.-30.09.), paddling pool, children's playground. In the house: reception, restaurant. Laundry (extra). Linen change weekly. Room cleaning 4 times per week. Public parking on the road. Railway station "Alcantarilha" 10 km. Please note: There are more similar properties for rent in this same residence. Reception is open 16 hours (0800-2400 hrs). Lounge and reading room, games room. Daily entertainment for adults and children. Bar-swimming pool open in summer. Restaurant with Take Away service. Breakfast buffet, lunch and dinner(to be paid for separately, on site). Trips arranged, entrance to water parks. Car hire. Electric cafetiere to be requested in adavance. Beach football pitch. IMPORTANT: access to the internet in the computer room (extra). The closest beach (350 m) is the "Sehora da Rocha", Playa de Armacao de Pera 2.5 km. Please note: the urbanisation comprises of eight 4 storey buildings, no lift, with a total of 185 apartments. Bus station in Armacao de Pera 4 km.

Potential tags:

(a) Stanford 'english.conll.4class.distsim.crf.ser' model.

2-room apartment 55 m2: living/dining room with 1 sofa bed and satellite- TV, exit to the balcony. 1 room with 2 beds (90 cm, length 190 cm). Open kitchen (4 hotplates, freezer). Bath/bidet/WC. Electric heating. Balcony 8 m2. Facilities: telephone, safe (extra). Terrace Club: Holiday complex, 3 storeys, built in 1995 2.5 km from the centre of Armacao de Pera, in a quiet position. For shared use: garden, swimming pool (25 x 12 m, 01.04.-30.09.), paddling pool, children's playground. In the house: reception, restaurant. Laundry (extra). Linen change weekly. Room cleaning 4 times per week. Public parking on the road. Railway station "Alcantarilha" 10 km. Please note: There are more similar properties for rent in this same residence. Reception is open 16 hours (0800-2400 hrs). Lounge and reading room, games room. Daily entertainment for adults and children. Bar-swimming pool open in summer. Restaurant with Take Away service. Breakfast buffet, lunch and dinner(to be paid for separately, on site). Trips arranged, entrance to water parks. Car hire. Electric cafetiere to be requested in adavance. Beach football pitch. IMPORTANT: access to the internet in the computer room (extra). The closest beach (350 m) is the "Sehora da Rocha", Playa de Armacao de Pera 2.5 km. Please note: the urbanisation comprises of eight 4 storey buildings, no lift, with a total of 185 apartments. Bus station in Armacao de Pera 4 km.

Potential tags:

(b) Stanford 'english.muc.7class.distsim.crf.ser' model.

2-room apartment 55 m2: living/dining room with 1 sofa bed and satellite- TV, exit to the balcony. 1 room with 2 beds (90 cm, length 190 cm). Open kitchen (4 hotplates, freezer). Bath/bidet/WC. Electric heating. Balcony 8 m2. Facilities: telephone, safe (extra). Terrace Club: Holiday complex, 3 storeys, built in 1995 2.5 km from the centre of Armacao de Pera, in a quiet position. For shared use: garden, swimming pool (25 x 12 m, 01.04.-30.09.), paddling pool, children's playground. In the house: reception, restaurant. Laundry (extra). Linen change weekly. Room cleaning 4 times per week. Public parking on the road. Railway station "Alcantarilha" 10 km. Please note: There are more similar properties for rent in this same residence. Reception is open 16 hours (0800-2400 hrs). Lounge and reading room, games room. Daily entertainment for adults and children. Bar-swimming pool open in summer. Restaurant with Take Away service. Breakfast buffet, lunch and dinner(to be paid for separately, on site). Trips arranged, entrance to water parks. Car hire. Electric cafetiere to be requested in adavance. Beach football pitch. IMPORTANT: access to the internet in the computer room (extra). The closest beach (350 m) is the "Sehora da Rocha", Playa de Armacao de Pera 2.5 km. Please note: the urbanisation comprises of eight 4 storey buildings, no lift, with a total of 185 apartments. Bus station in Armacao de Pera 4 km.

Potential tags:

(c) Stanford 'english.all.3class.distsim.crf.ser' model.

Fig. 2. Results of Stanford NER models applied on semi-formal text of holiday property description.

Let us first formalize the NEE and NED problems. Given a sequence of words (tokens) $\{w\} = \{w_1, w_2, ..w_n\}$, NEE is the process of identifying sub-lists of words that represents mentions of NEs where mention $\{m\} = \{w_i, w_{i+1}, ..w_j\}$. The process of NED is to assign m to one of its possible entities $\{e\} = \{e_1, e_2, ..e_n\}$. The final output of the two processes is list of pairs (m, e). Figure 4 shows the formalization of the two problems.

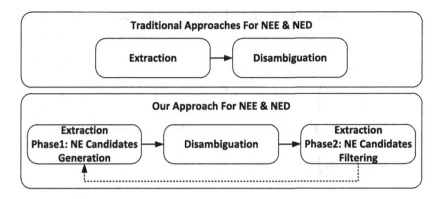

Fig. 3. Traditional approaches versus our approach for NEE and NED.

We claim that feedback derived from disambiguation would help in improving the extraction and hence the disambiguation. This is the same way we as humans understand text. The capability to successfully understand language requires one to acquire a range of skills including syntax, semantics, and an extensive vocabulary. We try to mimic a human's way of reasoning to solve the NEE and NED problems. Consider the tweet in Table 1 case 1. One would use syntax knowledge to recognize '10/13/2010' as a date. Furthermore, prior knowledge enables one to recognize '*Lady Gaga*' and '*Helsinki*' as a singer name and location name respectively or at least as names if one doesn't know exactly what they refer to. However, the term '*Speechless*' involves some ambiguity as it could be an adjective and also could be a name. A feedback clue from '*Lady Gaga*' would increase one's certainty that it refers to a song. Even without knowing that '*Speechless*' is a song of '*Lady Gaga*', there are sufficient clues to guess with quite high probability that it is a song. The pattern 'live @' in association with disambiguating '*Lady Gaga*' as a singer name and '*Helsinki*' as a location name, leads to infer '*Speechless*' as a song.

Although the logical order for a traditional Information Extraction (IE) system is to complete the extraction process before commencing the disambiguation, we start with an initial phase of extraction which aims to achieve high recall (find as many reasonable mention candidates as possible) then we apply the disambiguation for all the extracted possible mentions. Finally we filter those extracted mention candidates into true positives and false positives using features (clues) derived from the results of the disambiguation phase such as KB information and entity coherency. Figure 3 illustrates our general approach.

Unlike NER systems which extract entities mentions and assign them to one of the predefined categories (like location, person, organization), we focus first on extracting mentions regardless of their categories. We leave this classification to the disambiguation step which links the mention to its real entity.

The potential of this order is that the disambiguation step can give extra clues (such as entity-context similarity and entity-entity coherency) about each

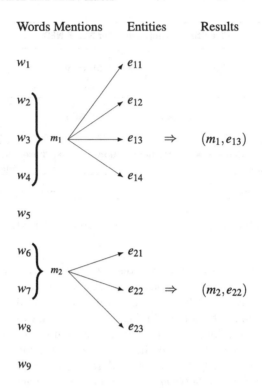

Fig. 4. Formalization of NEE and NED problems

NE candidate. This information can help in the decision whether the candidate is a true NE or not.

The general principal we claim is that NED could be very helpful in improving the NEE process. For example, consider the tweet in case 1 in Table 1. It is uncertain, even for humans, to recognize '*Speechless*' as a song name without having prior information about songs of '*Lady Gaga*'. Our approach is able to solve such problematic cases of named entities.

5 Case Study 1: Toponym Extraction and Disambiguation in Semi-formal Text

The task we focus on is to extract toponyms from EuroCottage holiday home descriptions[6] (an example is shown in Fig. 1) and use them to infer the country where the holiday property is located. We use this country inference task as a representative example of disambiguating extracted toponyms.

We propose an entity extraction and disambiguation approach based on uncertain annotations. The general approach illustrated in Fig. 5 has the following steps:

---

[6] www.eurocottage.com

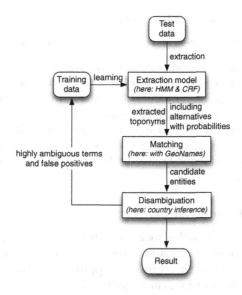

Fig. 5. Extraction and disambiguation approach

1. Prepare training data by manually annotating named entities.
2. Use the training data to build a statistical extraction model.
3. Apply the extraction model on test data and training data.
4. Match the extracted named entities against one or more gazetteers.
5. Use the toponym entity candidates for the disambiguation process.
6. Evaluate the extraction and disambiguation results for the training data. Automatically find a list of highly ambiguous named entities and false positives that affect the disambiguation results and use it to re-train the extraction model.
7. The steps from 2 to 6 are repeated automatically until there is no improvement any more in either the extraction or the disambiguation.

5.1 Toponym Extraction

For toponym extraction, we developed two statistical named entity extraction modules[7], one based on Hidden Markov Models (HMM) and one based on Conditional Ramdom Fields (CRF).

The goal of HMM [8] is to find the optimal tag sequence (in our case, whether the word is assigned to toponym tag or not) $T = t_1, t_2, t_3, ..., t_n$ for a given word sequence $W = w_1, w_2, w_3..., w_n$ that maximizes $P(T \mid W)$.

Conditional Random Fields (CRF) can model overlapping, non-independent features [9]. Here we used a linear chain CRF, the simplest model of CRF.

---

[7] We made use of the *lingpipe* toolkit for development: http://alias-i.com/lingpipe.

5.2   Extraction Modes of Operation

We used the extraction models to retrieve sets of annotations in two ways:

- **First-Best:** In this method, we only consider the first most likely set of annotations that maximize the probability $P(T \mid W)$ for the whole text. This method does not assign a probability for each individual annotation, but only to the whole retrieved set of annotations.
- **N-Best:** This method returns a top-25 of possible alternative hypotheses for terms annotations in order of their estimated likelihoods $p(t_i|w_i)$. The confidence scores are assumed to be conditional probabilities of the annotation given an input token.

5.3   Toponym Disambiguation

For the toponym disambiguation task, we only select those toponyms annotated by the extraction models that match a reference in GeoNames. We furthermore use an adapted version of the clustering approach of [10] to disambiguate to which entity an extracted toponym actually refers.

5.4   Handling Uncertainty of Annotations

Instead of giving equal contribution to all toponyms, we take the uncertainty in the extraction process into account to include the confidence of the extracted toponyms. In this way terms which are more likely to be toponyms have a higher contribution in determining the country of the document than less likely ones.

5.5   Improving Certainty of Extraction

In despite of the abovementioned improvement, the extraction probabilities are not accurate and reliable all the time. Some extraction models retrieve some false positive toponyms with high confidence probabilities. This is where we take advantage of the reinforcement effect. To be more precise. We introduce another class in the extraction model called 'highly ambiguous' and annotate those terms in the training set with this class that the disambiguation process finds more than τ countries for documents that contain this term.

The extraction model is subsequently re-trained and the whole process is repeated without any human interference as long as there is improvement in extraction and disambiguation process for the training set. The intention is that the extraction model learns to avoid prediction of terms to be toponyms when they appear to confuse the disambiguation process.

5.6   Experimental Results

Here we present the results of experiments with the presented methods of extraction and disambiguation applied to a collection of holiday properties descriptions. The data set consists of 1579 property descriptions for which we constructed a ground truth by manually annotating all toponyms.

Experiment 1: Effect of Extraction with Confidence Probabilities.
Table 3 shows the percentage of holiday home descriptions for which the correct country was successfully inferred. We can see that the **N-Best** method outperforms the **First-Best** method for both HMM and CRF models. This supports our claim that dealing with alternatives along with their confidences yields better results.

Table 3. Effectiveness of the disambiguation process for First-Best and N-Best methods in the extraction phase.

	HMM	CRF
No Filtering	68.95 %	68.19 %
1st Iteration	73.28 %	68.44 %

Table 4. Effectiveness of the disambiguation after iteration of refinement.

	HMM	CRF
No Filtering	68.95 %	68.19 %
1st Iteration	73.28 %	68.44 %

Experiment 2: Effect of Extraction Certainty Enhancement. Tables 4 and 5 show the effectiveness of the disambiguation and the extraction processes respectively before and after one iteration of refinement. We can see an improvement in HMM extraction and disambiguation results. The initial HMM results showed a high recall rate with a low precision. In spite of this, our approach managed to improve precision through iteration of refinement. The refinement process is based on removing highly ambiguous toponyms resulting in a slight decrease in recall and an increase in precision. In contrast, CRF started with high precision which could not be improved by the refinement process.

6   Case Study 2: Named Entity Extraction and Disambiguation Approach for Tweets

In this case study, we present a combined approach for NEE and NEL for tweets with an application on #Microposts 2014 challenge [11]. Although the logical order for such system is to do extraction first then the disambiguation, we start with an extraction phase which aims to achieve high recall (find as much NE candidates as possible). Then we apply disambiguation for all the extracted mentions. Finally, we filter those extracted NE candidates into true positives and false positives using features derived from the disambiguation phase in addition to other word shape and KB features. The potential of this order is that the disambiguation step gives extra information about each NE candidate that may help in the decision whether or not this candidate is a true NE. Figure 3 shows our system architecture versus traditional one.

Table 5. Effectiveness of the extraction process after iteration of refinement.

	HMM				CRF		
	Pre.	Rec.	F1		Pre.	Rec.	F1
No Filtering	0.3584	0.8517	0.5045	No Filtering	0.6969	0.7136	0.7051
1st Iteration	0.7667	0.5987	0.6724	1st Iteration	0.6989	0.7131	0.7059

6.1   NE Candidates Generation

For this task, we unionize the output of the following candidates generation methods:

- **Tweet Segmentation**: Tweet text is segmented using the segmentation algorithm described in [12]. Each segment is considered a NE candidate.
- **KB Lookup**: We scan all possible n-grams of the tweet against the mentions-entities table of DBpedia. N-grams that matches a DBpedia mention are considered NE candidates.
- **Regular Expressions**: We used regular expressions to extract numbers, dates and URLs from the tweet text.

6.2   NE Linking

Our NEL approach is composed of three steps; matcher, feature extractor, and SVM ranker.

- **Matcher**: This module takes each extracted mention candidate and looks for its Wikipedia reference candidates on DBpedia. Furthmore, for those mention candidates which don't have reference candidates in DBpedia, we use Google Search API to find possible Wikipedia pages for these mentions. This search helps to find references for misspelled or concatenated mentions like '*justinbieber*' and '*106andpark*'.
- **Feature Extractor**: This module is responsible for extracting a set of contextual and URL features for each candidate Wikipedia page as described in [13]. These features give indicators on how likely the candidate Wikipedia page could be a representative to the mention.
- **SVM Ranker**: After extracting the aforementioned set of features, SVM classifier is trained to rank candidate Wikipedia pages of a mention. For the challenge, we pick the page on the 1st order as a reference for the mention. The DBpedia URI is then generated from the selected Wikipedia URL.

6.3   NE Candidates Filtering

After generating the candidates list of NE, we apply our NE linking approach to disambiguate each extracted NE candidate. After the linking phase, we use SVM classifier to predict which candidates are true positives and which ones are not. We use the following set of features for each NE candidate to train the SVM:

- **Shape Features**: If the NE candidate is initially or fully capitalized and if it contains digits.
- **Probabilistic Features**:
  - The joint and the conditional probability of the candidate obtained from Microsoft Web N-Gram services.
  - The stickiness of the candidate as described in [12].
  - The candidate's frequency over around 5 million tweets[8].
- **KB Features**:
  - If the candidate appears in WordNet.
  - If the candidate appears as a mention in DBpedia KB.
- **Disambiguation Features**:
  - All the features used in the linking phase as described in [13]. We used only the feature set for the first top ranked entity page selected for the given NE candidate.

6.4   Final NE Set Generation

Beside the SVM, we also train a CRF model for NEE. We used the CRF model described in [14]. To generate the final NE set, we take the union of the CRF annotation set and SVM results, after removing duplicate extractions, to get the final set of annotations. We tried two methods to resolve overlapped mentions. In the first method (used in UTwente_Run1.tsv), we select the mention that appears in Yago KB [6]. If both mentions appear in Yago or both don't, we select the one with the longer length. In the second method (used in UTwente_Run2.tsv), we select only the mention with the longer length among the two overlapped mentions. The results shown in the next section are the results of the first method.

The idea behind this unionization is that SVM and CRF work in a different way. The former is a distance based classifier that uses numeric features for classification which CRF can not handle, while the latter is a probabilistic model that can naturally consider state-to-state dependencies and feature-to-state dependencies. On the other hand, SVM does not consider such dependencies. The hybrid approach of both makes use of the strength of each.

6.5   Experimental Results

In this section we show our experimental results of the proposed approaches on the challenge training data [11] in contrast with other competitors. All our experiments are done through a 4-fold cross validation approach for training and testing. Table 6 shows the results of '**Our Linking Approach**' presented in Sect. 6.2, in comparison with two modes of operation of AIDA [15]. The first mode is '**AIDA Cocktail**' which makes use of several ingredients: the prior probability of an entity being mentioned, the similarity between the context of the mention in the text and an entity, as well as the coherence among the entities. While the second mode is '**AIDA Prior**' which makes use only of the

---

[8] http://wis.ewi.tudelft.nl/umap2011/ + TREC 2011 Microblog track collection.

Table 6. Linking Results

	Percentage
Our Linking Approach	70.98 %
AIDA Cocktail	56.16 %
AIDA Prior	55.63 %

Table 7. Extraction Results

	Pre.	Rec.	F1
Candidates Generation	0.120	**0.945**	0.214
Candidates Filtering (SVM)	**0.722**	0.544	0.621
CRF	0.660	0.568	0.611
Final Set Generation	0.709	0.706	**0.708**
Stanford NER	0.716	0.392	0.507

Table 8. Extraction and Linking Results

	Pre.	Rec.	F1
Extraction + Linking	**0.533**	**0.534**	**0.534**
Stanford + AIDA	0.509	0.279	0.360

prior probability. The results show the percentage of finding the correct entity of the ground truth mentions. Table 7 shows the NEE results along the extraction process phases in contrast with '**Stanford NER**' [16]. Finally, Table 8 shows our final results of both extraction and entity linking in comparison with our competitor ('**Stanford + AIDA**') where '**Stanford NER**' is used for NEE and '**AIDA Cocktail**' is used for NEL.

7   Future Research Directions

Although many machine learning and fuzzy techniques abound, some aspects often remain absolute: extraction rules absolutely recognize and annotate a phrase or not, only a top item from a ranking is chosen for a next phase, etc. We envision an approach that *fundamentally* treats annotations and extracted information as uncertain throughout the process. We humans happily deal with doubt and misinterpretation every day, why shouldn't computers?

We envision developing information extractors 'Sherlock Holmes style' — *"when you have eliminated the impossible, whatever remains, however improbable, must be the truth"* — by adopting the principles and requirements below.

– Annotations are uncertain, hence we process both annotations as well as information about the uncertainty surrounding them.
– We have an unconventional conceptual starting point, namely not "no annotations" but "there is no knowledge hence anything is possible". Figure 6a shows all possible annotations for an example sentence for one entity type.
– A developer gradually and interactively defines an ontology with positive and negative knowledge about the correctness of certain (combinations of) annotations. At each iteration, added knowledge is immediately applied improving the extraction result until the result is good enough (see also [17]).
– Storage, querying and manipulation of annotations should be scalable. Probabilistic databases are an attractive technology for this.

Basic forms of knowledge are the entity types one is interested in and declarations like τ_1 —*dnc*— τ_2 (no subphrase of a τ_1-phrase should be interpreted as τ_2, e.g., Person—*dnc*—City). See Fig. 6b for a small example. We also envision application of background probability distributions, uncertain rules, etc. We hope these principles and forms of knowledge also allow for more effective handling of common problems (e.g., "you" is also the name of a place; should "Lake Como" or "Como" be annotated as a toponym).

7.1   Uncertain Annotation Model

An *annotation* $a = (b, e, \tau)$ declares a phrase φ_e^b from b to e to be interpreted as entity type τ. For example, a_8 in Fig. 6a declares $\varphi =$ "Paris Hilton" from $b = 1$ to $e = 2$ to be interpreted as type $\tau =$ Person. An *interpretation* $I = (A, \mathcal{U})$ of a sentence s consists of an annotation set A and a structure $\mathcal{U}$ representing the uncertainty among the annotations. In the sequel, we discuss what $\mathcal{U}$ should be, but for now view it as a set of random variables (RVs) R with their dependencies.

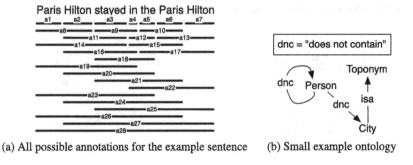

(a) All possible annotations for the example sentence   (b) Small example ontology

Fig. 6. Example sentence and NEE ontology

Rather unconventionally, we don't start with an empty A, but with a 'no knowledge' point-of-view where any phrase can have any interpretation. So our initial A is $\{a \mid a = (b, e, \tau) \wedge \tau \in T \wedge \varphi_e^b$ is a phrase of $s\}$ where T is the set of possible types.

With T finite, A is also finite. More importantly, $|A| = O(klt)$ where $k = |s|$ is the length of s, l is the maximum length phrases considered, and $t = |T|$. Hence, A grows linearly in size with each. In the example of Fig. 6a, $T = \{$Person, Toponym, City$\}$ and we have $28 \cdot |T| = 84$ annotations. Even though we envision a more ingenious implementation, no probabilistic database would be severely challenged by a complete annotation set for a typical text field.

7.2   Knowledge Application Is Conditioning

We explain how to 'apply knowledge' in our approach by means of the example of Fig. 6, i.e., with our A with 84 (possible) annotations and an ontology only containing Person, Toponym, and City. Suppose we like to add the knowledge Person—*dnc*—City. The effect should be the removal of some annotations and adjustment of the probabilities of the remaining ones.

An initial promising idea is to store the annotations in an uncertain relation in a probabilistic database, such as MayBMS [18]. In MayBMS, the existence of each tuple is determined by an associated world set descriptor (wsd) containing a set of RV assignments from a world set table (see Fig. 7). RVs are assumed independent. For example, the 3rd annotation tuple only exists when $x_8^1 = 1$ which is the case with a probability of 0.8. Each annotation can be seen as a probabilistic event, which are all independent in our starting point. Hence, we can store A by associating each annotation tuple a_i^j with one boolean RV x_i^j. Consequently, the database size is linear with $|A|$.

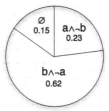

annotations

	b	e	type	...	wsd
a_1^1	1	1	Person	...	$\{x_1^1 = 1\}$
a_1^2	1	1	City	...	$\{x_1^2 = 1\}$
a_8^1	1	2	Person	...	$\{x_8^1 = 1\}$
...	...	...	...		...

world_set

x	v	P
x_1^1	0	0.4
x_1^1	1	0.6
x_1^2	0	0.7
x_1^2	1	0.3
x_8^1	0	0.2
x_8^1	1	0.8
...	...	...

Fig. 7. Initial annotation set stored in a probabilistic database (MayBMS-style)

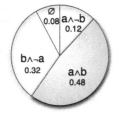
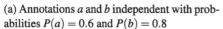

(a) Annotations a and b independent with probabilities $P(a) = 0.6$ and $P(b) = 0.8$

(b) a and b conditioned to be mutually exclusive ($a \wedge b$ not possible)

Fig. 8. Defining a and b to be mutually exclusive means conditioning the probabilities.

Adding knowledge such as Person—*dnc*—City means that certain RVs become dependent and that certain combinations of RV assignments become impossible. Let us focus on two individual annotations a_1^2 ("Paris" is a City) and a_8^1 ("Paris Hilton" is a Person). These two annotations become mutually exclusive. The process of adjusting the probabilities is called *conditioning* [19]. It boils down to redistributing the remaining probability mass. Figure 8 illustrates this for $a = a_1^2$ and $b = a_8^1$. The remaining probability mass is $1 - 0.48 = 0.52$. Hence, the distribution of this mass over the remaining possibilities is $P(a \wedge \neg b) = \frac{0.12}{0.52} \approx 0.23$, $P(b \wedge \neg a) = \frac{0.32}{0.52} \approx 0.62$, and $P(\emptyset) = P(\neg a \wedge \neg b) = \frac{0.08}{0.52} \approx 0.15$.

A first attempt is to replace x_1^2 and x_8^1 with one fresh three-valued RV x' with the probabilities just calculated, i.e., wsd(a_1^2) = $\{x' = 1\}$ and wsd(a_8^1) = $\{x' = 2\}$ with $P(x' = 0) = 0.15$, $P(x' = 1) = 0.23$, and $P(x' = 2) = 0.62$. Unfortunately, since annotations massively overlap, we face a combinatorial explosion. For this rule, we end up with one RV with up to $2^{2 \cdot 28} = 2^{56} \approx 7 \cdot 10^{16}$ cases.

Solution directions. What we are looking for in this paper is a structure that is expressive enough to capture all dependencies between RVs and at the same time allowing for scalable processing of conditioning operations. The work of [19] represents dependencies resulting from queries with a tree of RV assignments. We are also investigating the shared correlations work of [20].

References

1. Social networking reaches nearly one in four around the world
2. Chinchor, N.A.: Proceedings of the Seventh Message Understanding Conference (MUC-7) named entity task definition, Fairfax, VA, 21 p., April 1998. http://www.itl.nist.gov/iaui/894.02/related_projects/muc (version 3.5.)
3. Abbasi, M.-A., Chai, S.-K., Liu, H., Sagoo, K.: Real-world behavior analysis through a social media lens. In: Yang, S.J., Greenberg, A.M., Endsley, M. (eds.) SBP 2012. LNCS, vol. 7227, pp. 18–26. Springer, Heidelberg (2012)
4. Yu, S., Kak, S.: A survey of prediction using social media. CoRR, abs/1203.1647 (2012)
5. Lin, T., Mausam, Etzioni, O.: Entity linking at web scale. In: Proceedings of the Joint Workshop on Automatic Knowledge Base Construction and Web-scale Knowledge Extraction (AKBC-WEKEX), pp. 84–88 (2012)
6. Hoffart, J., Suchanek, F., Berberich, K., Kelham, E., de Melo, G., Weikum, G.: Yago2: Exploring and querying world knowledge in time, space, context, and many languages. In: Proceedings of WWW 2011, pp. 229–232 (2011)
7. Basave, A.E.C., Varga, A., Rowe, M., Stankovic, M., Dadzie, A.-S.: Making sense of microposts (#msm2013) concept extraction challenge. In: Making Sense of Microposts (#MSM2013) Concept Extraction Challenge, pp. 1–15 (2013)
8. Ekbal, A., Bandyopadhyay, S.: A hidden Markov model based named entity recognition system: Bengali and Hindi as case studies. In: Ghosh, A., De, R.K., Pal, S.K. (eds.) PReMI 2007. LNCS, vol. 4815, pp. 545–552. Springer, Heidelberg (2007)
9. Wallach, H.: Conditional random fields: An introduction. Technical Report MS-CIS-04-21, Department of Computer and Information Science, University of Pennsylvania (2004)
10. Habib, M.B., van Keulen, M.: Named entity extraction and disambiguation: The reinforcement effect. In: Proceedings of MUD 2011, Seatle, USA, pp. 9–16 (2011)
11. Cano, A.E., Rizzo, G., Varga, A., Rowe, M., Stankovic, M., Dadzie, A.-S.: #microposts2014 neel challenge: Measuring the performance of entity linking systems in social streams. In: Proceedings of the #Microposts2014 NEEL Challenge (2014)
12. Li, C., Weng, J., He, Q., Yao, Y., Datta, A., Sun, A., Lee, B.-S.: Twiner: named entity recognition in targeted twitter stream. In: SIGIR, pp. 721–730 (2012)
13. Habib, M.B., van Keulen, M.: A generic open world named entity disambiguation approach for tweets. In: Proceedings of the 5th International Conference on Knowledge Discovery and Information Retrieval, KDIR 2013, Vilamoura, Portugal, pp. 267–276, September 2013. SciTePress, Portugal (2013)
14. Habib, M., Van Keulen, M., Zhu, Z.: Concept extraction challenge: University of Twente at #msm2013. In: Making Sense of Microposts (#MSM2013) Concept Extraction Challenge, pp. 17–20 (2013)
15. Yosef, M.A., Hoffart, J., Bordino, I., Spaniol, M., Weikum, G.: Aida: An online tool for accurate disambiguation of named entities in text and tables. In: PVLDB, pp. 1450–1453 (2011)

16. Finkel, J.R., Grenager, T., Manning, C.: Incorporating non-local information into information extraction systems by gibbs sampling. In: ACL, pp. 363–370 (2005)
17. van Keulen, M., de Keijzer, A.: Qualitative effects of knowledge rules and user feedback in probabilistic data integration. VLDB J. **18**(5), 1191–1217 (2009)
18. Huang, J., Antova, L., Koch, C., Olteanu, D.: MayBMS: A probabilistic database management system. In: Proceedings of the 35th SIGMOD International Conference on Management of Data, Providence, Rhode Island, pp. 1071–1074 (2009)
19. Koch, C., Olteanu, D.: Conditioning probabilistic databases. Proc. VLDB Endow. **1**(1), 313–325 (2008)
20. Sen, P., Deshpande, A., Getoor, L.: Exploiting shared correlations in probabilistic databases. Proc. VLDB Endow. **1**(1), 809–820 (2008)

Author Index

Printed in the United States
By Bookmasters